Green Energy and Technology

For further volumes:
http://www.springer.com/series/8059

Fabian Wagner
Faculty of Business Administration
and Economics
University of Duisburg-Essen
Essen
Germany

Dissertation in fulfillment of the requirements for a Doctoral degree in Business Sciences (Dr. rer. pol.) from the Faculty for Business and Economics of Duisburg- Essen University (Fakultät für Wirtschaftswissenschaften, Universität Duisburg- Essen).
Submitted by: Fabian Wagner, born in Villingen
Date of the oral examination: August 7, 2013
First reviewer: Prof. Dr. Christoph Weber; Second reviewer: Prof. Dr. Volker Clausen

ISSN 1865-3529 ISSN 1865-3537 (electronic)
ISBN 978-3-319-05779-8 ISBN 978-3-319-05780-4 (eBook)
DOI 10.1007/978-3-319-05780-4
Springer Cham Heidelberg New York Dordrecht London

Library of Congress Control Number: 2014935085

Printed on acid-free paper

Springer is part of Springer Science+Business Media (www.springer.com)

Foreword

The world energy markets have undergone profound changes over the last decade. The record high prices for oil and other fossil energy carriers in 2008 were followed by a dramatic decline as a consequence of the global financial and economic crisis. For oil, a rapid recovery of the prices could be observed, whereas shale gas then turned out to be a game changer for the U.S. gas market and beyond. In parallel, the discussions on climate change intensified, yet a global long term agreement has not yet been concluded nor seems to be within immediate reach. Despite this lack of international accord, the relevance of renewables has increased considerably around the world. A major impediment for a global climate change mitigation strategy is, however, the still considerable uncertainty with respect to the future technical and economic potentials of renewable energies.

This is the starting point for Fabian Wagner's excellent research work. He focused on the future role of renewables in global energy supply and developed a comprehensive approach to identify optimal strategies for this. He tackles the three major challenges for renewable energy development within one integrated model: the site-specific potentials of renewables in different regions of the world, the implications of intermittency for solar and wind and the already mentioned uncertainty on future technological learning.

This is not just a book about a large numerical model. It also gives detailed insights in the potentials of renewables and in the analytics of optimal strategy choice under uncertainty. Fabian Wagner gathered here all available knowledge to sketch a broad picture of the path forward towards an economically efficient greenhouse gas reduction. He combined rigorous mathematical model formulations with his striking ability to describe key developments in a clear language and with appealing graphs. He supported his results with impressive sensitivity analyses leaving almost no questions unanswered.

This book offers answers to the readers. The most important one is about the role renewables will play in the future - despite the challenges they will need to face. In a carbon-constrained world there is a huge need for renewable, which are expected to dominate world energy supply by the end of the century, though this may not happen as rapidly as one would argue in Germany or in other part of the world. The reasons for this need can be found in Fabian's analysis.

It has been a great pleasure for me to work together with Fabian at his PhD. I full-heartily hope that this book - containing the result of his efforts - will find many readers who do agree with me on the intellectual rigor and the high quality of the research.

Essen, February 2014 Prof. Christoph Weber

Preface

Renewable power generation technologies, apart from hydro, account for less than 5% of global production today. From a cost perspective they can, barring a few exceptions, not compete with fossil- or fissile-based generation technologies. However, several future developments are likely to result in improved relative cost competitiveness for renewables. They include system and operating cost reductions, increasing costs for carbon dioxide emissions and increasing fossil fuel prices. Yet, whether the magnitude of those trends will be sufficient to render renewable technologies more competitive than conventional ones and, if so, which technologies will be the winners, is highly uncertain.

I experienced first-hand how relevant this topic is for players in the power industry when working for a wind turbine component manufacturer. Both owners and management were aware that in case governmental support for wind power ceased to exist, the thriving market would collapse. Although support schemes are fairly predictable in the short-term, in the mid- to long-term subsidies can be expected to be directed towards those technologies with the best prospects for reaching competitiveness at comparatively low costs to society.

The high degree of complexity associated with efforts to assess the future prospects of renewable technologies based on fundamental drivers were the main motivations to approach this topic using methods from Operations Research in a level of detail, which is generally possible in an academic rather than in an industry setting. Apart from its methodological contributions, this work therefore also aspires to produce practically relevant insights regarding the long-term competitiveness of renewable power.

The content of this book derives from my doctoral thesis work at the Chair for Management Science and Energy Economics (EWL) of Duisburg-Essen University between 2008 and 2013.

I would like to thank two persons in particular. My advisor, Prof. Dr. Christoph Weber, and my wife, Alesia Wagner. Without their support, inspiration and encouragement this book would have never seen the light of day. I would also like to thank a large group of people, who supported me in various ways, including Prof. Dr. Volker Clausen, my colleagues at the EWL and my parents, Rita and Otmar Wagner, as well as my parents-in-law, Margarita and Vasili Staravoitov.

Essen, 2013 Fabian Wagner

Contents

Foreword V

Preface VII

List of Abbreviations XV

1 Introduction 1

2 Renewables in Power Generation: Status Quo 3
2.1 Global Power Generation Landscape 3
2.2 Properties of Renewable Technologies 5
2.2.1 Generation Costs 7
2.2.2 Potential 10
2.2.3 Intermittency 11
2.3 Drivers of Future Technology Competitiveness 14
2.4 Research Question 16
References 18

3 Methods for Energy System Modeling 21
3.1 General Aspects 21
3.2 Equilibrium Models 23
3.2.1 Background 23
3.2.2 Energy System Applications 24
3.3 Mathematical Programming 26
3.3.1 Linear and Integer Programming 27
3.3.2 Nonlinear Programming 28
3.4 Solution Methods for Mathematical Programs 29
3.4.1 Analytical Approaches 29
3.4.2 Numerical Approaches 32
3.5 Risk and Uncertainty 34
3.5.1 General Aspects 35
3.5.2 Model-Endogenous Representation of Uncertainty ... 37
References 40

4 **Technological Change and the Experience Curve** **43**
4.1 Technological Change 43
4.2 Experience Curves 46
4.2.1 Basic Concept: The One-Factor Experience Curve ... 46
4.2.2 Drivers of Cost Reductions 49
4.2.3 Reasons for Differences in Empirical Learning Rates 52
4.2.4 Concepts with Several Explanatory Variables 54
4.2.5 Experience Curves as Forecasting Tool 55
4.3 Learning Rates for Power Generation Technologies 57
4.4 Endogenous Technological Change in E3 Models 62
4.4.1 Implementation Methods 63
4.4.2 Dealing with Non-convexity in Bottom-Up Models ... 66
4.5 Implications 68
References 69

5 **Optimal Investment Strategy for Competing Learning Technologies: An Analytical Approach** **75**
5.1 Objective and Approach 75
5.2 Basic Deterministic Model 79
5.2.1 Setup and Description 79
5.2.2 Solution 83
5.2.3 Investment Decision Rule 87
5.3 Deterministic Model with Capacity Endowment 88
5.3.1 Setup and Description 88
5.3.2 Solution 90
5.3.3 Investment Decision Rule 91
5.4 Dynamic Model with Uncertainty 93
5.4.1 Setup and Description 93
5.4.2 Solution 96
5.4.3 Investment Decision Rule 106
5.5 Pure versus Mixed Investment Strategies 111
5.5.1 Solution for Mixed Strategies 112
5.5.2 Global Optimality of Pure Strategies 115
5.6 Model Application: Offshore Wind and PV 124
5.7 Key Findings and Implications 127
References 128

6 **Optimal Future Deployment of Renewable Power Technologies: A System Model Approach** **131**
6.1 Objective and Approach 131
6.2 Model Description 134
6.2.1 Objective Function 137
6.2.2 Power Generation, Consumption and Trade 140
6.2.3 Energy Storage 144

6.2.4 Capacity Additions and Divestments ... 145
6.2.5 Renewable and Nuclear Resources ... 149
6.2.6 Cost Components ... 152
6.3 Critical Assumptions ... 160
6.4 Results ... 165
6.4.1 Base Case ... 166
6.4.2 Technology Substitution and the Impact of Cost Reductions ... 174
6.4.3 Value of Differentiated Renewable Potentials ... 177
6.4.4 Impact of Fluctuating Renewable Energy Carrier Availability ... 179
6.4.5 Impact of Fossil Fuel Prices and the Carbon Cap ... 180
6.4.6 Nuclear Power as Carbon-Free Substitute for Renewables ... 181
6.4.7 Influence of Interest Rate and Growth Constraint ... 184
6.5 Key Findings and Implications ... 184
References ... 187

7 Implications of Uncertainty for Renewable Power Deployment ... 191
7.1 Objective and Approach ... 191
7.2 Model Description ... 195
7.2.1 Non Learning-Related Modifications ... 197
7.2.2 Learning-Related Modifications ... 198
7.3 Results ... 202
7.3.1 Uncertain Learning Rates ... 203
7.3.2 Uncertain Carbon Bound ... 205
7.3.3 Uncertain Fossil Fuel Prices ... 207
7.3.4 Uncertain Renewable Potential ... 209
7.3.5 Multiple Uncertainties ... 211
7.3.6 Vertical Shift Omission ... 213
7.4 Key Findings and Implications ... 214
References ... 215

8 Summary and Outlook ... 217

A Renewable Power Technologies: Description, Generation Cost and Potential ... 223
A.1 Hydro Power ... 224
A.1.1 Overview ... 224
A.1.2 Generation Costs ... 226
A.1.3 Potential ... 229
A.2 Wind Power ... 230
A.2.1 Overview ... 230
A.2.2 Generation Costs ... 232

A.2.3 Potential 235
A.3 Photovoltaic Power 237
A.3.1 Overview 237
A.3.2 Generation Costs 239
A.3.3 Potential 242
A.4 Solarthermal Power 243
A.4.1 Overview 243
A.4.2 Generation Costs 244
A.4.3 Potential 247
A.5 Power from Biomass 249
A.5.1 Overview 249
A.5.2 Generation Costs 250
A.5.3 Potential 251
References 254

B Analytical Model Backup 257
B.1 Basic Model with Perfect Foresight 257
B.1.1 Extensive Model Formulation 257
B.1.2 First-Order Derivatives with Respect to Decision Variables 258
B.2 Deterministic Model with Capacity Endowment 258
B.2.1 Lagrangian Function 258
B.3 Dynamic Model with Uncertainty 259
B.3.1 Lagrangian Function 259
B.3.2 First-Order Derivatives with Respect to Decision Variables 259
B.3.3 Solutions for Even-Numbered Cases 261
B.3.4 Decision Rules for Even-Numbered Cases 263
B.3.5 First-Order Derivative with Respect to Learning Index 266
B.4 Pure versus Mixed Investment Strategies: Sample Solutions 267
B.4.1 No Change in Learning Rates 267
B.4.2 Learning Rate Increase 267
B.4.3 Learning Rate Decrease 269
B.4.4 Learning Rate Increase and Decrease 270

C Numerical Model Results 273
C.1 Deterministic Model Results 274
C.1.1 Base Case — Global Results 274
C.1.2 Base Case — Regional Results 275
C.2 Stochastic Model Results 283
C.2.1 Uncertain Learning Rates for Renewable Technologies 283
C.2.2 Uncertain Learning Rate for PV 284

List of Abbreviations

1FEC	One factor experience curve
1FLC/OFLC	One factor learning curve
2FEC	Two factor experience curve
AC	Alternating current
AEEI	Autonomous Energy Efficiency Improvement
BGR	Bundesanstalt für Geowissenschaften und Rohstoffe
BOS	Balance of System (Auxiliary components for PV generators)
CAGR	Compounded annual growth rate
CCS	Carbon Capture and Storage
CHP	Combined heat and power
CORE	Competitiveness of Renewable Electricity (Model name)
CSP	Concentrating Solar Power
DC	Direct current
DLR	Deutsches Zentrum für Luft- und Raumfahrt
DSC	Dedicated Steam Cycle
E3	Energy-Economy-Environment
EEV	Expected result of using the expected value solution
EIA	Energy Information Administration
EMF	Energy Modeling Forum
ENTSO-E	European Network of Transmission System Operators for Electricity
EPIA	European Photovoltaic Industry Association
EU	European Union
EUR	Euro
EWEA	European Wind Energy Association
EWL	Lehrstuhl für Energiewirtschaft (Chair for Energy Economics)
ff.	and the following
FLH/a	Full load hours per year
GAMS	General Algebraic Modeling System

i.e.	id est
IEA	International Energy Agency
IGCC	Integrated Gasification Combined Cycle
IIASA	International Institute for Applied Systems Analysis
IPCC	Intergovernmental Panel on Climate Change
IPS/UPS	Integrated Power System/Unified Power System of Russia
KKT	Karush-Kuhn-Tucker
LP	Linear Program(ming)
LR	Learning Rate
MINLP	Mixed Integer Nonlinear Program(ming)
MIP	Mixed Integer Program(ming)
MOX	Mixed oxide
NLP	Nonlinear Program(ming)
NREL	National Renewable Energy Laboratory
OECD	Organization for Economic Co-operation and Development
p.	page
pp.	pages
ppm	Parts per million
PR	Progress Ratio
PV	Photovoltaic
RD&D	Research, Development & Deployment
REN21	Renewable Energy Policy Network for the 21st century
RP	Recourse Problem
t	metric ton
UNFCCC	United Nations Framework Convention on Climate Change
US or USA	United States of America
USD	United States Dollar
VSS	Value of the Stochastic Solution
W	Watt (Unit for rate of energy conversion)
Wh	Watt hour (Unit for energy)

Chapter 1
Introduction

Global wind power capacity has increased more than tenfold over the last decade. Photovoltaic capacity additions have recently enjoyed annual growth rates in excess of 50%. Europe plans to satisfy a large share of its demand from solar thermal power generated in the North African desert. Such news may convey the message that a radical transformation of the global power sector away from conventional technologies towards renewable ones is taking place.

Fact is that investments into conventional technologies still outstrip investments into renewables by far. Renewable technologies, apart from Hydro, currently account for only three percent of global power production. The bulk of renewable capacity additions is concentrated in countries with sufficiently generous subsidy schemes, compensating investors for the comparatively high costs associated with these technologies.[1]

Yet, fundamental changes in the drivers determining the relative competitiveness among generation technologies are expected. The energy crises of the 1970s have already raised awareness that fossil and nuclear fuels are scarce and that their depletion goes along with fuel price increases in the long-term. More recently, mitigation of global warming has moved into focus, which requires greenhouse gas emission reductions. A large share of these emissions stem from power generation based on the combustion of fossil fuels. Finally, new renewable power technologies have experienced considerable cost reductions over the last two decades. As they can still be considered comparatively immature, further reductions are expected. When, if at all, those fundamental developments will lead to cost-competitive power production from renewables and which renewable technologies will provide this power, is, however, highly uncertain.

In this book the implications of those fundamental but highly uncertain developments on the future competitiveness of renewable power generation are

[1] Depending on the circumstances, hydro and a small number of biomass and onshore wind projects may already be cost-competitive today.

F. Wagner, *Renewables in Future Power Systems*,
Green Energy and Technology,
DOI: 10.1007/978-3-319-05780-4_1,

examined. Apart from its methodological contributions, this work therefore also aspires to produce practically relevant insights regarding the long-term competitiveness of renewable power generation. From a methodological perspective, the main contributions relate to the field of endogenous learning and uncertainty in optimizing energy system models. The primary objective here is closing the gap between the strand of literature covering renewable potential analyses on the one side and energy system modeling with endogenous technological change on the other side. The numerical model developed for this purpose provides a framework for technology assessment, which explicitly takes renewable-specific attributes into account. Also, particular model properties that require consideration when formulating models with endogenized experience curves and time-induced changes in the rate of cost reduction are first examined in an analytical setting and subsequently applied to a large-scale numerical model.

The book is divided into two main parts. In chapters 2 through 4 the point of departure as well as methods and concepts applied for the subsequent analyses are described. Chapter 2 covers the current power generation landscape as well as specific attributes of renewable generation technologies. From these discussions the research question is developed. Chapters 3 and 4 review concepts and methods from the fields of energy system modeling and technological change, respectively, which are then applied in the following chapters. The second part, made up of chapters 5 through 7, contains the main contributions of this book. In chapter 5 a small-scale power system model featuring endogenous cost reductions and uncertainty is formulated and solved analytically. Its main objective is methodological in nature but the derived solutions may also prove useful for rough-cut technology assessment carried out by industry players. In chapter 6 a large-scale power sector model is developed and numerically solved to assess the long-term prospects of renewable power technologies based on their fundamentals. Finally, uncertainty is introduced to this large-scale model in chapter 7, drawing upon the methodological findings from chapter 5.

Chapter 2
Renewables in Power Generation: Status Quo

Starting point for analyses on the future role of renewable power generation is the situation today. It is the objective of this chapter to describe this starting point as well as to provide additional background information required for the development of the research question. It begins with a description of today's power generation landscape from a global perspective. The focus hereby lies on the current role played by renewable technologies. In the following section, specific properties of renewable technologies, setting them apart from conventional ones, are covered. This includes their current cost situation, the notion of the *renewable potential* and the challenges associated with fluctuating energy carrier availability. Afterwards important trends impacting the future competitiveness of renewable technologies are introduced. Finally, drawing upon the discussions from the preceding sections, the research question is developed and the general approach to address it is outlined.

2.1 Global Power Generation Landscape

Fossil fuel based technologies dominate today's power generation landscape. They currently account for almost two thirds of worldwide generation (IEA, 2010a). The dominant fossil technology is, with a share of roughly 40% coal, followed by natural gas, which has a share of approximately 20%. Oil, although on the decline since the oil crises of the 1970s, makes up another 5%. The non-fossil third of total power generation is split almost equally between nuclear fission (14%) and renewables (18%).[1]

Hydropower represents the lion's share of the 18% coming from renewables, as illustrated in Fig. 2.1. Although considerable media coverage of non-hydro, often also referred to as "new", renewable technologies over the last years may have created the impression that those technologies make a considerable contribution to global production, their actual share stands at a mere 2–3%. Wind and biomass are the two largest contributors from this group, each

[1] Figures refer to global power production in 2008.

F. Wagner, *Renewables in Future Power Systems*,
Green Energy and Technology,
DOI: 10.1007/978-3-319-05780-4_2,

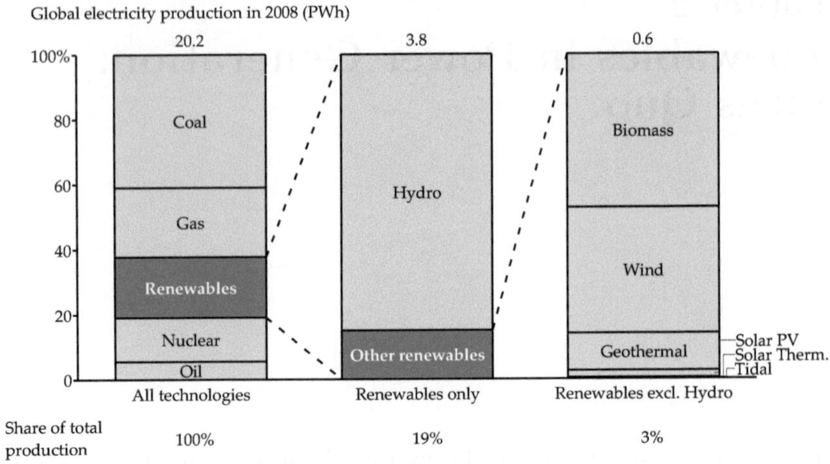

Fig. 2.1 Global electricity production 2008 by energy carrier[3]
Source: IEA (2010b), own representation

accounting for approximately 1% of worldwide production. Solar technologies taken together have contributed only about 0.1% of global electricity in 2008.

From a regional perspective the situation looks similar with renewable shares generally around the 20% mark. The four world regions with highest electricity production, Asia, North America, Europe and the former Soviet Union states[2] are characterized by renewable shares between 14% and 21% (see Fig. 2.2). Only two regions with lower power production clearly fall outside this range: Latin America with a 66% renewables share and the Middle East with virtually no renewable generation. Comparatively favorable conditions for hydro power are a key driver behind the high renewable share in Latin America, whereas relatively low costs for natural gas and oil constitute an important factor for the very low renewable penetration in the Middle East.

Both global and regional figures illustrate that non-hydro renewable technologies currently play a minor role for power generation. If there is indeed a power sector transformation towards high renewable shares, this process still stands at its very beginning. As the potential for additional hydro capacity at competitive costs is limited, a manifold of the non-hydro renewable capacity in place today would be required to replace the majority of fossil generation.

[2] Following IEA's classification, this region is further referred to Eastern Europe/ Eurasia. It also includes some Eastern European states that did not belong to the former Soviet Union.

[3] Hydro includes pumped storage; biomass includes waste.

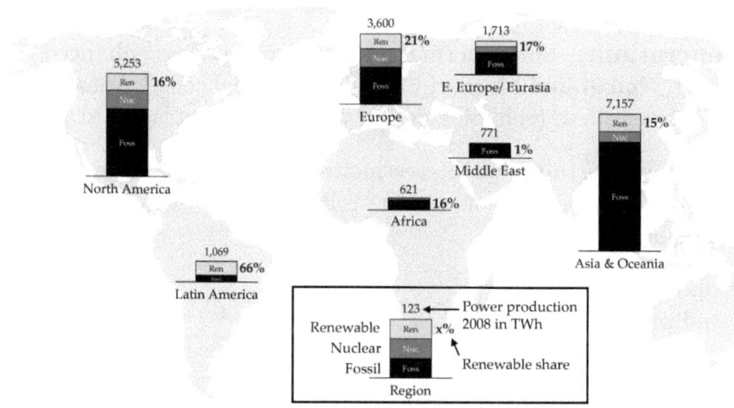

Fig. 2.2 Regional electricity production 2008
Source: IEA (2010b), own representation

2.2 Properties of Renewable Technologies

The European Union aspires to increase its share of renewable energy from less than 10% today to 20% of total primary energy usage over a decade (cf. European Commission, 2011), a large portion of this increase coming from the power sector.[4] But what are the reasons behind the introduction of this target and the related policy measures? On the one hand, renewable technologies possess certain properties, which are generally viewed as positive and not inherent to conventional technologies. On the other hand, other renewable properties prevent the market from autonomously moving towards higher renewable shares, forcing governments to take action.

The two main benefits from renewable power arise due to energy carriers being *renewable* and production generally being more *sustainable* than from conventional technologies. Primary energy carriers are referred to as *renewable*, if they are virtually inexhaustible from a human perspective (cf. Kaltschmitt et al., 2006, p. 4). This is clearly not the case for fossil fuels or Uranium, where resources are depleted faster than they can be replaced by natural processes. Renewable power generation technologies available today can be grouped based on technological maturity (cf. IEA, 2007):[5]

First generation: Hydro, biomass and geothermal power. First implementations date back as far as a century ago.

Second generation: Wind, modern biomass and solar photovoltaics. Demonstration started after the oil crises, with commercial deployment

[4] High increases of renewable shares in other energy sectors, such as transportation, may be more costly or simply infeasible in the mid-term.

[5] This list does not aspire to be exhaustive.

commencing approximately two decades ago and accelerating over the last decade.

Third generation: Solar thermal power,[6] ocean power, enhanced geothermal power,[7] integrated bioenergy systems and fuel cells.[8] Apart from solar thermal power these technologies are still in demonstration stage.

Although it would certainly be interesting to consider all of these technologies, the analyses in this book focus on the following five technology groups for practical reasons.[9]

i. Hydropower
ii. On- and offshore wind power
iii. Photovoltaic solar power
iv. Solarthermal power
v. Power from biomass

For background information on these technologies the interested reader is referred to Appendix A of this document.

Renewable energy carrier availability has been a key incentive for the first wave of government spending on renewable RD&D after the oil price shocks of the 1970s. When realizing that cost-competitive supply of fossil fuels may come to an end at some point in time, renewable technologies represented an appealing alternative. In today's efforts to increase renewable shares, the generally higher level of *sustainability* associated with renewable over conventional power generation also plays an important role. Sustainable power

[6] Often referred to as concentrating solar power (CSP). Strictly speaking, the category solar thermal also includes other technologies, e.g. solar updraft plants.

[7] Sometimes also referred to as Hot Dry Rock.

[8] IEA (2007) does not list fuel cells.

[9] Geothermal power, ocean power and fuel cells are not included in the analyses of this book. Traditional geothermal power installations have been stable at approximately 200 MW per year over the last three decades (IEA, 2008) and a significant step-up in capacity additions therefore appears unlikely. Technical potential is also lower than for any other of the renewable technologies considered (cf. Krewitt et al., 2009). The potential for enhanced geothermal power systems may be much higher than the one for traditional geothermal technology, yet costs of the currently existing demonstration plants put these systems far outside competitive ranges. The latter argument also applies to ocean power systems, for which considerable technical potential can be expected. Due to the early developmental stage of these technologies no empirical data on potential future cost reduction rates is available, rendering quantitative investigations not meaningful. Fuel cells are excluded for much the same reasons. An additional difficulty for quantitative analysis here emerges from the focus of this book on the power sector, whereas prospects for fuel cells can only be meaningfully assessed within a context covering the entire energy sector. It needs to be emphasized that omission of these technologies occurs for practical reasons and should not imply that these technologies will not play a role in future power systems.

generation can be defined as one that "meets the needs of the present without compromising the ability of future generations to meet their own needs" (Brundtland, 1987). Yet, sustainability is not necessarily always given for renewable technology deployment. The construction of large hydro stations, for example, may negatively impact entire ecosystems. Whether such a project meets the sustainability criterion needs to be assessed on a case-by-case basis (cf., for example, Frey and Linke, 2002). With respect to greenhouse gas emissions or other pollutants, however, renewable technologies outperform fossil ones by far.

Despite those benefits, renewable capacity additions are small when compared to conventional ones (IEA, 2010b, p. 226). Higher power generation costs for renewable than for conventional technologies render investment into renewables unattractive for profit maximizing industry players in absence of sufficiently attractive governmental support schemes. This fact is illustrated in the following subsection (sec. 2.2.1), where generation costs of renewable technologies are compared to conventional ones. In contrast to conventional plants, the output of renewable plants also relies on the availability of the primary energy carrier at the plant's site and energy carrier availability differs vastly across locations. The *renewable potential* is a concept that can be used to describe expected energy carrier availability for a particular region. It is introduced in subsection 2.2.2. Finally, even given favorable expected availability of a renewable energy carrier at a site, its supply is subject to fluctuations, further increasing opportunity costs for the respective technology and covered in sec. 2.2.3.

2.2.1 Generation Costs

A popular concept to compare economic viability among power generation technologies is the *levelized generation cost* approach. Its basic principle is allocation of capital and operating costs to a unit of production. The main advantage of this approach is the low degree of required efforts when compared to more sophisticated methods. Obviously such an approach is characterized by numerous drawbacks, largely related to the fact that the approach does not take a systemic perspective. Although electricity may be a homogeneous good from a technical point of view, it is not from an economic one. Depending on the time of availability, the value of power differs — peak-load is of higher value than base-load. Consequently comparing costs of a peak-load technology, such as gas turbines, to the one of a baseload technology, such as lignite, may be grossly misleading.[10] Moreover, the approach generally neglects externalities. It is certainly possible to factor quantifiable

[10] Arguably a comparison on the basis of levelized profit per unit of output would remedy this particular problem. However, obtaining representative electricity price data for different electricity load segments is challenging, e.g. due to electricity price fluctuations or differences in regional technology mixes.

externalities into levelized generation costs, such as CO_2 costs in countries with a corresponding trading scheme. Yet, other costs can hardly be considered under the approach, for example, the costs associated with providing backup capacity or grid-strengthening required in cases of large penetrations of technologies with fluctuating output. Despite these shortfalls, the levelized generation cost approach enables illustration of the order-of-magnitude in cost differences among technologies.

To illustrate these cost differences between renewable and conventional generation, levelized costs are shown in Fig. 2.3.[11] New nuclear, hard coal and gas turbine technologies are used as reference for base-, mid-, and peak-load power generation costs.[12] As power output from renewable technologies varies widely across locations, sample sites have been used. These include sites with particularly favorable conditions but also sites in countries, where large portions of current, subsidy-induced deployment of renewable technologies takes place. For details on the renewable cost calculation the interested reader is referred to Appendix A, for conventional technology assumptions to Table 6.4.

The most striking observation from Fig. 2.3 is the large cost gap between solar technologies and all other technologies. Even at favorable locations, generation costs are more than twice as high as the ones for the conventional reference plants. For the case of Germany, where most of current deployment takes place, power produced using PV is roughly four times as expensive as conventional one, even when emission costs are included.[13] Costs for biomass and wind are considerable closer to the costs of conventional technologies. Yet, even in case feedstock for biomass plants is available at very favorable prices, generation costs of the appropriate conventional reference technology, hard coal, cannot be matched.[14] The situation is slightly different for wind power. Although in the vast majority of cases costs for wind are also higher than costs for conventional base load power, given an excellent wind regime

[11] Generation costs include overnight capital, construction financing, operating and CO_2 costs. The latter are assumed at 25 $/t $_{CO_2}$. More detailed information on calculation method and underlying assumptions can be found on pp. 223ff in Appendix A.

[12] These costs lie above current wholesale prices, presumably since some of current power generation takes place at written-off facilities and a certain degree of overcapacity exists. Moreover, carbon emissions are free of charge in most countries.

[13] As a large share of PV implementation takes place in a distributed manner, it could be argued that, in case of cost-competitive local storage opportunities, PV generation costs should be compared to conventional retail instead of wholesale costs.

[14] A stylized biomass plant with solid fuel combustion is used for reference here. In reality, numerous biomass energy conversion technologies and paths exist, not only limited to power generation (see also Appendix A.5).

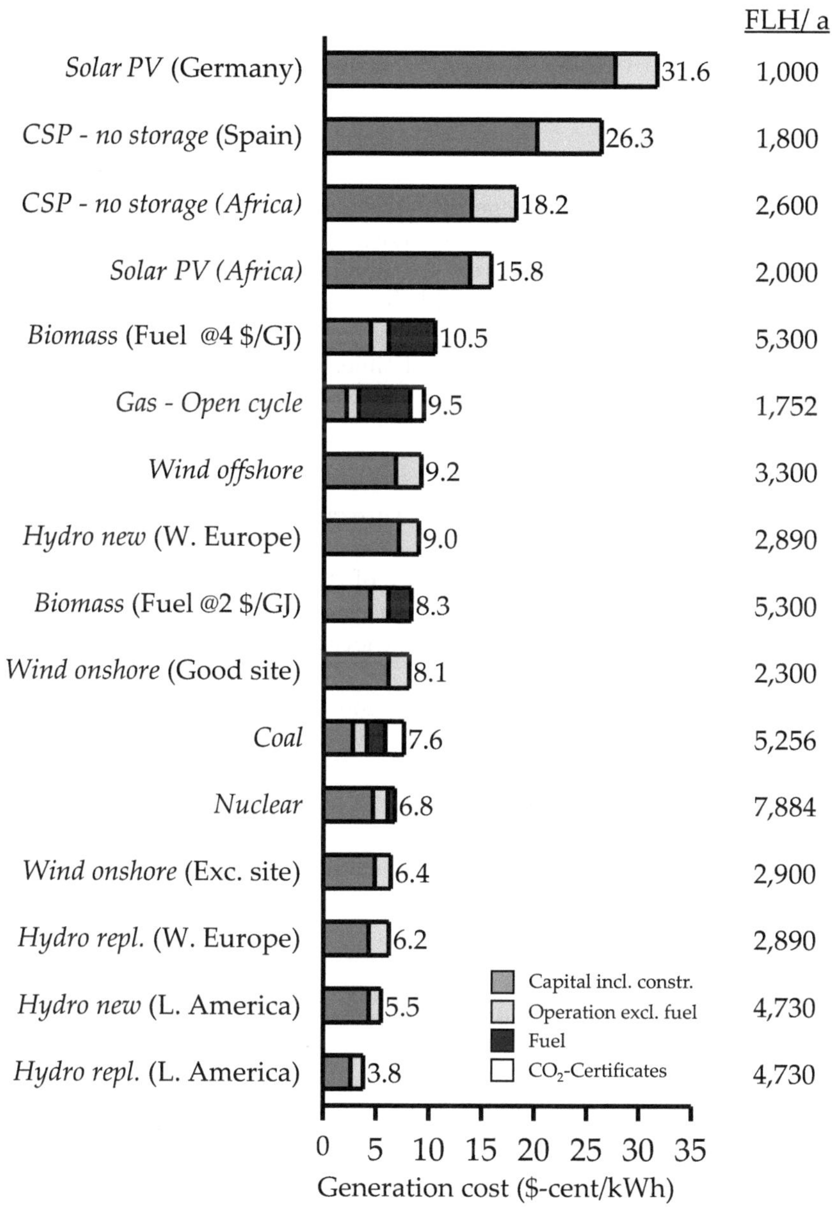

Fig. 2.3 Generation costs for selected technologies and locations

at the site of implementation wind power may reach cost-competitiveness.[15] Unsurprisingly hydro power is characterized by similar or even lower costs than conventional technologies.[16] The figures illustrate the reason why current non-hydro renewable capacity additions are limited to regions, where sufficiently attractive support schemes are in place.

2.2.2 Potential

From the fact that renewable energy carriers, such as wind or solar radiation, are both available free of charge and not depleted, one may be tempted to infer that these renewable resources are unlimited. A concept illustrating that the opposite is true, i.e. that renewable resources are limited, is the *renewable potential*. It describes the total amount of energy available from a renewable energy carrier over a certain period of time, generally a year, for a specific geographic area. The five categories of renewable potential introduced by van Wijk and Coelingh (1993) are:

Theoretical potential covers the theoretically available energy from a particular primary energy carrier. An example is the total wind energy available over a certain period.

Geographical potential is the share of theoretical potential available at locations suitable for installation of the appropriate conversion plants, for example, the wind energy available at locations, where wind turbines can be erected.

Technical potential represents the share of geographical potential that can be converted into electricity using an appropriate conversion technology. The Betz factor (Betz, 1926), for example, puts an upper limit on the share of total kinetic wind energy that a horizontal-axis wind turbine can extract from the wind and subsequently convert into electricity.

Economic potential is the share of technical potential exploitable in an economically viable manner, e.g. energy from sites with average wind speeds above a certain threshold.

Implementation potential represents the actually achievable contribution of the technology to the energy mix. It can be lower, e.g. due to an alternative use of the location, or higher, e.g. in presence of governmental subsidies, than the economic potential.

The categories with greatest relevance for this book are the technical, the economic and, to a lesser degree, the implementation potential. The technical

[15] Furthermore, the costs from output fluctuation must be small. This is given in a system with low penetration of technologies with uncontrollably fluctuating output.

[16] The value of hydro electricity depends on the type of hydro station. Run-of-river stations produce baseload power, whereas storage stations can be dispatched as needed.

potential obviously depends on the current state of technology. Increases in the efficiency of photovoltaic modules, for example, will lead to a corresponding increase in the technical potential of PV. Global technical potentials vary considerably across renewable technologies. Whereas technical potential for hydro and offshore wind power have been estimated at 10–20 PWh p.a., the technical potentials for solar technologies exceed these values by a factor of 10–100, depending on the source. The reference potentials for onshore wind and biomass power used in this book lie, with approximately 100 and 50 PWh p.a., respectively, above the ones for hydro or offshore wind but far below the potentials for solar technologies. Current global power production, for comparison, is roughly 20 PWh p.a.

Unfortunately it is difficult to make inferences on the economic potential from the figures above. Locations included in the technical potential category are characterized by large differences in energy carrier availability. In fact, a large share of technical onshore wind potential comes from sites with comparatively low average wind speeds (see p. 236). Presuming that specific capital and operating costs are similar for renewable plants across different locations, locations with higher energy carrier availability are characterized by lower generation costs. A breakdown of technical potentials based on energy carrier availability is thus required for meaningful assessment of economic potential. The economic potential for a technology is then the amount of power that can be generated at locations with energy carrier availability above the threshold corresponding to break-even costs. This implies that there already exists an, albeit small, positive economic potential for onshore wind power, whereas economic potential for grid-connected solar technologies are virtually zero.[17] An additional challenge in the derivation of economic potentials are large differences in technical potentials across regions. Whereas Northern Africa and Oceania, for example, are characterized by large amounts of high-productivity sites for solar plants,[18] locations in Central Europe are characterized by significantly lower productivity. For the analyses in this book, technical potentials differentiated by productivity and region are used to produce meaningful results. Once more, the interested reader is referred to appendix A for a more detailed discussion of the potentials for the covered renewable technologies.

2.2.3 Intermittency

Availability of some renewable energy carriers, notably wind and solar, is characterized by fluctuations over time. These naturally also lead to fluctuations in the power output of the corresponding conversion technologies,

[17] The implementation potential for solar technologies, however, is positive owed to the presence of governmental support schemes.

[18] Productivity in this context is used to express the amount of an energy carrier available at the site.

meaning that output can neither be fully predicted ex-ante nor controlled by the power system operator. This clearly undesirable property is often referred to as *intermittency* in literature. Two main types of fluctuation can be distinguished: Cyclical and stochastic ones. *Cyclical fluctuations* mainly relate to the diurnal or annual period. Fluctuations across the diurnal cycle are particularly relevant for solar technologies — at nighttime solar radiation drops to zero. Across the annual cycle, season-specific conditions impact energy carrier availability. Wind speeds, for example, are on average higher in winter than summer, whereas the opposite is true for solar radiation. In contrast to cyclical fluctuations, *stochastic fluctuations* cannot be predicted in advance based on historical data. Meteorological conditions are the source of most stochastic energy carrier supply fluctuations. They are particularly pronounced for wind and concentrating solar technologies but also relevant for PV.[19] In a system of geographically dispersed plants with intermittent output, fluctuations at single locations are, to a certain degree, smoothed as a result of portfolio effects.[20] Consequently there is generally a value associated with geographic dispersion of power generators characterized by intermittent output (cf., for example, Roques et al. 2010).[21]

An important consequence of intermittency is that power system operators cannot rely on the availability of the average amount of electricity across a particular time period, also referred to as *capacity factor*,[22] in times of peak demand. To address this limitation the notion of the *capacity credit* is often employed. "It [...] represents the contribution of [a generating] unit to the generation adequacy of a power system" (Amelin, 2009).[23] Beyond a certain intermittent technology penetration in a power system, further increase of the intermittent generation share will also require addition of backup capacity to achieve the desired system adequacy. This backup can be provided by dispatchable power generation as well as energy storage technologies. Another issue associated with intermittency is overproduction. In times of comparatively low demand and high output, a situation may arise, where more power is available than demanded, potentially endangering grid stability.

[19] Stochastic fluctuations are more pronounced for CSP than PV, since CSP relies on direct solar radiation only whereas PV technology is capable of converting indirect as well as direct radiation into electricity.

[20] Assuming that energy carrier availabilities at the different sites are not perfectly correlated.

[21] Portfolio theory has been introduced by Markowitz (1952) in the field of finance but recently the methodology has also been applied to problems of energy economics.

[22] "The ratio of the electrical energy produced by a generating unit for the period of time considered to the electrical energy that could have been produced at continuous full power operation during the same period." (US Energy Information Administration, 2012).

[23] For a detailed discussion of capacity credits and appropriate calculation methods the interested reader is referred to Amelin (2009).

Several studies aim to identify maximum possible solar and wind system penetration levels beyond which storage addition becomes necessary or excess production occurs. From their research, Stodola and Modi (2009) identify a threshold penetration of 8% for PV.[24] Similarly, Denholm and Margolis (2007) find that excess power is produced in a system with a PV penetration of 10–20% and Hoogwijk et al. (2007) conclude that excess production will occur for wind penetrations in excess of approximately 20%. Apart from these issues relating to generation capacity, integration of intermittent technologies into a system at large scale may also require strengthening of the transmission grid and lead to additional efforts in power system operations.

From the discussion above it becomes clear that integration of intermittent technologies into power systems entails costs beyond the ones considered in section 2.2.1. These additional costs can be explicit, such as the costs for adding storage or grid capacity, or implicit, such as the opportunity costs from having to "discard" energy available in excess of demand. Quantification of these "costs of intermittency" has been attempted in several pieces of research. Assessment methodologies and assumptions regarding the penetration of the intermittent technology, however, differ across those studies.[25] Unsurprisingly results also differ widely — wind integration costs from different studies span a broad range from 1 to 20 \$/MWh.[26] Swider and Weber (2007) identify costs of 5–13 \$/MWh[27] for the German system with wind penetrations of up to 20%, DeCarolis and Keith (2006) find values in the 10–20 \$/MWh range for the US system with a comparatively high penetration of 50%. Rosen (2008) arrives at 5 \$/MWh[28] for the EU-15 and penetration levels corresponding to the EU 2020 renewables target. The values derived by Holttinen et al. (2009) are, with 1–5 \$/MWh,[29] at the lower end of the overall range. Despite their difference in magnitude, these figures highlight that costs apart from straightforward capital and operating ones require consideration when assessing the competitiveness of renewable power generation technologies. Renewable generation costs presented in section 2.2.1 are therefore likely to be too optimistic.

[24] Their analysis is based on a selected number of US regions.

[25] Costs of intermittency depend on the penetration of the intermittent technology in the system. Ceteris paribus, the marginal value of installing intermittent capacity will decrease with increasing penetration levels (Lamont, 2008).

[26] Not including costs for grid reinforcement or increased operational efforts.

[27] 4–10 €/MWh.

[28] 4 €/MWh.

[29] 1–4 €/MWh increase in system operating costs from "variability and uncertainty".

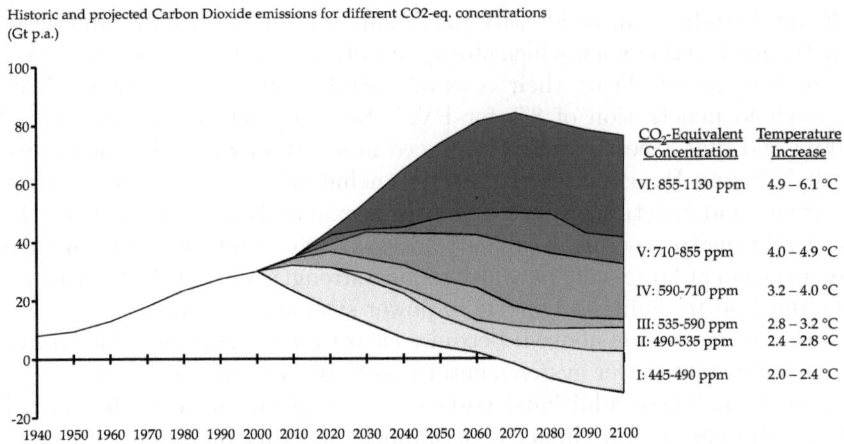

Fig. 2.4 Greenhouse gas emissions and global temperature increase
Source: Based on IPCC (2007a, p. 21)

2.3 Drivers of Future Technology Competitiveness

The previous sections have illustrated that costs for renewables are currently higher than for conventional power. Yet, several developments over the coming decades are expected to improve the relative cost position of renewable technologies. *Climate change*, higher *cost reductions* for renewable than conventional technologies and *increases in fossil and fissile fuel prices* count among those developments.

Climate change, more precisely global warming, and the need to react to it may lead to profound alterations of the energy sector over time. At the heart of global warming stands the *greenhouse effect*. Greenhouse gases[30] in the atmosphere absorb thermal infrared radiation. Some of this absorbed radiation is then re-emitted back towards the surface, leading to the greenhouse effect. The resulting change in the net irradiance is called radiative forcing, meaning that positive radiative forcing goes in line with an increase in atmospheric temperature. In the context of anthropogenic climate change, radiative forcing is generally calculated against a baseline value from the year 1750, i.e. before the industrial revolution (IPCC, 2007b, pp. 946, 951). Higher emission than removal rates of greenhouse gases will thus lead to a temperature increase. There is general agreement that an atmospheric temperature increase is likely to induce undesirable effects, such as a rise in sea level. Yet damages are expected to vary across world regions (cf. Baker, 2005). It is therefore desirable to limit the atmospheric

[30] Gases that "absorb and emit radiation at specific wavelengths within the spectrum of thermal infrared radiation." The gases covered under the Kyoto protocol are CO_2, N_2O, CH_4, SF_6 as well as the groups of hydrofluorocarbons and perfluorocarbons (IPCC, 2007b, p. 947).

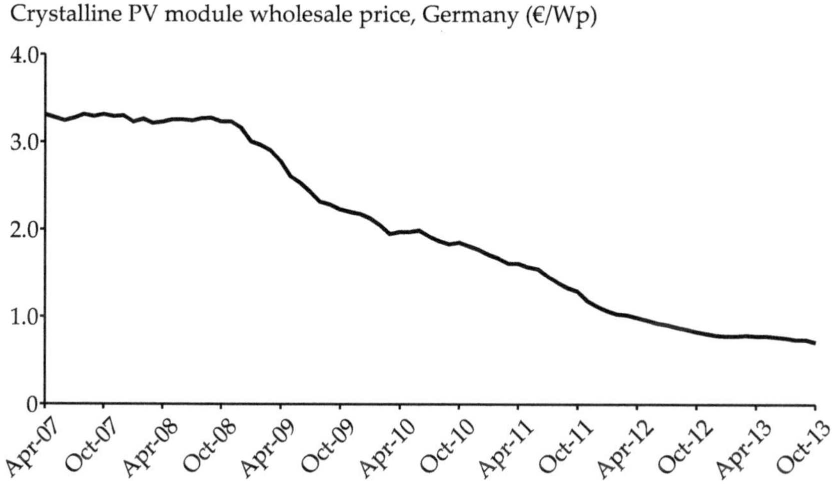

Fig. 2.5 Cost reductions for PV modules
Sources: pvXchange (2010), pvXchange (2013)

temperature increase — the international community has agreed on a target of 2°C in comparison to pre-industrial times (cf. UNFCCC, 2010). A reasonably high probability to meet this goal is achieved if radiative forcing is kept below 3 W/m^2, corresponding to atmospheric CO_2-equivalent concentrations around 450ppm (cf., for example, IPCC 2007c, p. 15). This target implies a massive reduction of emissions compared to today's level. In fact, greenhouse gas emissions will need to peak before the year 2020 (den Elzen et al., 2007) and drop to virtually zero towards the end of the century (see Fig. 2.4).[31] The carbon budget proposed by Meinshausen et al. (2009) illustrates how challenging this required emission reduction is: Cumulative CO_2 emissions 2000–50 of 1,000 Gt are associated with a 25% probability of exceeding the 2°C target — the cumulative emissions 2000–06, however, already amounted to approximately 234 Gt. The situation is further aggravated by the fact that global energy demand is growing. As the largest share of emission reductions must come from the energy sector, this implies that costs for power from CO_2-emitting, fossil-fuel based technologies are liable to increase, given that the costs of emission mitigation or abatement are allocated to the emitter. This in turn, would increase the relative competitiveness of renewable and nuclear power technologies over fossil ones.

[31] Different emission pathways are possible to reach a specific atmospheric concentration in a particular year (Manne and Richels, 2004). Since the publication of the fourth IPCC report in 2007, several studies have derived emission pathways and envelopes for low radiative forcing targets, presented, for example, in den Elzen et al. (2007) or Clarke et al. (2009). The latter publication provides an overview of ten studies.

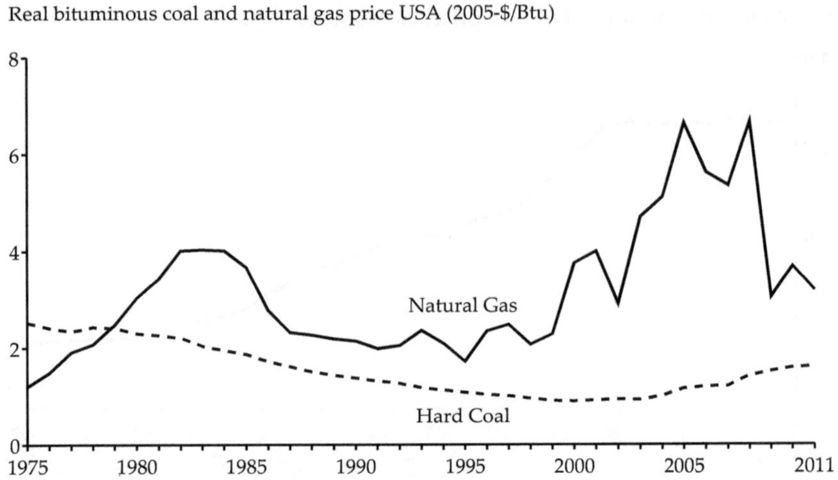

Fig. 2.6 US coal and natural gas prices 1975–2011
Source: EIA (2012)

The impact of *cost reductions* on renewable competitiveness is more obvious than the indirect one from climate change. Apart from hydro, renewable technologies are still comparatively immature and considerable cost reduction potentials remain to be exploited. Cost reductions largely relate to capital costs but may also be possible for operating cost. Figure 2.5 illustrates historical cost reductions for PV modules in Germany.[32] Cost reductions within the context of technological change are covered in more detail in chapter 4.

Coal and natural gas are the two *fossil primary energy carriers* with highest relevance for power generation. After fuel price decreases following the peaks of the 70s and early 80s, prices for both energy carriers have again risen continuously over the last decade. Figure 2.6 illustrates this development for US prices. Any statement on the short- to mid-term future price development for these fuels would be highly speculative but price increases can reasonably be expected over the course of the 21st century. For a review of concepts relating to fossil fuel price formation and a perspective on future fuel price developments based on fundamental drivers the interested reader is referred to Walther (2009).

2.4 Research Question

Different stakeholders strive to increase transparency on the future economic prospects of renewable power technologies. Policy makers aspire to allocate subsidies to those technologies, which can reach cost-competitiveness at

[32] Please note that the figures refer to modules only. Total system costs are more than twice the amount shown (pvXchange, 2013).

lowest costs to society. Potential investors into firms, whose business models depend upon renewable energy technologies, such as renewable plant manufacturers or maintenance service providers, must avoid betting on "the wrong horse." Even if a particular technology currently enjoys sufficiently attractive governmental support to motivate profit maximizing agents to invest, subsidy schemes may be revised upon realization that allocation of funds to competing technologies have a higher value to society.

Besides methodological contributions, it is the objective of this work to generate a perspective on the future economic prospects of renewable power generation technologies based on their fundamental drivers of competitiveness. The previous sections have illustrated that this undertaking is not trivial. On the one hand, a systemic point of view must be taken in order to capture the multitude of interactions among different renewable and conventional technologies within the power sector, e.g. their relative merits in presence of non-uniform demand. On the other hand, specific properties of renewables, such as available renewable potential and intermittency, must be considered as well as implications from cost reductions, CO_2 emission constraints or non-renewable fuel price developments. A particular challenge associated with the latter three factors is that their magnitude over the coming decades is highly uncertain. The analysis must therefore explicitly take this uncertainty into account. The research question is then formulated as:

> *What would be the cost-optimal investment strategy into renewable power generation technologies over the 21st century on a global scale taking into account renewable-specific properties, technological interdependencies within power systems and uncertainty in the key factors impacting future technology competitiveness?* [33]

A normative over a descriptive perspective is deliberately taken across the ensuing analyses to emphasize the importance of fundamental cost drivers. This way, governmental support for renewables can be excluded as relevant input to the analysis. In fact, under this general approach government support is rather a result of the analyses, playing a key role to enable implementation of the cost-optimal strategy given potential market failures in the context of increasing returns to scale and initially uncompetitive technologies.[34]

A global power system model is used to address the research question. The required methodological background is presented in the following two chapters. Chapter 3 introduces concepts of economic modeling, with a particular focus on energy system models and mathematical programming techniques.

[33] In the context of renewable potentials (see 2.2.2) the "economic potential of renewable power generation technologies" would be an equivalent expression for the "cost-optimal investment strategy into renewable power generation technologies" here.

[34] Increasing returns to scale are brought about by the potential to reduce cost by investing into the respective technologies. This topic will be further discussed in chapter 4.

Chapter 4 then covers the field of technological change. Here the focus lies on the experience curve concept and its consideration within energy system models. A small-scale power system model featuring endogenized experience curves and uncertain learning rates is formulated and analytically solved in chapter 5. Its main contribution is the examination of specific methodological properties in presence of experience curves, subject to uncertain development of learning rates. Moreover, the analytically derived model solution can be used as tool for quick competitiveness assessment by industry participants without having to formulate and solve a mathematical program. In chapter 6 a global, regionalized power sector model is developed to address the research question, yet still with a deterministic setup. The model's key objective is, on the one hand, to bridge the gap between renewable potential literature and, on the other hand, research based on integrated energy system modeling incorporating endogenous technological change. Finally, uncertainty is introduced to this model in chapter 7, drawing upon the methodological findings from the analytical model in chapter 5. Using the model, the impact of uncertain learning rates, a CO_2 cap, fossil fuel prices and renewable potential on the optimal investment strategy into renewable technologies are evaluated.

References

Amelin, M.: Comparison of Capacity Credit Calculation Methods for Conventional Power Plants and Wind Power. IEEE Transactions on Power Systems 24(2), 685–691 (2009)

Baker, E.: Uncertainty and learning in a strategic environment: global climate change. Resource and Energy Economics 27(1), 19–40 (2005)

Betz, A.: Windenergie und ihre Ausnutzung durch Windmühlen (Reprint 1994 ed.). Ökobuch Verlag, Staufen (1926)

Brundtland, G.H.: Our common future. Oxford Univ. Press, Oxford (1987)

Clarke, L., Edmonds, J., Krey, V., Richels, R., Rose, S., Tavoni, M.: International climate policy architectures: Overview of the EMF 22 International Scenarios. Energy Economics 31(suppl. 2), S64–S81 (2009)

DeCarolis, J.F., Keith, D.W.: The economics of large-scale wind power in a carbon constrained world. Energy Policy 34(4), 395–410 (2006)

den Elzen, M.G.J., Meinshausen, M., van Vuuren, D.P.: Multi-gas emission envelopes to meet greenhouse gas concentration targets: Costs versus certainty of limiting temperature increase. Global Environmental Change 17(2), 260–280 (2007)

Denholm, P., Margolis, R.M.: Evaluating the limits of solar photovoltaics (PV) in traditional electric power systems. Energy Policy 35(5), 2852–2861 (2007)

EIA. Annual Energy Review 2011, Washington D.C (2012)

European Commission. Citizens' summary: EU climate and energy package (2011)

Frey, G.W., Linke, D.M.: Hydropower as a renewable and sustainable energy resource meeting global energy challenges in a reasonable way. Energy Policy 30(14), 1261–1265 (2002)

Holttinen, H., Meibom, P., Orths, A., van Hulle, F., Lange, B., O'Malley, M., Pierik, J., Ummels, B., Tande, J.O., Estanqueiro, A., Matos, M., Gomez, E., Söder, L., Strbac, G., Shakoor, A., Ricardo, J., Smith, J.C., Milligan, M., Ela, E.: Design and operation of power systems with large amounts of wind power: IEA WIND Task 25. Final report, Phase one 2006–08. VTT, Espoo (2009)

Hoogwijk, M., van Vuuren, D.P., de Vries, B., Turkenburg, W.: Exploring the impact on cost and electricity production of high penetration levels of intermittent electricity in OECD Europe and the USA, results for wind energy. Energy 32(8), 1381–1402 (2007)

IEA. Renewables in Global Energy Supply: An IEA Fact Sheet, Paris (2007)

IEA. Energy Technology Perspectives: Scenarios & Strategies to 2050. IEA Publications, Paris (2008)

IEA. Electricity/Heat in World in 2007 (2010a)

IEA. World Energy Outlook 2010. IEA Publications, Paris (2010b)

IPCC. Climate Change 2007: Synthesis Report: Summary for Policymakers, Valencia (2007a)

IPCC. Contribution of Working Group I to the Fourth Assessment Report of the Intergovernmental Panel on Climate Change. Cambridge Univ. Press, Cambridge & New York (2007b)

IPCC. Contribution of Working Group III to the Fourth Assessment Report of the Intergovernmental Panel on Climate Change. Cambridge Univ. Press, Cambridge & New York (2007c)

Kaltschmitt, M., Streicher, W., Wiese, A. (eds.): Erneuerbare Energien: Systemtechnik, Wirtschaftlichkeit, Umweltaspekte. Springer, Heidelberg (2006)

Krewitt, W., Nienhaus, K., Kleßmann, C., Capone, C., Stricker, E., Graus, W., Hoogwijk, M., Supersberger, N., Winterfeld, U., Samadi, S.: Role and Potential of Renewable Energy and Energy Efficiency for Global Energy Supply. Dessau-Roßlau (2009)

Lamont, A.D.: Assessing the long-term system value of intermittent electric generation technologies. Energy Economics 30(3), 1208–1231 (2008)

Manne, A.S., Richels, R.: The impact of learning-by-doing on the timing and costs of CO2 abatement. Energy Economics 26(4), 603–619 (2004)

Markowitz, H.: Portfolio Selection. Journal of Finance 7(1), 77–91 (1952)

Meinshausen, M., Meinshausen, N., Hare, W., Raper, S.C.B., Frieler, K., Knutti, R., Frame, D.J., Allen, M.R.: Greenhouse-gas emission targets for limiting global warming to 2C. Nature 458(7242), 1158–1162 (2009)

pvXchange, PV price index (2010)

pvXchange. PV price index (2013)

Roques, F., Hiroux, C., Saguan, M.: Optimal wind power deployment in Europe - A portfolio approach. Energy Policy 38(7), 3245–3256 (2010)

Rosen, J.: The future role of renewable energy sources in European electricity supply: A model-based analysis for the EU-15. Universitätsverlag Karlsruhe, Karlsruhe (2008)

Stodola, N., Modi, V.: Penetration of solar power without storage. Energy Policy 37(11), 4730–4736 (2009)

Swider, D.J., Weber, C.: The costs of wind's intermittency in Germany: application of a stochastic electricity market model. European Transactions on Electrical Power 17(2), 151–172 (2007)

UNFCCC. Outcome of the work of the Ad Hoc Working group on long-term Cooperative Action under the Convention: Draft decision -/CP.16 (2010)
US Energy Information Administration. Glossary (2012)
van Wijk, A., Coelingh, J.: Wind power potential in the OECD Countries, Utrecht (1993)
Walther, W.: Long-term price uncertainities of fossil primary fuels and implications for the electricity industry. Cuvillier, Göttingen (2009)

Chapter 3
Methods for Energy System Modeling

In this book techno-economic models are used to examine the cost-optimal future deployment of renewable power generation technologies as well as to gain methodological insights. This chapter introduces modeling concepts and methodologies required to construct these models, with main focus on economic models and methodologies. The first section covers general aspects and offers a brief introduction to decision theory, followed by an overview of equilibrium modeling approaches and their application to energy system modeling. The model class of mathematical programs and corresponding solution methods are then discussed in sections 3.3 and 3.4, respectively. Finally, section 3.5 is dedicated to uncertainty and stochastic modeling.

3.1 General Aspects

An economic model can be defined as a "simplified description of the economy, or the part of the economy, relevant for the analysis" (Black et al., 2009). Such a model may take on different forms, its most simple one being a verbal description of relationships observed in the real world. A common form applied in economics is the mathematical model. The structure of such mathematical models is described by a set of equations. These equations relate "a number of variables to one another in certain ways" (Chiang and Wainwright, 2005, p. 5), thereby expressing analytical assumptions about real world relationships in mathematical form. Variables can broadly be classified as *endogenous*, for which solution values are identified within the model, and *exogenous*, which are an input to the model. In the remainder of this book the term *parameters* for the latter as well as for constants will be used.

In contrast to other sciences, it is usually impossible or at least difficult to conduct controllable experiments in order to support or refute hypotheses in economics. This is the main reason for the widespread use of models in this field. Modeling objectives include but are not limited to illumination and description of systems, forecasting of behavior based on past observations,

F. Wagner, *Renewables in Future Power Systems*,
Green Energy and Technology,
DOI: 10.1007/978-3-319-05780-4_3,

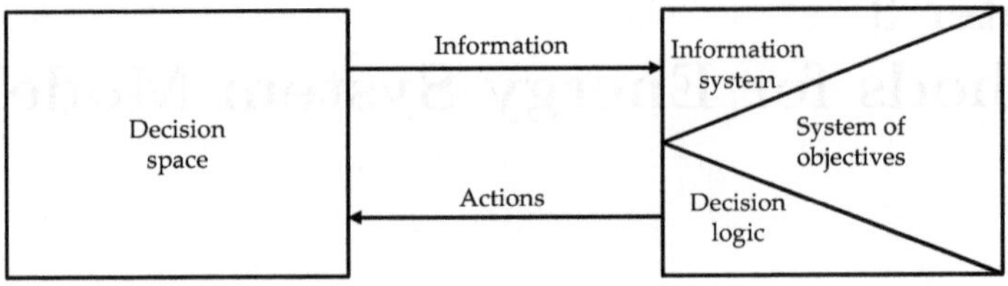

Fig. 3.1 Decision making process
Source: Bamberg and Coenenberg (2004, p. 1), own representation and translation

simulation of consequences from different actions or the testing of theories (Bannock et al., 2003).

A model's structure and complexity heavily depends upon the role played by time. *Static* models focus on a particular point in time. *Dynamic* models, in contrast, cover a time period. Variables in such models may influence each other across time periods. As the primary focus of the models in the following chapters is long-term technology assessment, dynamic models are applied here.

Many economic models strive to reflect the decision making process of one or more agents. Decision theory offers a formal framework describing this process. Bamberg and Coenenberg (2004) describe decision making as an interactive process (Fig. 3.1). On the one hand, the decision space covers all objects or persons influenced by actions of the decision maker. On the other hand, it includes the full set of possible states of the world. The latter are realized independently of the actions taken but influence the consequences of those actions. The decision space includes the set of all possible actions, $A = \{a_1, a_2, \ldots, a_m\}$, the set of all possible states of the world, $Z = \{z_1, z_2, \ldots, z_n\}$, and the results function g. The latter describes the consequences of all possible actions depending on the realized state of the world and can be expressed as $(a, z) \xrightarrow{g} x = g(a, z)$, x denoting the consequences of action a. In case this relationship is of deterministic nature, the consequences assigned to each ordered pair (a, z) are known with certainty[1] (Bamberg and Coenenberg, 2004, pp. 13ff.).

The degree of knowledge a decision maker has on the consequences of his actions is determined by the information system. It is characterized by a set of possible messages $Y = \{y_1, y_2, \ldots, y_k\}$ on the possible states of the world Z and a matrix containing the conditional probabilities, $p_{ij} = P(y_j|z_i)$, of receiving message y_j for the realization of state of the world z_i. Perfect information exists, if every message is assigned to a distinct state of the world with a probability of one. If this is not the case, decisions have to be made under imperfect information (Bamberg and Coenenberg, 2004, pp. 20ff.).

[1] This is the case for the deterministic models presented in section 5.2 and chapter 6.

The system of objectives is required to establish a ranking of possible actions A. It requires definition of one or more indicators capturing the consequences of actions as well as of decision maker preferences. Attitude towards risk/uncertainty counts among the four dimensions of preference mentioned by Bamberg and Coenenberg (2004, pp. 28ff.). For the models applied in this book, uncertainty will be explicitly considered, yet risk-neutrality of the decision maker is assumed and the optimality of decisions is evaluated based on discounted global power system costs.

3.2 Equilibrium Models

3.2.1 Background

An equilibrium can be defined as "a position of balance in the economy or, equivalently, as a situation in which no agent in the economy has an incentive to modify their chosen strategy." (Black et al., 2009). Although the term equilibrium is already mentioned in the early economic literature of the 18th century (Gervaise, 1720), it only became a central element of neoclassical teachings around 1900. Advocates of the neoclassical school propose a market economy characterized by decentralized planning as the superior system for the Pareto-efficient allocation of resources. In such a system, prices coordinate supply and demand (cf. Böhringer, 1996, p. 33).

Walras is credited for formulating the first mathematical general equilibrium model in 1874, yet was not yet able to prove the existence of an equilibrium in this set of equations. Its existence was proven much latter by Wald (1936), for the special case of a Cournot duopoly and, in a generally valid manner, by Arrow and Debreu (1954).

General equilibrium models consider all markets for different goods simultaneously. As substitutes and complements for virtually any good exist, a change in demand or supply of one market will, in the vast majority of cases, not only affect equilibrium prices and quantities in this particular but also other markets. Sometimes, however, detailed representation of a specific market may be of higher interest than its interactions with other ones. In this case, a *partial* equilibrium approach can be chosen, where only one or few markets are considered within the model. Developments in other markets remain exogenous to the model.[2] For the models presented in chapters 5 through 7 a partial equilibrium approach is taken. This approach allows for a more detailed representation of the power sector, for example, by explicitly modeling different technologies with endogenous technological change. The

[2] The definition of general versus partial equilibrium models is closely connected to the one of open versus closed models. The first term, respectively, describes a model, where all sectors of the economy are modeled simultaneously whereas only a portion of the economy is considered in the latter.

drawback of this approach is that interactions of the power sector with other sectors of the economy are ignored.

3.2.2 Energy System Applications

Growing public interest for energy policy in response to the oil crises alongside availability of modern computer technology motivated application of mathematical models to energy-related problems during the 1970s. The models presented by Nordhaus et al. (1973) and the MARKAL model developed and maintained by the International Energy Agency are examples of such early, numerically solved energy system models. These early applications were largely set up as partial equilibrium models covering the energy sector (cf. Bergman, 1988). Although technology-rich models of the energy sector are well-suited for planning purposes they may not be the method-of-choice for other questions, for example, the examination of economy-wide impacts of energy policy measures. For this reasons computable general equilibrium models appeared on the scene not much later (e.g. Hudson and Jorgenson, 1974, or Manne, 1977). This development marked the advent of Energy-Economy models, as interactions of the energy with other sectors of the economy were considered.[3]

The terms often used to describe these fundamentally different approaches in contemporary energy modeling literature are *bottom-up* and *top-down*. Bottom-up models focus on a detailed representation of the energy sector or even its sub-sectors and usually take a partial equilibrium approach. In contrast, top-down models aim to represent the energy sector as part of the entire economy, trading off the detailed representation of different technologies available in bottom-up approaches for a more consistent description of the overall economy. Many top-down applications follow a general equilibrium approach but other approaches also exist. Köhler et al. (2006), for example, use a Keynesian approach, where no general equilibrium state is attained. According to Hourcade et al. (2006), three methodological dimensions can be used to characterize energy-economy models: technological explicitness, microeconomic realism and macroeconomic completeness. Top-down approaches outperform bottom-up approaches on the dimensions of microeconomic realism and macroeconomic completeness, whereas bottom-up approaches score highest in terms of technological explicitness.

Increasing public and political interest in the effects of climate change and the need to examine the potential consequences of policy measures aimed at mitigating damages, led to the appearance of another factor with high relevance to the energy sector: the environment. Since decisions made in the energy sector impact not only the remainder of the economy but also the environment, most current energy system models fall within the

[3] Yet not all energy models considering interactions with other sectors necessarily follow a general equilibrium approach.

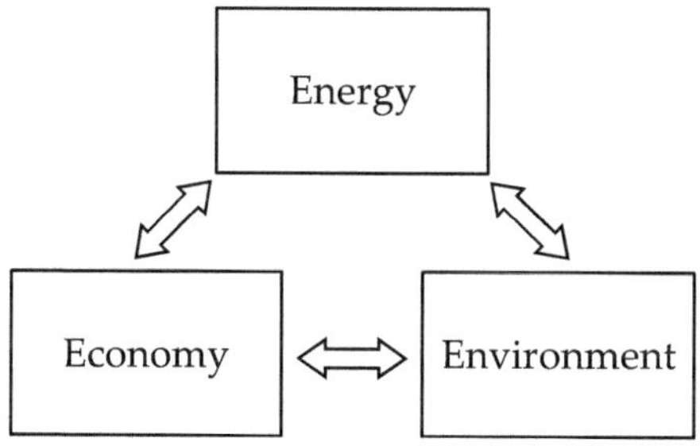

Fig. 3.2 Components of E3 models
Source: Own representation

Energy-Economy-Environment (E3) category (see Fig. 3.2). However, the extent to which those three elements are considered varies significantly from model to model, largely depending on what questions are to be answered and the methodological approach chosen.[4]

The need to investigate the impact of environmental policy measure has also motivated the development of *hybrid* models, which aim to combine the advantages of both top-down and bottom-up approaches. Böhringer and Rutherford (2008) distinguish three broad hybrid model categories:

I. Independently developed bottom-up and top-down models linked in some manner, often referred to as soft-linked in literature. These model combinations, however, are often characterized by "inconsistencies in behavioral assumptions and accounting concepts" (Böhringer and Rutherford, 2008).
II. Models focusing on either a top-down or bottom-up approach, representing the other in a reduced form. Examples include MESSAGE-MACRO (Messner and Schrattenholzer, 2000) or WITCH (Bosetti and Tavoni, 2009).
III. Completely integrated models based on mixed complementary programming to combine bottom-up and top-down approaches in a fully consistent manner. Although convincing in terms of coherence, practical application is likely limited due to dimensionality.

Böhringer (1996) distinguishes two generic approaches to formulating equilibrium problems: either as a system of simultaneous equations or as a mathematical program. Whereas mathematical programs allow for restriction of prices or volumes to specified ranges by means of inequalities, this is not possible for problems formulated as system of equations. Moreover, the system of equations approach is problematic when choice among discrete technological alternatives for the production of the same good is desired. In contrast,

[4] The models for renewable technology assessment presented in this book focus on the technologically explicit representation of the power generation sector. Ties to the general economy and the environment are represented exogenously.

concurrent constraints on prices and volumes are possible under the system of equations approach, whereas such constraints may lead to infeasibility under a mathematical program setup. (cf. Böhringer, 1996, pp. 58f.).

Apart from technical constraints, the choice between a system of equation versus mathematical program approach depends on the questions to be answered. Mathematical programs will produce *normative* or *prescriptive* solutions, i.e. expressing what should be be done. In contrast, solutions from systems of equations are typically *descriptive*, expressing what could happen. Those solutions, however, are in many cases not optimal. In practice, the strict division into descriptive versus normative models does not always hold. If a descriptive model, for example, describes the behavior of efficient markets, it could as well be set up as an optimization (Weber, 2005, p. 27). The models developed and applied in this book are set up as mathematical programs. On the one hand, choice among different technologies is required for technology assessment, on the other hand, a normative model setup is desired to ensure that selection of technologies takes place in a cost-optimal manner.

3.3 Mathematical Programming

Mathematical programming can be viewed as a special case of equilibrium analysis. In this case, the equilibrium state is defined as the optimal position for a given economic unit (Chiang and Wainwright, 2005, p. 220). The two main elements of a basic optimization problem are an objective function and one or more decision variables. A particular objective function value, often a maximum or minimum, is considered optimal according to the preferences of a particular decision maker. One or more sets of decision variable values may result in the desired objective function value, if it is attainable at all. This, however, also points to a problem encountered in optimization problems: Minor differences in exogenous model variable values, which induce small differences in the objective function value, may lead to vastly different solutions. This property is often referred to as *penny-switching.*

In most optimization problems the identification of a global extremum (maximum or minimum) of the objective function is desired. Given a continuous objective function, local extrema can be identified by means of differential calculus.[5] Nonlinear objective functions may have several local extrema among which the global one(s) must be identified.

Mathematical programming as constrained optimization emerged in the late 1930's and 1940's to solve production and organizational planning problems. Kantorovich is credited for the first published formulation and solution

[5] For an unconstrained objective function $y = f(x)$ the necessary first order condition for both minimization and maximization is $f'(x) = 0$, whereas the sufficient second order condition is $f''(x) < 0$ for a local maximum and $f''(x) > 0$ for a minimum.

of linear mathematical programming models in 1939.[6] The term *programming* goes back to the work of Dantzig, who introduced the standard formulation for a linear program still in use today along with the Simplex solution algorithm in 1948 (cf. Dantzig, 1949a).

3.3.1 Linear and Integer Programming

> "The basic assumptions of the model lead to a fundamental set of linear equations expressing the conditions which must be satisfied by the various levels of activity, X_i, in the dynamic system. These variables are subject to the further restriction $X_i > 0$. The determination of the "best" choice of X_i is made to depend on the maximization (or minimization) of linear form in X_i." (Dantzig, 1949a)

Dantzig's initial description of a linear program (LP) notes its key elements: An objective function, one or more decision variables and a number of inequality constraints. The latter ensure that decision variables can only take on economically meaningful values. Equations and inequalities must be in linear form. More formally, a linear program aiming at minimization of the objective function value can be expressed as given in Eq. (3.1) (cf., for example, Weber, 2005, p. 33, or Birge and Louveaux, 1997, p. 73).

$$\begin{aligned} Min. \quad & c^T x \\ s.t. \quad & Ax \geq b \\ & x \geq 0, \end{aligned} \tag{3.1}$$

where vector x denotes the decision variables and c its corresponding coefficients,[7] A the matrix of decision variable coefficients and b the corresponding right hand side vectors within the constraint. To ensure that decision variable values are meaningful from an economic perspective, they are customarily restricted to nonnegative values. An optimal solution x^* is characterized by at least one binding constraint, i.e. the left side of the constraint will equal the right one. A linear problem can have no, one or an infinite number of feasible optimal solutions and if an optimal solution exists, it must be a global one.

If the initial, also referred to as *primal* programming problem, is a minimization, its corresponding *dual* problem, will be a maximization. The dual problem is obtained by transposing the primal matrix. For the problem from Eq. (3.1), the dual is:

---

[6] His work was initially not known in the Western World and was published in English much later (Kantorovich, 1960).

[7] c^T is the transpose of vector c.

$$\begin{aligned} Max. \quad & b^T \pi \\ s.t. \quad & A^T \pi \leq c \\ & \pi \geq 0, \end{aligned} \tag{3.2}$$

where π denotes the set of dual variables, which are associated with the constraining resources b. The optimal objective function value is the same for both the primal and the dual problem. Closely connected to the property of duality is the one of complementary slackness: The value of π, corresponding to a non-binding constraint in the primal problem, must be zero in the optimum. This property is exploited in numerous solution algorithms. In economic applications the dual variable π is also often referred to as *shadow price* and expresses the impact an incremental unit of corresponding resource b would have on the objective function value. Thus, in many economic settings, it can be interpreted as the marginal value of the constraining resource b.

Integer programs are linear programs, where the decision variables have to assume integer values. The standard integer minimization problem, as shown in Eq. (3.3), looks similar to its linear counterpart from Eq. (3.1)

$$\begin{aligned} Min. \quad & c^T x \\ s.t. \quad & Ax \geq b \\ & x \geq 0 \\ & x \in \mathbb{Z}, \end{aligned} \tag{3.3}$$

In most applications, mixed integer programming (MIP) rather than pure integer programming is used. MIPs include decision variables assuming real as well as integer values. Finding solutions to mixed integer programs generally requires considerably more computational efforts than to their LP counterparts.

3.3.2 Nonlinear Programming

If objective function or constraints are nonlinear, the problem falls into the category of nonlinear programs. This leads to relationships of nonproportional nature among the decision variables and the objective function (Hillier and Hillier, 2003) as opposed to the proportional relationships found in linear programs. Eq. (3.4) describes a nonlinear program more generally (Chiang and Wainwright, 2005, p. 404).

$$\begin{aligned} Min. \quad & f(x) \\ s.t. \quad & g(x) \geq c \\ & x \geq 0, \end{aligned} \tag{3.4}$$

where $f(x)$ is a nonlinear objective function, $g(x)$ a vector-valued nonlinear function associated with the program's constraints and limiting resources c.

The "shape" of the feasible region plays an important role for the solution process of a nonlinear model. A convex feasible region is generally desirable. A key problem associated with nonlinear programs, in case their feasible regions are not strictly convex, may be the presence of more than one local optimum. In complex models this often renders identification of the global optimum among the local ones difficult, if not impossible (cf. Hillier and Hillier, 2003). Since most economic activities are characterized by decreasing marginal returns, a convex formulation can be achieved (cf. Chiang and Wainwright, 2005, pp. 372ff.)

3.4 Solution Methods for Mathematical Programs

Different approaches can be taken to identify the optimal solution(s) of a mathematical program. Analytical methods draw upon methods from differential calculus. One of their advantages is that often insights, which are valid across many, potentially even all, permissible input parameter configurations, can be produced and general inferences on model properties can be made. Their major drawback is that high manual solution efforts are already required for small scale models, usually rendering analytical solution methods unfit for real world problems. In contrast, numerical solution methods, generally making use of modern computer technology, are capable of identifying solutions in large-scale models. Their major drawback, however, is that the obtained solutions are only valid for a particular set of parameter values.

3.4.1 Analytical Approaches

Analytical solution of models is typically based on differential calculus. For unconstrained optimization problems the local extrema can be identified by setting all first-order partial derivatives to zero and solving the resulting equation system. For optimization problems with equality constraints, the same condition for extrema identification holds true — yet a method is needed to consider objective function and constraints simultaneously.

The objective of the *Lagrange-multiplier method* "is to convert a constrained-extremum problem into a form such that the first-order condition of the free-extremum problem can still be applied" (Chiang and Wainwright, 2005, p. 350). The method is applicable to non-linear programs with equality constraints. In presence of inequality constraints a modified approach with different first-order conditions needs to be applied, which will be discussed further below.

The generalized Lagrange-multiplier method with n variables and one equality constraint is described in the following paragraphs, which are, to a large extent, based on Chiang and Wainwright (2005, pp. 350ff).

Let z denote the objective function and g the constraint with corresponding resource c, given by Equations (3.5) and (3.6), respectively.

$$z = f(x_1, x_2, \ldots, x_n) \tag{3.5}$$

$$g(x_1, x_2, \ldots, x_n) = c \tag{3.6}$$

The Lagrangian function, Z, which brings together the elements from the objective function as well as the constraint in one equation, is given by Eq. (3.7).

$$Z = f(x_1, x_2, \ldots, x_n) + \lambda[c - g(x_1, x_2, \ldots, x_n)] \tag{3.7}$$

The corresponding first-order condition for a local extremum,[8] consisting of $n+1$ simultaneous equations, is given by the set of Equations (3.8). Solving this equation system will yield the local extrema of the program specified in Eqs. (3.5) and (3.6), if an extremum exists at all.

$$\begin{aligned} Z_\lambda &= c - g(x_1, x_2, \ldots, x_n) = 0 \\ Z_1 &= f_1 - \lambda g_1 \ = 0 \\ Z_2 &= f_2 - \lambda g_2 \ = 0 \\ &\vdots \qquad \vdots \qquad \vdots \\ Z_n &= f_n - \lambda g_n = 0 \end{aligned} \tag{3.8}$$

The Lagrange-multiplier method can be applied to any program with m equality constraints. In this case, the product of each constraint and a corresponding Lagrangian multiplier λ_m are added to the initial objective function in order to arrive at the Lagrangian function Z. The resulting equation system consists of $(n+m)$ equations.

The second-order sufficient conditions for either a maximum or minimum of an equality-constraint program can be expressed in terms of the second-order total differential of z[9] (Chiang and Wainwright, 2005, p. 358):

---

[8] Derived through partial differentiation of the Lagrangian function with respect to λ and the model's decision variables.

[9] The second-order sufficient condition can also be expressed in determinantal form. For a maximum, the sign of the leading principal minors, $|\overline{H}_2|, |\overline{H}_3|, \ldots$, of the bordered Hessian matrix $|\overline{H}|$ must be alternating, for a minimum all principal minors must take a negative sign. This concept can also be extended to the multi-constraint case, where the bordered Hessian is characterized by multiple borders.

- For a maximum: d^2z must be negative definite, subject to $dg = 0$
- For a minimum: d^2z must be positive definite, subject to $dg = 0$

For a particular local maximum of a twice-differentiable function to also constitute a global one, the following sufficient conditions hold (cf. Chiang and Wainwright, 2005, p. 373):

a. z^* is an absolute constrained maximum if f is explicitly quasiconcave and the constraint set is convex
b. z^* is a unique absolute constrained maximum if f is strictly quasiconcave and the constraint set is convex[10]

Most applications of mathematical programs include inequality constraints. For this class of models the Lagrange-Multiplier method can no longer be applied to identify extrema. Instead, the so-called *Karush-Kuhn-Tucker (KKT) conditions* (cf. Kuhn and Tucker, 1951) must be used. This set of conditions is based on the property of complementary slackness. The first steps to stating the KKT conditions for a particular problem are the same as for the Lagrange-multiplier method: The problem needs to be formulated in Lagrangian form and the partial, first-order derivatives with respect to all decision and Lagrangian variables must be obtained. Based on the latter, the KKT conditions can be stated. Their general form for a minimization problem with m constraints and n decision variables is (cf. Chiang and Wainwright, 2005, p. 410):

$$Z_{x_j} \geq 0 \qquad x_j \geq 0 \quad \text{and} \quad x_j \perp Z_{x_j} \tag{3.9}$$

$$Z_{\lambda_i} \leq 0 \qquad \lambda_i \geq 0 \quad \text{and} \quad \lambda_i \perp Z_{\lambda_i} \tag{3.10}$$

where $i = \{1, 2, \ldots, m\}$ and $j = \{1, 2, \ldots, n\}$. These conditions are necessary for a minimum of a nonlinear mathematical program with inequality constraints.

The solution process based on KKT conditions is one of trial-and-error. Starting point is the assumption that one or more constraints are not binding at the extremum, implying that the associated Lagrangian multiplier λ_i will be zero. By requirement of complementary slackness, the inequality[11] corresponding to this variable can be transformed into an equality. Moreover, often guesses on whether particular decision variables x_j will be positive or zero can be made. Terms containing variables with zero-value will drop out, reducing model complexity. After these initial assumptions on variable signs

[10] Strict quasiconcavity precludes the existence of "horizontal plain segments," which is not the case for explicit quasiconcavity.
[11] First-order, partial derivative with respect to the Lagrangian variable under consideration.

have been made, the resulting equation system can be solved. The final step is then to check whether the obtained solution fulfills all KKT-conditions. If this is not the case, the described process has to be repeated with other assumptions on variable signs until the solution is found that fulfills KKT conditions.

In some cases, KKT-conditions are not valid necessary conditions for an extremum. This issue is connected to the shape of the feasible region. Yet, for any program with exclusively linear constraints, KKT conditions "will always hold at the optimal solution" (Chiang and Wainwright, 2005, p. 416).[12] Linearity in a model's constraints has an additional benefit: KKT-conditions for a minimum/ maximum are not only necessary but also sufficient conditions for a local extremum. The analytical model from chapter 5 is solved based on the KKT-condition approach. Although its objective function is nonlinear, linearity in its constraints is deliberately preserved to ensure that KKT conditions are valid for any parameter configuration.

3.4.2 Numerical Approaches

Although analytical solution methods are appealing for small-scale applications, they are no longer suitable for most contemporary models addressing real world problems. The latter are often characterized by thousands of variables and the manual efforts associated with an analytical solution would be prohibitive. Instead, numerical solution methods can effectively be applied. Two methods with relevance for this book[13] are (Dual) Simplex and branch & cut algorithms for LPs and MIPs, respectively.

The Simplex algorithm is used to solve linear programs and was first presented by Dantzig in the late 1940's (cf., for example, Dantzig 1963, p. 15).[14] Initially the solutions based on this algorithm were computed by hand,[15] whereas the advent of modern computer technology made automation of this process possible. The general principle of the algorithm is a movement along the edges of the polyhedron representing the feasible region of the problem. It identifies better solutions with each iteration until the global optimum is found. The simplex method can be divided into two main steps. In the first

---

[12] More generally, the KKT conditions will hold for problems where the so-called *constraint qualification* is satisfied (cf., for example, Chiang and Wainwright 2005, p. 415).

[13] Applied to solve the models in chapters 6 and 7 of this document.

[14] Another important class of LPs solver algorithms is based on the *Interior Point Method* (Karmarkar, 1984). It is not covered here in detail, since its relevance for models set up as MIP, such as the numerically solved ones in this book, is comparatively low.

[15] The objective of an early application was the cost-optimal diet for soldiers. A linear program considering 9 dietary elements and 77 foods required 120 manhours to compute by hand (cf. Dantzig, 1949b).

step, an initial corner solution[16] is identified as starting point. This is often the origin of the problem, where all decision variables are equal to zero. In case this point is not included in the feasible region, additional calculations are required to identify a feasible starting point. In the second step, the rate of solution improvement from moving to adjacent corner points[17] are calculated and compared. The algorithm then "moves" along the edge of the feasible region characterized by the highest rate of improvement to the corresponding adjacent corner solution. This step is repeated until a corner solution is found, where a movement towards any of the adjacent corner solutions will not improve the solution. This corner solution is the/a globally optimal solution[18] to the problem.[19] For a more formal description of the simplex algorithm the interested reader is referred to Dantzig (1963) or Sydsæter and Hammond (2009, pp. 737ff.).

Historically, two main methodologies for solving Integer problems could be distinguished: *cutting plane* and *branch & bound* algorithms. Both of these approaches rely on the *relaxation* of the integer problem. In this relaxation the integer requirement is dropped and the resulting LP can be solved using a (Dual) simplex algorithm. Under a *cutting plane* approach this LP relaxation is *tightened* by introducing additional constraints. These cut off incumbent solutions located outside the convex hull of feasible integer solutions, thus forcing the new solution to be closer to the set of feasible solutions. The first solution algorithm based on such cuts was proposed by Gomory (1958). Although he was able to prove that an optimal integer solution can be found after a finite number of cuts, the algorithm was not found suitable for practical applications as it "adds an exponential number of cuts to the relaxed problem" (cf. Achterberg, 2008, p. 18).

[16] A corner solution is a feasible solution that satisfies n constraints simultaneously for a problem with n decision variables (Hillier and Hillier, 2003, p. CD 17-38).

[17] Adjacent corner solutions share all but one constraint boundary (Hillier and Hillier, 2003, p. CD 17-38).

[18] If the rate of solution improvement to an adjacent corner solution is zero, an infinite number of globally optimal solutions exists along the corresponding edge of the feasible region.

[19] More technically, an initial LP problem featuring inequalities is transformed into an equation system with slack/surplus variables. For such a system, the *basic feasible solution* is a description of corner solutions in terms of all decision as well as slack/surplus variables. Non-zero variables are called *basic* variables, variables with zero value *non-basic variables* of the corner solution. For each iteration of the algorithm a formerly non-basic variable becomes a basic one (entering variable) and a formerly basic variable becomes a non-basic one (leaving variable). The entering variable chosen is the one leading to the highest improvement to the solution, the leaving variable the one with the least deterioration to the latter. After the exchange of basic and non-basic variables, the problem is brought into "Proper form for Gaussian Elimination" (cf. Hillier and Hillier, 2003, pp. CD 17s-14ff.).

At least for binary problems it would be theoretically possible to enumerate all possible variable combinations and solve the corresponding LPs. The binary variable combination associated with the lowest objective function value would then characterize the global optimum. The disproportionate increase in combinations for addition of binary variables, however, renders this approach unsuitable for large-scale applications. Nonetheless *branch & bound* algorithms are based on this property.[20] In this class of algorithms the initial problem is split into subproblems, which in turn are further divided. A branching tree is used to keep track of these different subproblems, which constitute the tree's nodes. At these nodes, LP relaxations of the corresponding subproblem are solved, which yield information on the bounds for the globally optimal integer solution.[21] Based upon this information those sections of the tree can be cut off, which cannot contain an integer solution outperforming the incumbent one. Following this approach, a full enumeration of the problem can be avoided (cf. Achterberg, 2008, pp. 15–18)

Contemporary solution algorithms for Mixed-Integer problems generally combine both of the methods from above as *branch & cut* algorithms, a concept going back to the work of Crowder et al. (1983). Such algorithms generally follow a branch & bound strategy but introduce cuts for both the problem as a whole as well as subproblems to strengthen the respective LP relaxations.[22] For further detail on functions of sophisticated MIP solvers the interested reader is referred to Achterberg (2008, pp. 57ff.). The large-scale power system model in chapters 6 and 7, which is set up as MIP, is solved using a branch & cut approach with the solver package CPLEX (cf. IBM ILOG, 2011)

3.5 Risk and Uncertainty

The power sector is characterized by comparatively high asset intensity and long useful lives of those assets. Nuclear or hydro power stations, for example, may well be in operation for half a century or longer. Consequently not all factors, which determine the relative future cost position of long-lived power technologies are known at the time, when an investment decision must be made — the decision is made under risk or uncertainty.

[20] Land and Doig (1960) are credited for first describing a branch & bound algorithm.

[21] Bound information is also generated from other model functionalities, e.g. heuristics.

[22] Cut & branch algorithms apply cuts to the root problem only and then continue the solution process following a branch & bound strategy (cf. Achterberg, 2008, p. 19).

3.5.1 General Aspects

Traditionally, risk and uncertainty have been used as distinct terms. Their definition in an economic context goes back to Knight, who defines risk as "measurable" uncertainty, whereas he restricts "true" uncertainty "to cases of the non-quantitative type" (Knight, 1921, p. I.I.26). In contemporary decision theory, a setting is characterized by risk, if a decision maker has knowledge of the probabilities associated with the possible future states of the world. In absence of such probabilistic information, the situation is characterized by uncertainty (cf. Bamberg and Coenenberg 2004, pp. 76, 127). This strict distinction between risk and uncertainty, however, is not generally made in literature. If not explicitly mentioned, the term uncertainty will be used in a broader sense, comprising the term risk — particularly since the following chapters describe models explicitly taking uncertainty into account but for which a clear distinction between risk and uncertainty may be difficult.[23] There have also been efforts to further subdivide the term uncertainty based on the degree to which future events are uncertain. Courtney et al. (1997) distinguish four different types in the context of a strategy development process:

I. *A clear-enough future*: A single forecast is precise enough for the development of a strategy
II. *Alternate futures*: The future can be described by a finite set of discrete scenarios
III. *A range of futures*: The future is characterized by a range of possible outcomes; no natural scenarios exist
IV. *True ambiguity*: No basis for forecasting the future exists

In this context Weber (2005) notes that these types of uncertainty in the power sector primarily apply in the long-term. The long-term uncertainties, which are considered in the models described in this book, can largely be associated with class III.[24]

The presence of uncertainty has an impact on optimal decision making. Reverting to the decision making process from section 3.1, under risk the consequences of combination (a, z)[25] are no longer determined in a deterministic but in a stochastic manner. Consequence $g(a, z)$ turns into a random variable. Preferences of the decision maker will impact the value of the objective function, which are used to compare different possible actions a against each other (Bamberg and Coenenberg, 2004). In this context, the decision maker's

[23] Probabilities for different states of the worlds are used but they are not based on measurable, empirical data.

[24] Short-term uncertainties also influence the relative cost position of renewable technologies, e.g. stochastic output fluctuations of intermittent technologies. In the system model described in chapters 6 and 7, this type of uncertainty is reflected in form of a deterministic approximation.

[25] Where a is an action taken by the decision maker and z a state of the world.

attitude towards risk plays an important role. Whereas the presence of risk will reduce the value of the objective variable for a risk averse decision maker, it will increase the value for a risk-seeking one. For risk-neutral decision makers the presence of risk will not alter the value of the objective variable. Methods applied to determine optimal actions differ depending upon whether a situation is characterized by risk or uncertainty in a narrow sense, i.e. whether probabilistic information on the realization of the different possible states of the world is available.

In presence of uncertainty in a narrow sense, i.e. in absence of probabilistic information, a number of rules have been devised to identify optimal decisions. They may yield different decision recommendations for the same problem, as they often differ with respect to their implicit assumptions on decision maker preferences. These include the widely known Maximax, Maximin, Minimax or Laplace[26] rules, yet numerous other rules exist. Since for the stochastic models in this book an assumption on probabilities is made, these rules are not discussed in further detail but the interested reader is referred to other publications instead, e.g. Kramer (1967), Hillier and Hillier (2003, pp. 520ff. and CD 12-1ff.) or Bamberg and Coenenberg (2004, pp. 128ff.).

In case probabilities can be attached to different states of the world, other methods to determine optimal decision making can be applied. Bayes' rule,[27] also often referred to as μ-rule,[28] assumes risk-neutrality of the decision maker. Possible actions are evaluated on the basis of their expected payoff $E(x)$, as given by Eq. 3.11.

$$E(x) = \mu = \sum_i p_i x_i \tag{3.11}$$

where i is the total set of possible events, p_i the probability associated with event i and x_i the corresponding payoff. Bayes' rule is particularly useful when probabilities for events are conditional upon the realization of other events. This is also the case, if decisions must be made at different points in time, where information updating takes place. In such a setup, decision trees are a useful tool to visualize these relationships. Fig. 3.3 illustrates a simple decision tree, where sets $i = \{A, B\}$ and $j = \{1, 2\}$ describe possible states of the world and x_{ij} the associated payoffs. The probability for the realization of a state of the world in model stage 2 from a stage 1 perspective is $p_{ij} = (p_j \cap p_i)$.

In reality, decision makers are rarely risk-neutral. Bernoulli already illustrated this point in the 18th century using the so-called St. Petersburg paradox.[29] He proposes to use a (concave) utility function in order to capture

[26] Also known as the equally likely criterion.
[27] Named after the 18th century mathematician Thomas Bayes.
[28] μ represents the expected payoff of an action.
[29] See Bernoulli (1954) for an English translation of Bernoulli's work on this subject.

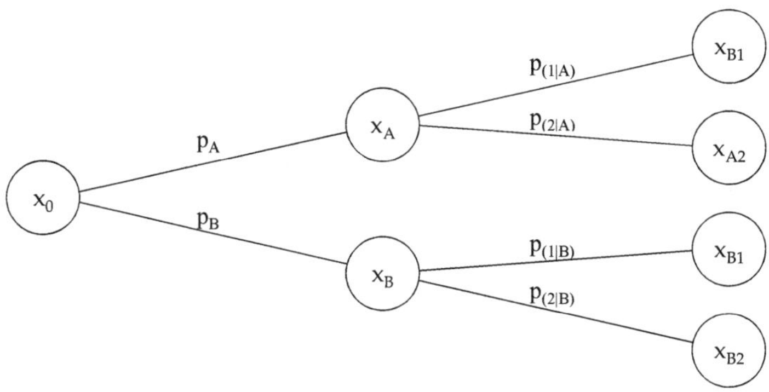

Fig. 3.3 Illustrative Decision Tree
Source: Own representation

risk-aversion, where marginal utility decreases with increasing risk.[30] Before an optimal action can be selected, the expected payoffs and associated risks must first be converted into *utility*, which is then used to compare alternative actions. This principle is today referred to as *expected utility hypothesis* (cf. Bamberg and Coenenberg, 2004, p. 86).

3.5.2 Model-Endogenous Representation of Uncertainty

In deterministic decision models uncertainties existing in the real world are ignored and expected future values for the model's parameters are used. For cases, where few changes in parameter values are expected over time or where these changes are reasonably predictable, deterministic models may yield similar results to stochastic ones, which explicitly take uncertainty into account. The development of the power sector over the 21st century, however, is characterized by a high degree of uncertainty. Apart from uncertain fossil fuel prices, possible cost reductions for new technologies as well as costs from mitigating or adapting to climate change cannot be predicted. Such uncertainties, paired with long asset lifetimes, may profoundly affect optimal decision making for electricity industry players and policy makers alike.[31]

[30] Conversely, a convex utility function describes a risk-seeking attitude. Such a utility function must exist for decision makers engaging in gambling activities, where the expected payoff is lower than the wager.

[31] It needs to be noted that not all uncertainties associated with model-based forecasting relate to the model's parameters. In addition to parameter uncertainty, van der Zwaan and Seebregts (2004) point out modeling and methodological uncertainty. Modeling uncertainty arises from differences in assumptions on processes within and outside the model and which of these processes should be considered endogenously. Methodological uncertainties relate to the inevitable simplifications in the model's structure when compared to the complexity of real world processes.

Stochastic models explicitly consider (parameter) uncertainty and may therefore be more appropriate for decision support on optimal investment strategies in the power sector than deterministic ones.

Kann and Weyant (2000) distinguish two main approaches to uncertainty representation in economic models: *sequential decision making under uncertainty* and *uncertainty propagation*. They differ mainly with regards to presence of information updating over the modeling horizon and corresponding decision adjustments — this property is only given in case of *sequential decision making under uncertainty*. In literature the process of information updating resulting in uncertainty reduction or even full resolution is often referred to as "learning" and the model setup as one described by the term act-learn-act.[32] In reality, information updating on the actual realization of random parameters and the corresponding decisions would occur continuously. Computational constraints render representation of this property in large-scale E3 models infeasible and uncertainty is (partially) resolved at a finite, discrete number of points in time. Different mechanisms can be implemented to trigger uncertainty resolution. *Active learning* requires certain activities, such as deployment of a particular technology, to reduce the degree of uncertainty. If information can be purchased, for example, through allocating funds to R&D, the mechanism is termed *purchased learning*. Finally, *autonomous learning* refers to information updating at a certain point in time (Kann and Weyant, 2000). The output from a multistage stochastic optimization model featuring an act-learn-act setup is a strategy, which is optimal considering all possible random variable realizations.

The modeling approach of *uncertainty propagation* does not allow for information updating and sequential decision making. Instead, uncertainty is resolved at the beginning of the model horizon. This approach often involves sampling from a probability distribution of the uncertain parameters.[33] A common implementation approach in large-scale models is Monte-Carlo simulation, where the optimization model is solved numerous times with parameter values drawn from their respective probability distributions. The output under such an approach is also a set of probability distributions describing the solution.

The stochastic power sector models developed in the following chapters are set up as optimization with sequential decision making under uncertainty. For the small-scale analytical model in chapter 5 uncertainty is resolved once, in the large-scale, numerically solved model in chapter 7 information on uncertain parameter values is updated twice. This corresponds to a two- and

[32] This meaning of learning is distinct from the one used to describe technological cost reductions, which will be touched upon in the next chapter. In case of ambiguity the term uncertainty resolution rather than learning will be used in this document.

[33] If more than one parameter is uncertain, it is necessary to specify correlations among probability distributions.

three-stage setup, respectively.[34] The general form of a two-stage stochastic minimization problem with fixed recourse is given by Eq. (3.12) (Birge and Louveaux, 1997, pp. 54ff.).[35]

$$\begin{aligned} Min. \quad & c^T x + E_\xi \left[min\ q(\omega)^T y(\omega)\right] \\ s.t. \quad & Ax = b \\ & T(\omega)x + Wy(\omega) = h(\omega) \\ & x \geq 0,\ y(\omega) \geq 0, \end{aligned} \tag{3.12}$$

The elements describing the first stage correspond to the ones in the deterministic problem given by Eq. (3.1). x is the first-stage decision variable, c the corresponding objective function coefficient, A the respective coefficient in the constraint and b the constraining resource. In the second stage a random event ω is realized, which is an element of the full set of possible events Ω and which impacts the optimal value of second-stage decision variable $y(\omega)$ in an indirect manner. The second-stage objective parameter $q(\omega)$ as well as parameters appearing in the corresponding constraint, $T(\omega)$ and $h(\omega)$ are also functions of the realized event ω.[36] For this program with fixed recourse,[37] both decision variables x from the first stage as well as $y(\omega)$ from the second stage appear in the constraint $T(\omega)x+Wy(\omega) = h(\omega)$, illustrating that ω will not only affect second-stage but also first-stage optimal decisions. The objective function is composed of the deterministic term $c^T x$, corresponding to the first stage, and the "expectation of the second-stage objective $q(\omega)^T y(\omega)$ taken over all realizations of the random event ω" (Birge and Louveaux, 1997, p. 54).

The stochastic program from (3.12) can be transformed into its *deterministic equivalent program*, given by

$$\begin{aligned} Min. \quad & c^T x + E_\xi Q(x, \xi(\omega)) \\ s.t. \quad & Ax = b \\ & x \geq 0, \end{aligned} \tag{3.13}$$

with

$$Q(x, \xi(\omega)) = \min y\{q(\omega)^T y | Wy = h(\omega) - T(\omega)x,\ y \geq 0\} \tag{3.14}$$

[34] In both models information updating takes place autonomously, i.e. at prespecified points in time.

[35] Dantzig (1955) and Beale (1955) are credited for first describing this class of problems.

[36] $T(\omega)$ is called technology matrix, $h(\omega)$ is the constraining resource.

[37] Recourse matrix W is a constant.

In case the number of possible events ω is finite, the program from above can also be transformed into an *extensive form* (cf. Birge and Louveaux, 1997, p. 8). In this form, the decision variables for the second stage are explicitly described for all possible realizations of ω. Such an approach is followed for the stochastic models in this book, as there is a maximum of four possible states of the world in a particular stage.

A stochastic program with recourse may yield an optimal "hedging" strategy against the possible realizations of uncertain parameters. In contrast, a solution based on a deterministic model, where the uncertain parameters have been replaced with their respective expected values, will not yield this "hedged" solution. The value from using a stochastic recourse problem over its deterministic counterpart based on expected values can be expressed as the *value of the stochastic solution (VSS)*. For a minimization it is defined as

$$VSS = EEV - RP \tag{3.15}$$

where EEV denotes the *expected result of using the expected value solution* and RP the solution of the corresponding stochastic recourse problem (Birge and Louveaux, 1997, p. 139). There exist cases, where there is a zero VSS despite presence of uncertainty. This can occur, if uncertainty relates to objective function coefficients only (cf. Birge and Louveaux, 1997, p. 57).

A major drawback of stochastic over deterministic models is the associated increase of computational efforts in the solution process. Even for a finite number of events, the addition of another problem dimension disproportionately increases the size of the problem. This property is referred to as "curse of dimensionality," an expression going back to Bellman (1957). It considerably limits the number of states of the world, which can be explicitly considered in energy system models.

References

Achterberg, T.: Constraint Integer Programming. Verlag Dr. Hut, Munich (2008)

Arrow, K.J., Debreu, G.: Existence of an Equilibrium for a Competitive Economy. Econometrica 22(3), 265–290 (1954)

Bamberg, G., Coenenberg, A.G.: Betriebswirtschaftliche Entscheidungslehre, 12th revised edn. Verlag Franz Vahlen, Munich (2004)

Bannock, G., Baxter, R.E., Davis, E.: Dictionary of Economics, 4th edn. Bloomberg Press, Princeton (2003)

Beale, E.M.L.: On Minizing A Convex Function Subject to Linear Inequalities. Journal of the Royal Statistical Society. Series B 17(2), 173–184 (1955)

Bellman, R.: Dynamic Programming. Princeton Univ. Press, Princeton (1957)

Bergman, L.: Energy Policy Modeling: A Survey of General Equilibrium Approaches. Journal of Policy Modeling 10(3), 377–399 (1988)

Bernoulli, D.: Specimen theoriae novae de mensura sortis. In: Commentarii Academiae Scientiarum Imperialis Petropolitanae, vol. 5, pp. 175–192. Petersburg (1738)

Bernoulli, D.: Exposition of a New Theory on the Measurement of Risk: Translated by L. Sommer. Econometrica 22(1), 23–36 (1954)
Birge, J.R., Louveaux, F.: Introduction to stochastic programming. Springer, New York (1997)
Black, J., Hashimzade, N., Miles, G.D.: A dictionary of economics. Oxford Univ. Press, Oxford (2009)
Böhringer, C.: Allgemeine Gleichgewichtsmodelle als Instrument der energie- und umweltpolitischen Analyse: Theoretische Grundlagen und empirische Anwendung. Peter Lang, Frankfurt a. M. (1996)
Böhringer, C., Rutherford, T.F.: Combining bottom-up and top-down. Energy Economics 30(2), 574–596 (2008)
Bosetti, V., Tavoni, M.: Uncertain R&D, backstop technology and GHGs stabilization. Energy Economics 31(suppl. 1), S18–S26 (2009)
Chiang, A.C., Wainwright, K.: Fundamental methods of mathematical economics, 4th international edn. McGraw-Hill, Boston (2005)
Courtney, H., Kirkland, J., Viguerie, P.: Strategy under uncertainty - What makes for a good strategy in highly uncertain business environments. Harvard Business Review 75(6), 66–81 (1997)
Crowder, H., Johnson, E.L., Padberg, M.: Solving Large-Scale Zero-One Linear Programming Problems. Operations Research 31(5), 803–834 (1983)
Dantzig, G.B.: Programming in a Linear Structure. Econometrica 17(1), 73–74 (1949a)
Dantzig, G.B.: Programming of interdependent activities II: Mathematical model. Econometrica 17(3-4), 200–211 (1949b)
Dantzig, G.B.: Linear Programming under Uncertainty. Management Science 1(3-4), 197–206 (1955)
Dantzig, G.B.: Linear Programming and Extensions. Princeton Univ. Press, Princeton (1963)
Gervaise, I.: The System or Theory of the Trade of the World. H. Woodfall, London (1720)
Gomory, R.E.: Outline of an algorithm for integer solutions to linear programs. Bulletin of the American Mathematical Society 64(5), 275–278 (1958)
Hillier, F.S., Hillier, M.S.: Introduction to Management Science, 2nd edn. McGraw-Hill/ Irwin, Boston (2003)
Hourcade, J.-C., Jaccard, M., Bataille, C., Ghersi, F.: Hybrid Modeling: New Answers to Old Challenges. The Energy Journal 27(special issue), 1–12 (2006)
Hudson, E.A., Jorgenson, D.W.: U. S. Energy Policy and Economic Growth, 1975-2000. The Bell Journal of Economics and Management Science 5(2), 461–514 (1974)
IBM ILOG. CPLEX Optimizer (2011)
Kann, A., Weyant, J.: Approaches for performing uncertainty analysis in large-scale energy/economic policy models. Environmental Modeling and Assessment 5(1), 29–46 (2000)
Kantorovich, L.V.: Mathematical Methods of Organizing and Planning Production: Translated by W. Campbell and W. H. Marlow. Management Science 6(4), 366–422 (1960)
Karmarkar, N.: A New Polynomial-Time Algorithm for Linear Programming. Combinatorica 4(4), 373–395 (1984)
Knight, F.H.: Risk, Uncertainty and Profit. Houghton Mifflin, Boston (1921)

Köhler, J., Barker, T., Anderson, D., Pan, H.: Combining Energy Technology Dynamics and Macroeconometrics: The E3MG Model. The Energy Journal 27(special issue 2), 113–134 (2006)
Kramer, G.: Entscheidungsproblem, Entscheidungskriterien bei völliger Ungewißheit und Chernoffsches Axiomensystem. Metrika 11(1), 15–38 (1967)
Kuhn, H.W., Tucker, A.W.: Nonlinear Programming. In: Neyman, J. (ed.) Proceedings of the Second Berkeley Symposium on Mathematical Statistics and Probability, pp. 481–492. Univ. of California Press, Berkeley (1951)
Land, A.H., Doig, A.G.: An Automatic Method of Solving Discrete Programming Problems. Econometrica 28(3), 497–520 (1960)
Manne, A.S.: ETA-MACRO: A Model of Energy-Economy Interactions. In: Hitch, C.J. (ed.) Modeling energy-economy interactions: Five approaches, pp. 1–45. Resources for the Future, Washington (1977)
Messner, S., Schrattenholzer, L.: MESSAGE-MACRO: linking an energy supply model with a macroeconomic module and solving it iteratively. Energy 25(3), 267–282 (2000)
Nordhaus, W.D., Houthakker, H., Solow, R.: The Allocation of Energy Resources. Brookings Papers on Economic Activity 1973(3), 529–576 (1973)
Sydsæter, K., Hammond, P.: Mathematik für Wirtschaftswissenschaftler: Basiswissen mit Praxisbezug: Translated by F. Böker, 3rd revised edn. Pearson, Munich (2009)
van der Zwaan, B., Seebregts, A.: Endogenous learning in climate-energy-economic models - an inventory of key uncertainties. International Journal for Energy Technology and Policy 2(1-2), 130–141 (2004)
Wald, A.: Über einige Gleichungssysteme der mathematischen Ökonomie. Zeitschrift für Nationalökonomie 7(5), 637–670 (1936)
Walras, L.: Eléments d'économie politique pure, ou théorie de la richesse sociale. L. Corbaz & Cie, Lausanne (1874)
Weber, C.: Uncertainty in the Electric Power Industry: Methods and Models for Decision Support. International Series in Operations Research & Management Science, vol. 77. Springer, New York (2005)

Chapter 4
Technological Change and the Experience Curve

Cost reductions through technological progress are a key driver of renewable competitiveness. The main objective of this chapter is therefore to provide background on technological change in general with a specific focus on experience curves. After a general introduction of the topic in section 4.1, the concept of experience curves is introduced in section 4.2. The section covers the basic experience curve concept, drivers behind the cost reductions described by the experience curve, reasons for differences in learning rates among technologies, advanced experience curve concepts as well as the benefits and drawbacks of using experience curves as forecasting tool. A literature review to identify appropriate learning rates for renewable power generation technologies is then carried out in section 4.3. Section 4.4 then covers methods available to implement technological change in economic models and discusses options for dealing with non-convexity in bottom-up models. The chapter ends with a short discussion on how technological change is reflected in the models covered by this book and the benefits and pitfalls of this approach.

4.1 Technological Change

Technological change is widely recognized as the most important driver of long term economic growth (Solow, 1956). Schumpeter (1934)[1] views this process of technological change as one of *creative destruction*, where technologies are subsequently replaced by newer ones. Different ideas and the technologies developed based on those ideas compete for adopters, whose number is limited by the size of the respective market. Unsurprisingly such a process is not a smooth one but rather characterized by barriers and subsequent technological breakthroughs. A technological breakthrough requires changing the physical configuration or methodology inherent to a technology. Technical progress is then generally highest immediately after overcoming a

[1] The original in German was already published in 1912.

F. Wagner, *Renewables in Future Power Systems*,
Green Energy and Technology,
DOI: 10.1007/978-3-319-05780-4_4,

barrier and slows down again when a new one is approached (Ayres, 1994). Grübler et al. (1999) identify three robust attributes to technological change: (1) cost reductions and performance improvements, (2) regular patterns of dynamic competition and (3) co-evolution of long-lived infrastructure. Those three attributes will be touched upon in the following paragraphs.

According to Schumpeter (1934), the process of technological change can be divided into three main phases: *invention*, *innovation* and *diffusion*.[2] Invention is the creation of an idea. During the innovation phase the invention is then applied to real world problems. During diffusion the new technology is adopted in the market.

The diffusion phase can further be divided into four substages to describe the role played by a technology in a market: *niche market commercialization*, *pervasive diffusion*, *saturation* and *senescence* (Grübler et al., 1999). Cost level and market share can serve as indicators for a technology's current life cycle stage. Costs are high in the invention and innovation phases but decrease rapidly during pervasive diffusion. Reasons for these reductions are, among others, economies of scale and learning effects.[3] Cost increases incurred to improve performance may offset some of the cost reductions realized under the technology's previous configuration. Although cost reduction trajectories of complex technologies, such as wind turbines, may look smooth when regarding the technology as a whole, the cost reductions for its components move in steps of technological breakthroughs (cf. Kobos et al., 2006). The development of commercial market shares usually mirrors the cost trajectory. Adoption starts in the phase of niche market commercialization. Decreasing cost during the phase of pervasive diffusion then led to higher demand for the technology, translating into a fast increase in market share.[4] Market share growth then slows down during maturity and turns negative in senescence, when the substitution of the technology by a new one begins. The development of costs and market shares are often illustrated using S-curves: market share development follows an *S*, cost development a *reversed S*.

Prices need not necessarily move in line with costs. A popular hypothesis on the generic development of prices across a technology's life cycle was proposed by Henderson (1968). In pre-commercial and niche market stages, prices may be set lower than costs in expectation of large profits during the phase of pervasive diffusion. If market structure permits, prices are not reduced at the same rate as costs fall during pervasive diffusion and potentially early maturity. This *price umbrella* allows competitors with higher

[2] For some technologies, the process of *diffusion* is rather one of *substitution*. This is the case, if technology adoption takes place in an existing market by replacing the incumbent technology. *Diffusion* in a narrower sense then only refers to adoption in a new market. In practice, however, the distinction between diffusion and substitution is difficult as both processes affect supply and demand and are hence difficult to disentangle (Grübler et al., 1999).

[3] For a more detailed discussion on the drivers of cost reductions see sec. 4.2.

[4] Both developments may reinforce each other in presence of increasing returns.

production costs than the market leader's to enter the market, which ultimately increases competitive intensity. In this environment, prices decline rapidly, often resulting in a *shakeout phase*, during which players with uncompetitive cost structures are eliminated from the market. After this phase, a mature industry structure emerges and prices follow costs.

The neoclassical school emphasizes the central role of prices for the allocation of resources. Yet, explaining the technological adoption process merely from a price or cost perspective ignores other factors that may play a vital role in technological change. According to Jacobsson and Johnson (2000), an innovation system perspective better captures the different factors influencing the technology adoption process. They group these factors into the three categories actors, networks and institutions. These factors "reinforce one another" and can either induce technological change or obstruct it. Among the barriers faced by new technologies, asides from potentially high cost levels, count the need for additional infrastructure, a slow turnover of the existing capacity stock, unfavorable market organization or information and financing constraints (cf. Sagar and van der Zwaan, 2006; IEA, 2000). Barriers arising from network effects also impact the transformation of the power generation sector. The need for a reinforced transmission grid to incorporate large capacities of intermittent technologies, for example, constitutes an important barrier to the substitution of fossil generation technologies.

New technologies evolving in a synergetic manner around the same infrastructure raise barriers for potential substitutes (Grübler et al., 1999). This is one of the reasons for the emergence of *path-dependence* or *lock-in* of interdependent technology systems. Investment into a particular technology makes investment into a related rather than an unrelated one less costly, which will in turn further decrease costs for the related technologies. Correspondingly, opportunity costs associated with investments into unrelated technologies increase as the investment process into the related technologies advances. Self-reinforcement of this process is a necessary condition for path dependence. If this property is given, the technology or technology cluster is characterized by *increasing returns* and the degree of *lock-in* will increase as adoption of the technology proceeds. In presence of increasing returns, small events at the beginning of a technology's life cycle may be decisive for the diffusion of competing technologies. The increasing returns property reinforces initial adoption at small scale and makes adoption of the competing technology less attractive (cf. Arthur, 1989). This can ultimately even lead to large-scale adoption of a *suboptimal* technology. A prominent example for such a development is the QWERTY keyboard layout (David, 1985). The presence of increasing returns may not only be given in case of network effects but may also be a result of cost reductions due to learning or economies of scale. A concept relating cumulative investment into a technology with decreasing costs, thereby implying presence of increasing returns, is the experience curve.

4.2 Experience Curves

4.2.1 Basic Concept: The One-Factor Experience Curve

The experience curve is a concept describing the historical cost reductions observed in an *industry* as function of cumulative industry output. The concept implicitly assumes that activities of industry participants are at the origin of cost reductions and that time is a worse predictor of the latter. Numerous studies have confirmed that cumulative output is indeed a better predictor of cost reductions than time (cf., for example, Rapping, 1965; Lieberman, 1984). Yet it needs to be emphasized that the experience curve merely describes an empirical relationship between costs and cumulative capacity — it does not imply that experience or learning are the main *causes* of cost reductions.

As learning is subject to diminishing returns (Arrow, 1962), the experience curve takes the form of a potential function. Is is described by Equation (4.1).

$$C_t = C_0 \left(\frac{\kappa_t}{\kappa_0} \right)^{-b}, \tag{4.1}$$

where C_t denotes costs at time t, C_0 costs at an arbitrary time 0, κ the cumulative output of the industry and b the learning index.[5] For convenience, the learning index is usually transformed into the learning rate, LR, which denotes the relative cost reduction resulting from a doubling of cumulative output κ.

$$LR = 1 - 2^{-b} \tag{4.2}$$

Closely related is the progress ratio, PR, which indicates the relative specific cost level after a doubling of cumulative output. It is given by

$$PR = 2^{-b} = 1 - LR \tag{4.3}$$

Econometric estimation of the learning rate is usually carried out as linear, ordinary least squares regression on data in log-log format. This has the advantage that the fitted cost development can be graphed as a straight line.

Historically, three closely related concepts describing cost reductions have been distinguished: learning, progress and experience curves. They all share the same functional form, as given by Eq. (4.1), but differ with respect to cost

[5] An alternative formulation frequently seen in literature refers to the costs of the first unit, C_1. The formulation changes to $C_t = C_1 \kappa_t^{-b}$ in this case.

type and economic unit of reference. The *learning curve* is used to describe labor-related improvements at the assembly line or manufacturing process level. The observation that a potential function best describes the decreases in labor input was first reported in the 1930s for airframe production (Wright, 1936) and afterwards frequently applied in this field during World War II (cf., for example, Alchian 1963). In addition to labor, *progress functions* also include improvements for other cost types than labor, such as "in materials inputs, process or product technologies, or managerial technologies" (Dutton and Thomas, 1984). They apply to economic units up to the firm level. *Experience curves* are characterized by an even higher degree of aggregation and comprehensiveness, usually describing total cost reductions at the industry level.[6]

Aggregation of different cost types for progress and experience curves, however, renders distinguishing which cost reductions have occurred due to learning, i.e. an increase in knowledge within the system of reference, and which ones are attributable to other factors, difficult. For learning curves, on the other hand, it is fairly safe to attribute the labor cost savings to knowledge increases. Aggregation of costs across entire industries raises the question whether cost reductions take place at the regional or global level. Capital cost reductions for the renewable power technologies Wind and PV appear to be largely taking place at the global level (cf. Junginger et al., 2005; Nemet, 2006; Kobos et al., 2006; Ferioli et al., 2009). This implies a high level of international knowledge transfer. In the case of wind turbine manufacturing this appears plausible from an industry structure perspective, since the largest manufacturers are also highly internationalized and geographical borders do not impede information flows.[7]

The functional form of the experience curve implies the presence of increasing returns. From this property Henderson (1968) concluded that increasing accumulated experience would be a dominant strategy for the maximization of a firm's profit as well as for societal benefit.[8] However, for a technology that has not yet reached a competitive cost level, the presence of increasing returns will often not lead to a self-reinforcing process of investment. This is due to the fact that an investor may not be able to appropriate the full

[6] These definitions are not used consistently in literature. Krawiec (1983), for example, defines the learning curve as "a generalization about the sources of short-run cost reductions" whereas historically the learning curve has exclusively been referred to savings in labor.

[7] Vestas, General Electric, and Enercon, numbers 1,2, and 4 in market share in 2009, respectively, (cf. BTM Consult Aps, 2010) are each characterized by large shares of sales outsides their home countries and even continent.

[8] Increase of cumulative experience per se will not lead to cost reductions but rather provides a managerial opportunity to realize cost reductions (Abell and Hammond, 1979). For this reason Dutton and Thomas (1984) suggest that the learning rate not be treated as a constant but a dependent variable that is influenced by several policy variables.

benefits from his investments. As Arrow (1962) puts it: "The presence of learning means that an act of investment benefits future investors but this benefit is not paid for by the market."

Two important mechanisms to bring an initially uncompetitive technology to competitive cost levels are niche markets and policy measures. Higher prices can be realized in niche than in mass markets. The resulting deployments then create opportunities for cost reductions, as described by the experience curve. Off-grid PV installations in remote areas are an example of such a niche market. In many cases, however, niche markets are not sufficient to bridge the full gap to competitiveness. Policy measures can then be used to create incentives for technology deployment and thus creation of cost reduction opportunities. An example for such support measures are the feed-in tariffs for Wind and PV currently in place in Germany. These measures, however, also come at a cost. Apart from the opportunity cost of not using resources in other sectors of the economy, feed-in tariffs lower the competitive pressure from potential substitutes and thus incentives to lower costs. For example, based on the results from an econometric innovation-diffusion model, Söderholm and Klaassen (2007) suggest that the high wind power feed-in tariffs in Germany have resulted in a lower price-based learning rate[9] than in the competitive bidding system employed the UK.[10]

Apart from the obvious shortfall of not offering an explanation for the causes of cost reductions, a number of other issues arise when using the experience curve concept. Two important problems relate to improvements in product performance and the use of price instead of cost data. Initial applications of the learning curve were confined to highly standardized products with few design changes, such as airframes. Today, however, the experience curve concept is also applied to products that may be characterized by considerable performance improvements. Performance improvements generally result in a cost increase. When an experience curve is derived from the resulting data, the additional costs from the performance improvement will mask some of the cost reductions actually experienced (cf. Abernathy and Wayne, 1974). An example for this property in the power sector are nuclear plants, for which virtually no decrease in unit investment costs has been observed over several decades. Yet, safety standards for new nuclear power plants have increased steadily over the last 50 years and the associated costs have offset cost reductions over the initial designs from the 1950's and 60's (cf. Neij, 2008). As cost data may not be available for the estimation of an experience curve, prices are frequently used as proxy instead. Yet, if price changes are unrelated to the development of costs, the implied learning rates will be biased (cf. Papineau, 2006). This can, for example, be the case in presence of monopolistic

[9] Learning rates derived on prices rather than costs.

[10] This should not be interpreted in a way that a competitive bidding system is generally superior. Feed-in tariffs have resulted in a more widespread deployment of wind power.

or oligopolistic market structures in the early diffusion stages of a technology or due to changes in governmental support.

Additional issues arise when using experience curves, initially a merely descriptive concept, as forecasting tool (see section 4.2.5). The following two subsections are dedicated to the causes of cost reductions and the potential reasons for differences in learning rates across and within industries.

4.2.2 Drivers of Cost Reductions

The strength of the experience curve concept lies in the description of historical cost reductions. Its explanatory power for those cost reductions is, however, low. The term experience curve is also somewhat misleading, since often only a minor share of cost reductions can be attributed to learning/ experience effects of some sort.

In the past numerous schemes have been suggested to distinguish the different sources of the aggregated cost reductions described by the experience curve. Ghemawat (1985) distinguishes three main categories: exogenous progress, "exploiting scale economies" and "basic improvements learnt from cumulated output." Dutton and Thomas (1984) propose four categories: "technological progress in capital goods," "Horndal-Plant effects",[11] "local industry and firm characteristics" and scale effects. They also distinguish different types of learning based on the dimensions exogenous versus endogenous and autonomous versus induced.[12] Clarke et al. (2008) distinguish R&D, learning-by-doing and spillovers as sources of technical change.[13] Those three examples demonstrate that different perspectives can be taken for the classification of cost reduction sources.

To structure the following discussion on the sources of the cost reductions described by experience curves, the classification system shown in Figure 4.1 will be used. In line with the definition of the experience curve, the industry will serve as reference economic unit to distinguish endogenous from exogenous cost reductions.

Exogenous sources of cost reductions are often referred to as spillovers, here from other industries. They can be subdivided into rent and knowledge ones. Rent spillovers are a direct transfer of economic benefits, whereas knowledge spillovers refer to the transfer of knowledge (Clarke et al., 2008). A typical example for rent spillovers are reductions in input costs, such as material and labor. These spillovers can directly lead to cost reductions without requiring action in the receiving industry. If a spillover is of indirect nature, it merely

[11] At the Horndal iron works labor productivity increases were observed despite the absence of investments over a 15-year-period.

[12] This classification is partly based on Levy (1965), who distinguishes between exogenous, planned and autonomous learning.

[13] They also mention economies of scale but note that those are not a driver of long-term technical change.

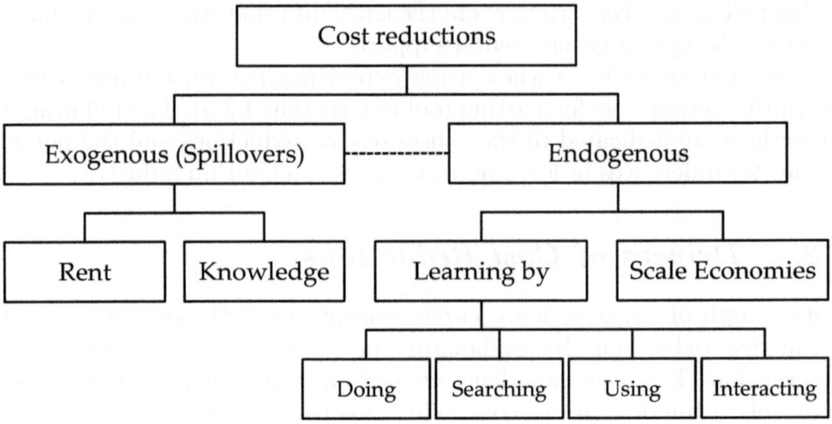

Fig. 4.1 Sources of cost reductions
Source: Own representation partly based on Clarke et al. (2008)

results in a "pool of technological opportunities that might be exploited by the receiving industry" (cf. Clarke et al., 2008; Cohen and Levinthal, 1989). In this case, the particular industry can actively exploit the emerging opportunities by means of the mechanisms summarized under the endogenous category from Fig. 4.1.

The key endogenous drivers of cost reductions are learning effects and economies of scale. The first are related to the accumulation of knowledge, except for direct knowledge spillovers, the latter to the current rate of production output. For this reason learning will result in a long-term cost reduction, whereas economies of scale will disappear again in case the rate of output is reduced.[14] Put differently, learning will shift the unit cost curve downwards, whereas economies of scale represent a movement along the unit cost curve (Isoard and Soria, 2001). In contemporary literature learning effects are usually classified into four main categories (cf., for example, Kahouli-Brahmi 2008; Junginger 2005; Junginger et al. 2005; Kamp et al. 2004; Schaeffer et al. 2004).[15]

Learning-by-doing is the acquisition of knowledge in the course of the production process. Costs are reduced due to increased operational efficiency (cf., for example, Rosenberg 1985, p. 121) and effectiveness. Knowledge

[14] Accumulated knowledge may also be subject to reductions over time due to obsolesce or organizational forgetting.

[15] Additional categories mentioned sometimes are Learning-by-learning, which relates to an improvement of the learning process itself (cf. Rotmans and Kemp, 2003) and Learning-by-expanding/ Learning-by-network-growth. The latter refers to a process in which additional "actors, organizational structures and industrial sectors become involved in, focused on, dependent on and adapted to the new technology." (Schaeffer et al., 2004).

accumulated is generally of tacit nature (Kamp et al., 2004). The concept was first introduced by Arrow (1962). In fact, learning-by-doing was the key source of cost reductions in early learning-curve literature related to airframe production (cf., for example, Wright, 1936; Alchian, 1963). Production experience increased with cumulative output resulting in reduced labor costs. Within the technology life-cycle framework, learning by doing comes into play in the diffusion phase. It is important to distinguish learning-by-doing as source of cost reductions from learning-by-doing, which is frequently used as synonym for one-factor experience curves in modeling literature. The latter describe cost reductions from a multitude of different sources, not limited to learning in a narrower sense.

Learning-by-(re)searching is the systematic and organized search for knowledge, often also referred to as Research and Development (R&D). Activities cover a wide range from basic research to the design of products (cf. Kamp et al., 2004; Garud, 1997). The knowledge generated is usually formalized. R&D plays an important role across the entire technology life-cycle with particular efforts required before the start of the diffusion stage.

Learning-by-using is the acquisition of knowledge from user feedback. The concept was introduced by Rosenberg (1985, p. 122ff.) and is particularly important for complex products, such as capital goods, where performance characteristics are only well understood after intensive user experience. This type of learning occurs during the technology diffusion stage.

Learning-by-interacting refers to the transfer of knowledge among different groups. Lundvall (1988, p. 349ff.) is credited with introducing this concept. Similar to learning-by-using, complex products require a higher degree of learning-by-interacting as manufacturers themselves may not possess all required capabilities for the continuous improvement of the technology. Learning by interacting may already become important before the start of the diffusion phase.

The described sources of cost reductions may also interact with one another. Learning and economies of scale, for example, may reinforce each other as one of the mechanisms reduces costs and thereby in turn generates the opportunity for the other mechanism to apply (cf. Isoard and Soria, 2001). Adler and Clark (1991) stress the complex relationship between first- and second order learning.[16]

The case of cost reductions for PV modules illustrates the fact that learning, in particular learning-by-doing, is often only a minor source of cost reductions. Over the period from 1980 to 2001, Nemet (2006) finds that 43% of reductions in module costs are attributable to plant size, 30% to efficiency

[16] First-order learning occurs autonomously, whereas second-order learning involves managerial action (Adler and Clark, 1991).

improvements[17] and 12% to lower silicon costs.[18] The factor plant size highlights the importance of economies of scale. Increases in efficiency can, to a high degree, be associated with R&D rather than learning-by-doing and silicon cost reductions can be considered a rent spillover from the semiconductor industry. None of the factors associated with learning-by-doing, such as yield improvement, more efficient use of raw materials and silicon wafer size increases, explain more than 3% of the overall cost reductions.

4.2.3 Reasons for Differences in Empirical Learning Rates

Empirical learning rates vary substantially across and within industries. This has been illustrated by the work of Dutton and Thomas (1984), who compiled progress ratios for 108 different learning curves. The largest share of the surveyed learning rate values lie in the 10–30% range with a mode of 20%. However, several values located outside the above mentioned range exist. Interestingly, a higher degree of variation can often be observed across "organizations or organizational units producing the same product than within organizations producing different products" (Argote and Epple, 1990). Empirical learning rate values for renewable power technologies are also characterized by large variations both for the same technology and across technologies. For example, empirical learning rates for wind power are, with approximately 10%, lower than the ones for PV, which are distributed around a value of 20% (see sec. 4.3).

Numerous reasons for these observed learning rate differences have been pointed out. To structure the following discussion, they are grouped into five categories: technology specifications, technology life-cycle stage, knowledge depreciation, impact of institutions, and estimation issues.

(1) Technologies or products are characterized by different specifications. Complex products are made up of many components. Those components are characterized by different technological maturity, i.e. have experienced different levels of cumulated output over their respective lifetimes (cf. Ferioli et al., 2009). If one of two products contains components, which are relatively mature (and have already experienced numerous doublings of output) whereas the other one is made up of relatively immature components, the second product will be characterized by a higher aggregate learning rate than the first. Wind turbines, for example, incorporate several relatively mature

[17] Strictly speaking, module efficiency increases are a performance improvement and should therefore not lead to cost reductions when the dependent variable is costs per rated W_p. The performance increase will, however, reduce power generation costs as higher power output per unit of investment can be achieved. Nemet directly relates efficiency improvements to module costs.

[18] Nemet carries out an econometric estimation with seven explanatory variables for cost reductions.

components, such as tower steel, gearboxes or generators. PV generators, on the other hand, include many components that have not experienced significant cost reductions outside the PV manufacturing industry. This could be a reason why empirical learning rates for wind turbines lie below the ones for PV generators. When considering intra-industry differences, product specification can also be a reason for learning rates differences. Although the implicit assumption of a standardized product is made when deriving experience curves, in practice products differ from firm to firm or even within firms.

(2) Learning rates may vary across a technology's life cycle. Especially at the beginning and towards the end of the life-cycle, departure from log-linearity has been witnessed (cf., for example, Asher 1956; Henderson 1968; Jamasb 2006). Based on these observations some authors have suggested to modify the derivation of experience curves in a manner that takes a technology's current life-cycle stage into account (e.g. Ayres and Martinàs, 1992). Dependence of the learning rate on life cycle stage, however, cannot be generalized and is only observable for some technologies (Alberth, 2008). An example supporting this statement is provided by PV technology. Although it has been common perception that learning rates decrease over the life of a technology, Schaeffer et al. (2004) find that observed learning rates for PV have, in fact, increased over time. Therefore, meaningful adjustment of experience curve estimation based on technology life-cycle stage will likely prove difficult, if not impossible.

(3) Knowledge depreciation constitutes another potential source of learning rate differences (Li and Rajagopalan, 1998). Two key factors leading to knowledge depreciation are organizational forgetting and employee turnover. Organizational forgetting can occur when production processes are interrupted, records are lost or previously acquired knowledge becomes obsolete (Argote and Epple, 1990). Different rates of organizational forgetting and turnover across or within industries may then result in different learning rates.

(4) Institutions, such as markets or governments, can also impact learning rates — in particular when prices are used as a proxy for costs. Prices may not fall in line with costs in a monopolistic or oligopolistic market. The same may be true for industries or particular firms benefiting from governmental subsidies.

(5) If a variable with positive regression coefficient is omitted in an econometric model and that variable shows a positive correlation with variables included in the model, the resulting regression coefficient will be biased (cf. Kahouli-Brahmi, 2008; Berndt, 2000). Consequently a learning rate solely based on cumulative output as independent variable will be biased, as numerous other explanatory variables for cost reductions exist. Examples are R&D spending or economies of scale. In case the relationship of these omitted variables to cumulative output as well as the relationships among the omitted variables differ, the estimation process itself can be a reason for learning rate differences.

4.2.4 Concepts with Several Explanatory Variables

In response to the criticism that the explanatory power of one-factor experience curves (1FEC),[19] where cumulative capacity is the single independent variable, concepts have been proposed that relate cost reductions to several variables.

Especially in the early stages of their life cycle, technologies require high levels of R&D activity, often in combination with other learning mechanisms, to reach cost levels enabling wider diffusion (cf., for example, Ferioli et al. 2009; Grübler et al. 1999). A concept explicitly considering learning-by-searching is the two factor experience curve (2FEC).[20] Its functional form is given by Equation (4.4) (cf., for example, Kahouli-Brahmi 2008; Barreto and Kypreos 2004; Miketa and Schrattenholzer 2004).

$$C_t = C_0 \left(\frac{\theta_t}{\theta_0}\right)^{-a} \left(\frac{\kappa_t}{\kappa_0}\right)^{-b} \tag{4.4}$$

where θ represents the knowledge stock and κ cumulative output. a and b denote the learning indexes corresponding to knowledge stock and cumulative output, respectively. Development of the knowledge stock occurs as described in Equation (4.5).

$$\theta_t = (1 - \rho)\theta_{t-1} + \delta_{t-g} \tag{4.5}$$

R&D expenditures, here denoted by δ, are commonly used as proxy for additions to the knowledge stock. There is a time lag, g, between the occurrence of R&D spending and the resulting additions to the knowledge stock. Following the approach of Li and Rajagopalan (1998), a constant knowledge depreciation factor, here ρ, is usually introduced to represent reductions in the knowledge stock, for example, from obsolescence.

Starting with Kouvaritakis et al. (2000), several studies based on 2FECs have been published (e.g. Kobos et al., 2006; Jamasb, 2006; Miketa and Schrattenholzer, 2004; Barreto and Kypreos, 2004; Klaassen et al., 2002), proposing that explicit representation of R&D, on the one hand, more accurately represents the process of technological learning but, on the other hand, also brings about several drawbacks. Reliable data on R&D expenditures is difficult to obtain, since much of R&D takes place in the private domain. Data for the knowledge depreciation rate and the time lag between

[19] In literature term one factor learning curve (1FLC or OFLC) is frequently extended to experience curves. In this document the expression 1FEC is employed to emphasize that the experience curves considered describe cost reductions beyond the ones for labor.

[20] The rationale for using this expression instead of the more widespread 2FLC is analogous to the one for 1FECs.

expenditure and addition to the knowledge stock are equally difficult to come by. Finally, multicollinearity is an issue, which arises when a change in one of the independent variables triggers a change in the other (cf. Barreto and Kypreos, 2004).

Despite the addition of another explanatory variable when compared with the 1FEC, the 2FEC still leaves numerous sources of cost reductions, such as spillovers and economies of scale unexplained. Other publications have attempted to uncover these effects by means of econometric models with more than two explanatory variables. Isoard and Soria (2001), for example, find that both economies of scale as well as learning effects were significant in the cost reductions for wind power and PV. Nemet (2006) highlights the importance of economies of scale, learning-by-searching and spillovers for PV cost reductions in an econometric analysis comprising seven potential cost reductions sources, whereas learning-by-doing only plays a minor role. These detailed analyses provide valuable insights into the reasons for cost reductions for particular technologies from an ex-post perspective. For forecasting purposes, however, they have the drawback that numerous assumptions on the future development of the considered sources of cost reductions must be made.

4.2.5 Experience Curves as Forecasting Tool

Although originally a purely descriptive concept, the learning curve was soon discovered as forecasting tool. Alchian (1963) reports the use of progress curves to forecast future direct labor requirements for different airframe models.[21] Analyzing the reliability of progress curves for forecasting purposes, he finds a forecasting error of 20–25% — interestingly the range of errors was the same no matter if the forecast had been based on individual models or the aggregation of all 20 aircraft models.

Opinions on the reliability of applying one-factor experience curves for forecasting purposes differ widely. Hall and Howell (1985), for example, find that "experience curves are partly spurious and of little practical value in forecasting or decision making," whereas Alberth (2008) finds the 1FEC a "very useful first order approximation with the distribution of the forecast error being both symmetrical and unbiased."[22]

One-factor experience curves are used in forecasting for several reasons. First, it is generally agreed that forecasts based on cumulative output are

[21] Although the research underlying this paper had been completed before 1950 and the data originated from World War II airframe production, publication took place much later as the data had been rated military classified.

[22] In case it is "considered in log-format".

superior to forecasts based on time.[23] Second, the use of 1FECs in practice requires far less efforts in comparison with more sophisticated approaches. Access to cumulative capacity data is generally much easier than access to the additional data required for more sophisticated approaches, such as R&D spending for 2FECs. Sensitivity analyses based on one instead of several variables are arguably also easier and faster to carry out. In most bottom-up energy models, cumulative output is treated as an endogenous variable, whereas assumptions must be made for other variables, e.g. the R&D budget at given points in time. Third, statistical analysis of historical data demonstrate that errors from 1FEC-based forecasts in log-format have generally produced unbiased and symmetric results (Alberth, 2008). This does not mean that there will not be a forecasting error but that using 1FEC forecasting in combination with stochastic methods may be appropriate.

Despite several advantages when compared to more sophisticated approaches, a long list of issues related to forecasting based on 1FECs exists. Forecasts are highly sensitive to the learning rate value. Due to the functional form of a power function, "technology maturing costs and break even cost grow faster than exponentially" for decreases in learning rates (McDonald and Schrattenholzer, 2001). Consequently, managerial decisions or policy based on overconfident learning rate estimates may lead to high costs without ever realizing the expected benefits, such as a cost competitive technology despite large up-front deployment support.

Two methodological issues further impact the accuracy of forecasts based on experience curves derived from historical data. The fitted experience curve will lead to an overestimate of future cost reductions in presence of spillovers or economies of scale, which are factors unexplained by the regression. Moreover, experience curve forecast errors will not show a bias in log-log format but will show one when viewed on a regular scale (cf. Nordhaus, 2009; Alberth, 2008). This overestimation of learning rates will result in lower marginal costs for the technologies under consideration and thus incorrect forecasts or energy system model results (cf. Nordhaus, 2009).

Performance improvements are generally not reflected in empirical learning rates. For example, experience curves describing the capital cost development for wind turbines in \$/kW will not capture improvements in blade design. This design change may result in a cost increase in terms of \$/kW but be overcompensated by increased power output given the same wind regime, leading to overall lower power generation costs in term of \$/kWh. In contrast to the previous concern of overestimated learning rates, the presence of performance improvements will thus lead to an underestimation of the learning rate. An approach circumventing this problem would be to use power generation costs instead of capital costs as dependent variable. This approach, however, has the serious drawback that generation costs are sensitive to the

[23] However, Papineau (2006) finds that cumulative output becomes insignificant if time is used as additional explanatory variable for historical PV and Wind cost development.

opportunity costs of capital. Changes in interest rate may then mask the cost developments for the technology under consideration.

For several technologies it is argued that learning rates for 1FECs at very early and late developmental stages are lower than in the stages in between. The log-log learning curve then takes the shape of a reversed S with kinks instead of a straight line (Hall and Howell, 1985). This may render the application of the experience curve concept to technologies in very early developmental stages, such as fuel cells, inappropriate.[24]

A closely related issue is the one of infinite cost reductions. Although the absolute amount of cost reductions over time diminishes, costs can still reach unrealistically low levels. Especially in presence of high learning rates, cost forecasts after several doublings of capacity may yield values even below the costs of material input. An approach to address this issue in modeling consists of specifying a *floor cost*, below which costs are not allowed to drop.

If prices are used as proxy for costs, market structure will exert considerable influence on estimated learning rates. For example, if learning rates are estimated based on data from a *price umbrella* and used to forecast costs in a *shakeout* phase, costs will likely be overestimated. In most cases, however, using price instead of cost data will result in an overestimation of learning rates (cf. Nemet, 2006), as profit margins are generally positive.

The discussion above highlights that forecasts based on experience curves have benefits, the key ones being higher accuracy than time based forecasts and good data availability. Yet, there are also numerous issues related to applying the descriptive concept of experience curves as forecasting tool.

It is therefore required to realize and appropriately address the inherent uncertainty associated with the use of experience curves as forecasting tool. For example, to verify learning rate plausibility, Neij (2008) uses complementary bottom-up engineering estimates and judgmental expert assessments. In isolated technology cost forecasts based on experience curves (e.g van der Zwaan and Rabl, 2004; Poponi, 2003; Neij, 1999), sensitivity analysis are used to test the robustness of recommendations for managerial decisions or policy. In more sophisticated energy system models with endogenous cost reduction mechanisms, methods to explicitly consider uncertainty can be employed (see sec. 3.5).

4.3 Learning Rates for Power Generation Technologies

Although reliable estimates for learning rates will not resolve many of the associated issues when experience curves are used as a forecasting tool, they will certainly improve the robustness of results. As for learning rates in general (cf. Dutton and Thomas, 1984), empirically derived learning rates differ significantly across energy technologies but also across studies covering the same

[24] For those technologies there generally also exists too little data to meaningfully fit an experience curve.

technology. Surveying a large number of studies, Kahouli-Brahmi (2008) finds energy technology learning-by-doing rates[25] ranging from 1–42%. A generic rate applied to all technologies would therefore be liable to producing highly unrealistic forecasts.

In response to the need for realistic learning rate assumptions to be used in models with endogenous learning, IEA (2000) and McDonald and Schrattenholzer (2001) have published an overview of empirically assessed learning rates for energy technologies. More recent overviews include Neij (2008) and Kahouli-Brahmi (2008). The comprehensive learning rate overview compiled by Junginger (2005) focuses on learning in wind energy. In addition to reviewing learning rates from literature, Neij's analysis also includes bottom-up cost assessments and judgmental expert opinions to validate the plausibility of her learning rate recommendations to be used for forecasting purposes.

A fundamental difference between most conventional and renewable[26] power plants is the size of a typical installation. Technologies that can be manufactured in "factory-based, automatic production" are characterized by different learning rates than technologies implemented as large scale plants in the field (Neij, 1997). Nuclear, coal or hydro stations have to be built as large units at the site of implementation, whereas most components of wind or PV generators can be built in factories and only require assembly work at the site of implementation. Solarthermal plants lie somewhat in between those extremes. Collectors can be produced in a factory-centered process but significant efforts relating to the power block and the installation of collectors, especially for tower plants, are still required in the field.

In the following, 1FEC learning rates for selected power generation technologies and the potential sources for the observed cost reductions will be discussed. Statistical distributions of learning rates found in literature are not shown for two reasons: The overall limited amount of studies and the differences in data sets employed for the estimation process. Numerous studies are based on different regional, others on global data.[27] Time periods considered and the reference unit for the dependent variable vary as well, i.e. some experience curves refer to capital costs in \$/kW, some to power generation costs in \$/kWh. Kahouli-Brahmi (2008) or McDonald and Schrattenholzer (2001) are suggested for overviews of learning rates with their corresponding data sources.[28]

Wind power: Wind power has experienced several doublings of cumulative capacity over the last two decades, fueled by governmental support.

[25] Learning rates associated with the explanatory variable cumulative output in both 1FECs and 2FECs.

[26] Except for hydro technology.

[27] On the perspective of global versus regional experience curves for Wind and PV see p. 47.

[28] Some PV experience curve studies not covered in these two publications are considered in Neij (2008), some additional Wind experience curve studies in Junginger (2005).

From 1995 to 2009 cumulative output[29] has doubled approximately five times.[30] As one would expect given the comparatively good data availability, a number of experience curve estimations exist. Learning rates based on capital costs for regional studies range from -17% (Neij et al., 2003) to +19% (Junginger, 2005). However, most studies arrive at values in the range of 5–15% (e.g. Coulomb and Neuhoff, 2006; Neij et al., 2003; Neij, 1999; Seebregts et al., 1998). Interestingly Junginger (2005) and Coulomb and Neuhoff (2006), which base their analysis on global capacity, obtain higher learning rates than the studies which derive regional experience curves.[31] A "consensus" learning rate suggestion for wind power based on capital costs is 5–15% (Neij, 2008). This is a low value when compared to the "generic" average across industries of approximately 20% proposed by Dutton and Thomas (1984).

Learning rates for wind power based on generation costs as dependent variable (e.g. in Neij et al., 2003; IEA, 2000) are generally higher than the ones based on capital costs. This can be attributed to turbine performance increases, which are reflected as capital cost escalations rather than reductions. Turbine efficiency, for example, has experienced considerable increases over the last two decades. Contemporary turbines reach efficiencies close to the maximum theoretical value, the Betz coefficient (Betz, 1926).

Historical cost reductions can, to a large degree, be attributed to the upscaling of turbines. From 1985 to 2005 rotor diameters increased from about 20m to more than 120m (European Wind Energy Association, 2009). This corresponds to an increase in rated turbine power by a factor of 36. In line with rotor diameters, hub heights have increased to take advantage of higher wind speeds at greater elevation. As reason for the comparatively low learning rate of wind turbines the maturity of many of its components is suggested (cf., for example, Neij, 1999). Tower steel, generators or gearboxes have been adapted from other applications. An exception are the rotor blades from fiber-reinforced plastics, which are an area where comparatively higher cost reductions can still be expected for the future.[32]

Photovoltaics: PV technology has been characterized by the highest learning rates among energy technologies over the last decades. In contrast to wind, PV generators do not include many mature components, which at least explains part of the difference in learning rates. Although extreme learning rate estimates have been reported, most analyses suggest a global value in the 15–25% range for PV modules (cf., for example,

[29] In terms of capacity installed.

[30] 5 GW in 1995 to 160 GW in 2009 (GWEC, 2007, 2010).

[31] The research by Coulomb and Neuhoff (2006) is based on data provided by Junginger.

[32] Today's blade manufacturing process is still very labor intense, leading to low factory throughput and high scrap or warranty costs (depending on whether defects are discovered before installation or not).

Schaeffer et al. 2004; Harmon 2000; Williams and Terzian 1993) and a similar value for balance of system (BOS) components. Interestingly the learning rate of PV observed since the mid-1980s has been higher than the one observed previously, which stands in contrast to common perceptions that learning rates always decrease as a technology matures (Schaeffer et al., 2004).[33] Prices for PV modules, which are customarily used as proxy for costs, have not declined steadily but there have been alternating periods of stable and decreasing prices. It is often suggested that this discontinuous nature of decreases may, at least partially, be driven by the piecewise modifications in support schemes rather than actual discontinuities in realized manufacturing cost reductions (cf., for example, Neij 2008).

A study using power generation instead of capital costs as dependent variable is IEA (2000), proposing a learning rate of approximately 35%. Even if this is only an approximate figure,[34] it indicates that the costs of power production from PV have decreased at a higher rate than capital costs. This in turn, similar to wind power, indicates presence of performance improvements. In fact, typical efficiency of commercial crystalline-silicon PV cells has increased from around 6–7% in 1975 (Nemet, 2006) to about 14–20% today (IEA, 2010).

Although 1FECs show a good fit for the cost development of PV modules, most of the capital cost reductions have not come about through learning effects. Economies of scale, a result from up-scaling module manufacturing plants and increasing automation, have been quoted as the largest source of historical cost reductions (cf. Nemet, 2006).[35] Another important factor has been silicon cost reduction, a rent spillover from the semiconductor industry. Learning-by-doing related savings, such as yield increases, lower material input requirements or increased wafer size have only had a comparatively minor impact on capital costs (Nemet, 2006).[36]

Future PV generation cost reductions are expected from module lifetime increases, higher efficiencies, lower silicon input requirements and higher yields (Schaeffer et al., 2004). Additional economies of scale may also be

[33] PV could count among the technologies that exhibit a "flat" beginning and end of the experience curve. The change observed by Schaeffer et al. might then coincide with a transition from a "flatter" (though by far not entirely flat) section of the curve to a somewhat steeper one.

[34] Data from EU-Atlas project. As this data is available in five-year steps only and the experience is thus based on 3 data points (1985 to 1995), the implied learning rate should be considered as a rough estimate only.

[35] Isoard and Soria (2001) also suggest that flexible returns to scale exist for PV but highlight that returns to scale may be constant in the long-run once plants have reached optimal scale.

[36] Nemet also mentions efficiency increases as significant source of cost reductions. Strictly speaking, this is a performance improvement when capital costs are used as dependent variable in the analysis.

exploited in case optimal plant size and throughput have not yet been attained. Improvements in module life and higher efficiencies, however, will not lead to capital cost reductions. Therefore, learning rates derived in the future by treating $/W as dependent variable might understate the actual cost reductions achieved in terms of generation costs.

Solarthermal technology: Solarthermal power can be considered an immature technology in comparison to Wind or PV. Yet, as for wind power, maturity of its components differ. Power block technology is similar to the one found in conventional plants, whereas collector field and thermal storage are specific to solarthermal technology. For this reason Viebahn et al. (2008) propose using a learning rate of only 5% for the power block and 12% for both collector field and thermal storage components. The overall learning rate estimated based on the empirical data of the first nine solar through plants in the Mojave desert[37] is 12% (Enermodal Engineering Ltd., 1999). Based on this figure and several bottom-up analyses, Neij (2008) suggests a 10% learning rate for use in forecasting, characterized by a high degree of uncertainty.

Coal and Biomass: Conversion technologies for power from biomass are similar to the ones used for coal (see sec. A.5), therefore a similar learning rate can be expected (cf. Neij, 2008).[38] For coal power plants learning rates in the 1–8% range have been proposed (cf. Rubin et al., 2007; Kouvaritakis et al., 2000; Joskow and Rose, 1985).

Combined cycle gas turbines: This technology is characterized by higher learning rates than coal, possibly since it is still in an earlier technology life cycle stage. During the 1980s prices for combined cycle gas turbines even increased, implying a learning rate of -11%, followed by a sharp decrease with an implied learning rate of +25% in the 1990s. The first period of price increases could correspond to the initial technology development stages, where performance improvements led to cost increases, as well as the price umbrella stage (see p. 45), where only few producers were in the market. The second period with high price reductions could then correspond to the shake-out phase (Claeson Colpier and Cornland, 2002). For the following phase of stable margins Claeson Colpier and Cornland propose a learning rate of 10%.

Other generation technologies: Several generation technologies are characterized by no capital cost reductions or even cost increases. Hydro power, one of the oldest power generation options, has only shown marginal cost improvements.[39] For nuclear power, specific investment costs have even increased over the course of the 1980s and 1990s (cf. Neij, 2008).[40]

[37] SEGS I–IX.

[38] Other bioenergy technologies, e.g. related to the production of biomass or conversion apart from power generation may well experience higher learning rates.

[39] Kouvaritakis et al. (2000) find a learning rate of 1.4%.

[40] Neij bases this assessment on unpublished data for France.

Overall 1FEC learning rates for power generation technologies appear to lie below the economy "average" of 20% put forth by Dutton and Thomas (1984), with PV modules constituting the only exception to this observation.

4.4 Endogenous Technological Change in E3 Models

Initially, Energy-Economy-Environment models treated technological change as time-dependent model input. An example for such an approach is *autonomous energy efficiency improvement* (AEEI), which reflects increasing energy efficiency of the economy over time. A consequence of such an approach, treating technological change as *exogenous*, is that investments into a technology will be postponed until the technology has reached a certain cost level. Once implementation begins, the technology will diffuse gradually. Up-front investments, apart from long-enough waiting, are not required and the technology is therefore available to society as a free good (Messner, 1997).

It is generally agreed that technical change arises from within an economic system (cf., for example, Grübler et al. 1999). Significant up-front investments, for example, into R&D, are required to bring costs of a particular technology to competitive levels. Once a competitive price level has been reached, positive feedback between increased output and costs may quickly reduce costs further and lead to more radical diffusion than exhibited by models treating technological change as exogenous.

Endogenous modeling of technological change offers the possibility to reflect the requirement for up-front investments as well as the self-reinforcing dynamics emerging during diffusion. *Endogenous* in this context means that the value of model variables will have an influence on the rate and direction of technological change.[41]

Energy models incorporating some kind of endogenous change mechanism go back to the 1970's (e.g. Nordhaus and van der Heyden, 1977) but the widespread inclusion of endogenous change in energy models didn't start before the 1990s. Several streams of literature have contributed to this development, which can be grouped into three main categories: endogenous growth, innovation and experience curve literature (Köhler et al., 2006).

Endogenous growth literature stresses the importance of knowledge spillovers for technological change. Technology is considered a *non-rival, partially excludable good.*[42] Non-rivalry implies the existence of increasing

[41] A closely connected notion is the one of *induced* technical change. Induced technical change refers to the alteration of the rate and direction of technological change by means of policy measures (Clarke et al., 2008). Models aiming at analyzing the impact of policy measures on technical change therefore require endogenous representation of the technical change progress.

[42] *Non-rival* goods can be used by more than one firm or person simultaneously. Owners can prevent others from using *excludable goods*. Excludability is a function of "the technology and the legal system." (Romer, 1990).

returns, whereas the property of only partial excludability implies existence of spillovers (cf. Romer, 1990).

Apart from its contributions to the topic of spillovers, *innovation literature*, as defined by Köhler et al. (2006), has also addressed the roles of uncertainty, path dependence and diffusion in the process of technical change. As the return on investment into new technologies, for example, in form of R&D or deployment subsidies, is uncertain, future cost development is uncertain. The existence of uncertainty impacts decision making at the firm and government level, for example, in the development of low-carbon technologies. Uncertainty on future damages from climate change may increase reluctance of policymakers to commit to long term support schemes. Uncertainty about future support, in turn, may prevent firms from allocating investments to such technologies (cf. Montgomery and Smith, 2007; Köhler et al., 2006). Increasing returns often accompany technological change and lead to path dependence and lock-in of particular technologies (cf., for example, Weyant and Olavson, 1999). An important implication of this property is that an initially inferior technology may end up dominating a market. Small events early in the life of a technology, such as a government subsidy, may provide an initial advantage and the effect of increasing returns may subsequently be strong enough to render following this path less costly than starting to pursue the cost reduction path of an alternative, initially superior technology (cf. Arthur, 1989). Moreover, market failures impact the diffusion of new technologies and can lead to a situation of sub-optimal underinvestment. Besides the presence of increasing and uncertain returns on investments, barriers from trade restrictions or transaction costs can hinder diffusion. Köhler et al. (2006) emphasize that addressing these issues through policy measures can generate large returns to society.

Experience curve literature addresses the topic of technological change from a microeconomic perspective. This topic has been discussed in the preceding two sections (4.2–4.3).

The previous paragraphs have illustrated that capturing all facets of technological change in an energy model is an impossible task. Yet, even simplified representation of the endogenous drivers of change in a model is still preferable to none at all, assuming that the resulting model closer resembles the real world. The next two sections provide an overview of the methods currently available for introducing endogenous change into models and how to deal with the issue of non-convexity resulting from the presence of increasing returns.

4.4.1 Implementation Methods

Methods available for implementing endogenous technical change in E3 models depend on whether a bottom-up or top-down approach is followed. The two approaches differ mainly with respect to the degree of technological detail (cf. Löschel, 2002). Whereas bottom-up models allow for inclusion of several competing technologies, top-down models represent the energy sector in an aggregate manner. The higher technological detail in bottom-up models comes at the cost of foregoing general economic coherence. For example, the opportunity costs of allocating R&D investment to the energy sector are not taken into account. Other issues include ignoring transaction and information related costs. On the other hand, bottom-up models allow for explicit tracking of capital stock vintage. This allows confinement of effects from technological change to newly installed capacity instead of application thereof to the entire capital stock.

The selection of the appropriate approach depends on the research question at hand. If the objective is technology assessment under consideration of technological improvement, a bottom-up approach will be more suitable than a top-down one. On the other hand, if the focus is on policy assessment considering effects of energy policy outside the energy sector or the analysis of the general technological diffusion and innovation process, a top-down model will be the method of choice. Hybrid models combine the strengths of both model types (for more details see p. 25), combining a bottom-up energy sector with a top-down description of the economy. Yet, these combinations will generally not perform as well in their particular "area of expertise" as the corresponding "pure" approach. For example, hybrid models are generally characterized by a lower degree of technological detail than a generic bottom-up model.

Sue Wing (2006) distinguishes four approaches to implement endogenous technological change in E3 models:

I. Backstop technology[43]
II. Price-induced input augmentation
III. Stock of knowledge
IV. Learning-by-doing[44]

More than one of these mechanisms can be introduced simultaneously. Löschel and Otto (2009), for example, include a backstop technology as well as improvements via a stock-of-knowledge. Castelnuovo et al. (2005) apply learning-by-doing and a stock-of-knowledge approach.

The *backstop technology* approach is available for both bottom-up and top-down approaches. In top-down models, an alternative production function is specified, which is generally available after an exogenously specified date. Rising prices may then trigger the use of the alternative production function. Transition usually occurs in a *flip-flop* manner, i.e. the entire production shifts

[43] A backstop technology is "a set of processes that is capable of meeting the demand requirements, and has a virtually infinite resource base" (Nordhaus et al., 1973).

[44] Learning-by-doing in this context refers to the dependence of costs on cumulative output and not to the particular source of cost reductions. For disambiguation of the expression learning-by-doing see p. 50.

to the new technology. In bottom-up models, the backstop can be implemented simply as another technology, available after a certain date or triggered by a model-endogenous activity (cf. Sue Wing, 2006). Sue Wing classifies the backstop approach as "semi-endogenous" since backstop production is triggered endogenously but the technology's parameter assumptions are independent of activity levels within the model.

Under the *price-induced input augmentation* approach, relative price changes lead to technical change. Instead of an AEEI coefficient, production function coefficients are a function of relative prices. Describing this function can be difficult, among others due to the "lack of empirical information about the shape of" this function and the difficulty of simultaneously describing "complementarities and tradeoffs among the augmentation of different inputs." Consequently the approach is rarely used, although implementations, e.g. Dowlatabadi (1998), exist (Sue Wing, 2006).

The *stock-of-knowledge* approach is a popular method in top-down modeling. A stock of knowledge relating to a particular production input is explicitly represented. Investment into R&D increases, obsolesce decreases the stock of knowledge. Input coefficients in the production function are then modeled as function of the respective knowledge stocks (cf. Sue Wing, 2006). For example, if the production function for energy is a function of both low- and high-carbon technologies, the knowledge stock associated with low carbon technologies will modify the production function coefficient for low carbon technologies. An advantage of this approach is that it captures the opportunity costs of redirecting R&D funds from other sectors of the economy. Unfortunately the approach is generally not applicable to partial equilibrium models.[45]

The *learning-by-doing* approach is available to both bottom-up and top-down models, yet implementation methods differ depending on model setup. In top-down models, learning-by-doing can be represented by means of a production function with increasing returns to a variable, such as physical capital (e.g. Castelnuovo et al. 2005). Goulder and Mathai (2000) introduce learning-by-doing into a CO_2 abatement function, where higher cumulative abatement decreases specific abatement costs. For bottom-up models, the use of experience curves is a popular approach to endogenize technical change. Generally, the capital costs of one or more technologies are formulated as function of cumulative capacity additions. Besides the conceptual issues arising from the use of experience curves as forecasting tool (see sec. 4.2), introduction of experience curves to cost-minimizing models will result in non-convexity. Moreover, presence of increasing returns in absence of sufficiently stringent constraints will lead to *flip-flop* behavior in perfect-foresight models, i.e. once deployment of a technology is cost-optimal, it is introduced as rapidly as possible, even if this may be unrealistic from a manufacturing capacity perspective.

[45] The two-factor experience curve contains elements of both the stock-of-knowledge approach as well as the experience curve approach.

Messner (1997) and Mattsson and Wene (1997) are credited for first implementing experience curves in the bottom-up energy system models MESSAGE and GENIE, respectively. Since then, several applications of the 1FEC to bottom-up or hybrid models have been carried out. Examples include POLES (Kouvaritakis et al., 2000), MARKAL (Seebregts et al., 1998; Barreto and Kypreos, 2002), ERIS (Seebregts et al., 2000), DNE21+ (Sano et al., 2006), GET-LFL (Hedenus et al., 2006), MERGE-ETL (Kypreos and Bahn, 2003) or TIMES (Rout et al., 2009).

To disentangle the impact of R&D efforts and other cost reduction sources, 2FECs have been endogenized in bottom-up models. Kouvaritakis et al. (2000) use a formulation, where R&D investments directly influenced capital costs. Later implementations (e.g. in Barreto and Kypreos, 2004 and Miketa and Schrattenholzer, 2004) have used a knowledge stock formulation, corresponding to the one illustrated in Eqs. (4.4) and (4.5). Although the use of 2FECs over 1FECs is conceptually more convincing, it has a number of drawbacks. The inclusion of an additional nonlinear factor into the objective function aggravates the issue of non-convexity. The estimation of learning rates is more difficult for 2FECs than 1FECs, due to nonavailability of coherent private R&D spending data and existence of multicollinearity between cumulative capacity and the variables driving the knowledge stock. Moreover, in partial equilibrium models, future R&D budgets must be exogenously assigned.

Under both the 1FEC and 2FEC approach, a popular technique to limit indefinite cost reductions is the introduction of *floor costs*, below which specific costs cannot fall. To limit unrealistically rapid deployment from a manufacturing capacity perspective, maximum growth constraints are also frequently introduced. As these assumptions, along with the learning rates, are exogenous, technical change is generally never entirely endogenous to a model.

4.4.2 Dealing with Non-convexity in Bottom-Up Models

The introduction of experience curves in any model will lead to non-linearity. If a model has the form of a linear program, endogenization of experience curves will convert it into a nonlinear one with increasing returns. In models with the objective of cost-minimization, the presence of increasing returns renders the objective function concave. The result is a non-convex problem, characterized by multiple local equilibria. The application of standard solver algorithms in such a setting unfortunately does not guarantee identification of the global optimum. To address this issue, three approaches can be taken:

I. Retain NLP, apply appropriate numerical solution methods and heuristics
II. Retain NLP, solve analytically
III. Convert into MIP, apply standard numerical solution methods

If the non-convex NLP is retained and numerical methods are employed, the attainment of a global optimum cannot be guaranteed with certainty. The probability of finding the global optimum, however, can be greatly improved. One possibility is the application of specialized algorithms, such as BARON (Tawarmalani and Sahinidis, 2005), which can identify global extrema in non-convex problems of limited complexity. Manne and Barreto (2004) apply this approach in a stylized model of three power generation technologies, two thereof capable of learning. Although they find that BARON effectively identifies the global optimum for this small-scale model, they emphasize that for large-scale, non-convex models the most effective way of identifying the global optimum is by means of employing "different starting points and terminal conditions." Yet, under this approach there is no certainty that the discovered optimum is indeed the global and not simply the best of the identified local ones. Mattsson and Wene (1997) and Sano et al. (2006) are other studies retaining non-convexity and applying numerical solution methods in energy system models with a limited amount of learning technologies.[46]

It is also possible to solve small-scale, non-convex NLPs analytically. By enumerating all possible local optima and comparing their corresponding objective function values, it is possible to identify the global optimum with certainty. In chapter 5 this method is applied in a two-period, two-technology model with uncertain learning. Yet, the considerable efforts required for the analytical solution process, even in small-scale models, render this approach unfit for system models characterized by numerous technologies and periods.

The third approach, conversion into a Mixed-Integer program, has been the most popular for bottom-up energy system models so far (e.g. Messner, 1997; Mattsson, 1997; Kram et al., 2000; Barreto and Kypreos, 2002; Rout et al., 2009). The concave cumulative cost function is approximated by a number of linear segments. Those linear segments generally lie below the nonlinear cumulative cost function and only touch it at each segment starting and ending point. The approximation consequently yields lower capital costs than the original, nonlinear function.[47] Binary variables indicate the active segment of the approximated cost function. For a detailed description of the approximation and implementation procedure the interested reader is referred to pp. 153ff. in chapter 6. The advantage of this piece-wise linear approach is that attaining a global optimum, even in large-scale models, can be assured. A severe disadvantage of this approach are the vastly higher computational requirements than in problems with continuous solution spaces. Specific algorithms, for example, branch & cut ones, need to be employed to exhaustively explore the non-continuous solution space and identify the global optimum.

In summary, there is no "silver bullet" approach to solving the issues related to the introduction of increasing returns into energy system models.

[46] Mattsson and Wene (1997) use two, Sano et al. (2006) three learning technologies.

[47] Except at the points, where the linear segments coincide with the original function.

For mid- to large-scale models the modeler must choose between potential non-global optimality under the non-convex NLP approach or increased computational complexity and a certain understatement of capital costs under the MIP approach.

4.5 Implications

The models applied to analyze optimal deployment strategies for renewable power generation technologies in this book follow a bottom-up, partial equilibrium approach. Therefore, only the experience curve and the backstop approach are available for representation of endogenous technological change. The more sophisticated experience curve approach, in form of a 1FEC, referring to turnkey capital costs, is chosen. The 2FEC approach is avoided, as it would require exogenous assumptions on the R&D budget until 2100 and cause computational difficulties when converting the non-convex problem into a MIP.[48]

The 1FEC approach requires assumptions on the future value of learning rates. Uncertainty associated with this value in combination with the high sensitivity of model results to the learning rate, have led to the widespread call for endogenous treatment of uncertainty in models with endogenous technical change (cf., for example, Alberth 2008; Berglund and Söderholm 2006; Grübler et al. 1999). Grübler et al. assert that characteristic S-curves, which have been observed in many historical technology diffusion processes, can only be witnessed in models treating technical change as well as the associated uncertainty endogenously. Consequently, the models presented in the following chapters will explicitly consider uncertainty by means of a stochastic programming approach.

Solution approaches differ for these models. For the small-scale model presented in chapter 5, non-convexity is retained and the model is solved analytically. The large-scale power sector model with eight learning technologies is converted into a MIP and solved numerically. It also includes floor costs and maximum growth constraints to avoid unrealistically low capital costs as well as unrealistically high diffusion rates from a manufacturing capacity perspective (see chapters 6 and 7).

When interpreting model results, several issues related to using the experience curve as forecasting tool need to be considered. On the one hand, empirical learning rates are upwards biased in presence of spillovers. Nordhaus (2009) argues that using these empirical rates to forecast future costs will understate marginal costs and "bias optimization models to tilt towards technologies that are incorrectly specified as having high learning coefficients." On the other hand, these empirical learning rates refer to capital cost reductions and therefore do not reflect the observed cost of energy reductions from efficiency improvements. Using empirical learning rates for capital costs

[48] The MIP would become a MINLP.

only will therefore overstate the cost of energy from the particular technology. Which of the two described effects is of greater magnitude in existing empirical studies would have to be determined for each technology and data set individually — a virtually impossible task. For the models presented hereafter, one thus cannot determine whether using empirical learning rates to forecast future developments will over- or understate marginal costs for the learning technologies but it is clear that they compensate each other to a certain extent.

Another limitation relates to the neoclassical approach followed in the models, where costs are the sole driver of technological change. This approach thus ignores non-cost related variables affecting the diffusion of technologies, such as network effects or transaction costs. Interactions with other sectors of the economy are also ignored due to the partial equilibrium approach.

Although limitations exist for the chosen approaches of endogenizing technological change, a model considering technological change as exogenous would be less suitable to investigating the optimal deployment of renewable power generation technologies. As new renewable technologies are generally still not competitive with conventional technologies, cost reductions are required so that large-scale deployment can take place. The endogenization of technological change yields an answer to two important questions: (1) *Which* technologies should be pursued and which ones do not merit the high up-front investment? (2) *When* should the investment into the particular technology take place?

References

Abell, D.F., Hammond, J.S.: Strategic market planning: Problems and analytical approaches. Prentice-Hall, Englewood Cliffs (1979)

Abernathy, W.J., Wayne, K.: Limits of the learning curve. Harvard Business Review 52(5), 109–119 (1974)

Adler, P.S., Clark, K.B.: Behind the Learning Curve: A Sketch of the Learning Process. Management Science 37(3), 267–281 (1991)

Alberth, S.: Forecasting technology costs via the experience curve - Myth or magic. Technological Forecasting and Social Change 75(7), 952–983 (2008)

Alchian, A.: Reliability of Progress Curves in Airframe Production. Econometrica 31(4), 679–693 (1963)

Argote, L., Epple, D.: Learning curves in manufacturing. Science 247(23), 920–924 (1990)

Arrow, K.J.: The Economic Implications of Learning by Doing. The Review of Economic Studies 29(3), 155–173 (1962)

Arthur, W.B.: Competing Technologies, Increasing Returns, and Lock-In by Historical Events. The Economic Journal 99(394), 116–131 (1989)

Asher, H.: Cost-Quantity Relationships in the Airframe Industry. The RAND Corporation, Santa Monica (1956)

Ayres, R.U.: Toward a non-linear dynamics of technological progress. Journal of Economic Behavior & Organization 24(1), 35–69 (1994)

Ayres, R.U., Martinàs, K.: Experience and the life cycle: Some analytic implications. Technovation 12(7), 465–486 (1992)

Barreto, L., Kypreos, S.: Multiregional technological learning in the energy systems MARKAL model. International Journal of Global Energy Issues 17(3) (2002)

Barreto, L., Kypreos, S.: Endogenizing R&D and market experience in the "bottom-up" energy-systems ERIS model. Technovation 24(8), 615–629 (2004)

Berglund, C., Söderholm, P.: Modeling technical change in energy system analysis: analyzing the introduction of learning-by-doing in bottom-up energy models. Energy Policy 34(12), 1344–1356 (2006)

Berndt, E.R.: The practice of econometrics: Classic and contemporary. Addison-Wesley, Reading (2000)

Betz, A.: Windenergie und ihre Ausnutzung durch Windmühlen (Reprint 1994 ed.). Ökobuch Verlag, Staufen (1926)

BTM Consult Aps. World Market Update 2009: Forecast 2010–2014 (2010)

Castelnuovo, E., Galeotti, M., Gambarelli, G., Vergalli, S.: Learning-by-Doing vs. Learning by Researching in a model of climate change policy analysis. Ecological Economics 54(2-3), 261–276 (2005)

Claeson Colpier, U., Cornland, D.: The economics of the combined cycle gas turbine — an experience curve analysis. Energy Policy 30(4), 309–316 (2002)

Clarke, L., Weyant, J., Edmonds, J.: On the source of technological change: What do models assume. Energy Economics 30(2), 409–424 (2008)

Cohen, W.M., Levinthal, D.A.: Innovation and learning: The two faces of R&D. The Economic Journal 99(397), 569–596 (1989)

Coulomb, L., Neuhoff, K.: Learning Curves and Changing Product Attributes: the Case of Wind Turbines: Cambridge Working Paper in Economics 0618, Cambridge (2006)

David, P.A.: Clio and the Economics of QWERTY. The American Economic Review 75(2), 332–337 (1985)

Dowlatabadi, H.: Sensitivity of climate change mitigation estimates to assumptions about technical change. Energy Economics 20(5-6), 473–493 (1998)

Dutton, J., Thomas, A.: Treating Progress Functions as a Managerial Opportunity. The Academy of Management Review 9(2), 235–247 (1984)

Enermodal Engineering Ltd. Cost reduction study for solar thermal power plants: Final Report (1999)

European Wind Energy Association. The economics of wind energy, Brussels (2009)

Ferioli, F., Schoots, K., van der Zwaan, B.: Use and limitations of learning curves for energy technology policy: A component-learning hypothesis. Energy Policy 37(7), 2525–2535 (2009)

Garud, R.: Technological innovation: Oversights and foresights. Cambridge University Press, Cambridge (1997)

Ghemawat, P.: Building Strategy on the experience curve. Harvard Business Review 63(2), 143–149 (1985)

Goulder, L.H., Mathai, K.: Optimal CO_2 Abatement in the Presence of Induced Technological Change. Journal of Environmental Economics and Management 39(1), 1–38 (2000)

Grübler, A., Nakićenović, N., Victor, D.G.: Dynamics of energy technologies and global change. Energy Policy 27(5), 247–280 (1999)

GWEC. Global Wind 2006 Report, Brussels (2007)

GWEC. Global Wind 2009 Report, Brussels (2010)

Hall, G., Howell, S.: The Experience Curve from the Economist's Perspective. Strategic Management Journal 6(3), 197–212 (1985)

Harmon, C.: Experience Curves of Photovoltaic Technology, Laxenburg (2000)

Hedenus, F., Azar, C., Lindgren, K.: Induced Technological Change in a Limited Foresight Optimization Model. The Energy Journal 27(special issue 1) (2006)

Henderson, B.: Perspectives on experience, Boston (1968)

IEA. Experience curves for energy technology policy. IEA Publications, Paris (2000)

IEA. Technology Roadmap: Solar photovoltaic energy, Paris (2010)

Isoard, S., Soria, A.: Technical change dynamics: evidence from the emerging renewable energy technologies. Energy Economics 23(6), 619–636 (2001)

Jacobsson, S., Johnson, A.: The diffusion of renewable energy technology: an analytical framework and key issues for research. Energy Policy 28(9), 625–640 (2000)

Jamasb, T.: Technical Change Theory and Learning Curves: Patterns of Progress in Energy Technologies: Cambridge Working Papers in Economics 0625, Cambridge (2006)

Joskow, P.L., Rose, N.L.: The effects of Technological Change, Experience, and Environmental Regulation on the Construction Cost of Coal-Burning Generating Units. The RAND Journal of Economics 16(1), 1–27 (1985)

Junginger, M.: Learning in Renewable Energy Technology Development: Doctoral thesis. Doctoral thesis, Utrecht University, Utrecht (2005)

Junginger, M., Faaij, A., Turkenburg, W.C.: Global experience curves for wind farms. Energy Policy 33(2), 133–150 (2005)

Kahouli-Brahmi, S.: Technological learning in energy–environment–economy modelling: A survey. Energy Policy 36(1), 138–162 (2008)

Kamp, L.M., Smits, R.E.H.M., Andriesse, C.D.: Notions on learning applied to wind turbine development in the Netherlands and Denmark. Energy Policy 32(14), 1625–1637 (2004)

Klaassen, G., Miketa, A., Riahi, K., Schrattenholzer, L.: Technological Progress Towards Sustainable Development. Energy and Environment 13(4-5), 553–577 (2002)

Kobos, P.H., Erickson, J.D., Drennen, T.E.: Technological learning and renewable energy costs: implications for US renewable energy policy. Energy Policy 34(13), 1645–1658 (2006)

Köhler, J., Grubb, M., Popp, D., Edenhofer, O.: The Transition to Endogenous Technical Change in Climate-Economy Models: A Technical Overview to the Innovation Modeling Comparison Project. The Energy Journal 27(special issue 1), 17–55 (2006)

Kouvaritakis, N., Soria, A., Isoard, S.: Modelling energy technology dynamics: methodology for adaptive expectations models with learning by doing and learning by searching. International Journal of Global Energy Issues 14(1-4), 104–115 (2000)

Kram, T., Seebregts, A., Schaeffer, G.J., Barreto, L., Kypreos, S., Messner, S., Schrattenholzer, L.: Technology dynamics in energy systems models with perfect foresight. International Journal of Global Energy Issues 14(1-4), 48–64 (2000)

Krawiec, F.: Concepts of Learning and Experience in Developing Solar Thermal Technologies. Technological Forecasting and Social Change 24(3), 207–246 (1983)

Kypreos, S., Bahn, O.: A MERGE Model with endogenous technological progress. Environmental Modeling and Assessment 8(3), 249–259 (2003)
Levy, F.K.: Adaptation in the Production Process. Management Science 11(6), B136–B154 (1965)
Li, G., Rajagopalan, S.: A learning curve model with knowledge depreciation. European Journal of Operational Research 105(1), 143–154 (1998)
Lieberman, M.: The Learning Curve and Pricing in the Chemical Processing Industries. The RAND Journal of Economics 15(2), 213–228 (1984)
Löschel, A.: Technological change in economic models of environmental policy: a survey. Ecological Economics 43(2-3), 105–126 (2002)
Löschel, A., Otto, V.M.: Technological uncertainty and cost effectiveness of CO_2 emission reduction. Energy Economics 31(suppl. 1), S4–S17 (2009)
Lundvall, B.A.: Innovation as an interactive process: from user–producer interaction to the national system of innovation. In: Dosi, G., Freeman, C., Nelson, R., Silverberg, G., Soete, L. (eds.) Technical Change and Economic Theory, pp. 349–369. Pinter, London (1988)
Manne, A.S., Barreto, L.: Learn-by-doing and carbon dioxide abatement. Energy Economics 26(4), 621–633 (2004)
Mattsson, N.: Internalizing technological development in energy system models: Licentiate. Chalmers University, Gothenburg (1997)
Mattsson, N., Wene, C.-O.: Assessing New Energy Technologies Using an Energy System Model With Endognized Experience Curves. International Journal of Energy Research 21(4), 385–393 (1997)
McDonald, A., Schrattenholzer, L.: Learning rates for energy technologies. Energy Policy 29(4), 255–261 (2001)
Messner, S.: Endogenized technological learning in an energy systems model. Journal of Evolutionary Economics 7(3), 291–313 (1997)
Miketa, A., Schrattenholzer, L.: Experiments with a methodology to model the role of R&D expenditures in energy technology learning processes; first results. Energy Policy 32(15), 1679–1692 (2004)
Montgomery, D.W., Smith, A.: Price, quantity and technology strategies for climate change policy. In: Schlensinger, M.E., Kheshgi, H., Smith, J.B., La Chesnaye, F.C., Reilly, J.M., Wilson, T., Kolstad, C. (eds.) Human-Induced Climate Change: An Interdisciplinary Assessment, pp. 328–342. Cambridge Univ. Press, New York (2007)
Neij, L.: Use of experience curves to analyse the prospects for diffusion and adoption of renewable energy technology. Energy Policy 23(13), 1099–1107 (1997)
Neij, L.: Cost dynamics of wind power. Energy 24(5), 375–389 (1999)
Neij, L.: Cost development of future technologies for power generation — A study based on experience curves and complementary bottom-up assessments. Energy Policy 36(6), 2200–2211 (2008)
Neij, L., Dannemand Andersen, P., Durstewitz, M., Helby, P., Hoppe-Kilpper, M., Morthorst, P.E.: Experience Curves: A Tool for Energy Policy Assessment, Lund (2003)
Nemet, G.F.: Beyond the learning curve: factors influencing cost reductions in photovoltaics. Energy Policy 34(17), 3218–3232 (2006)
Nordhaus, W.D.: The Perils of the Learning Model For Modeling Endogenous Technological Change: NBER Working Paper 14638, Cambridge (2009)
Nordhaus, W.D., Houthakker, H., Solow, R.: The Allocation of Energy Resources. Brookings Papers on Economic Activity 1973(3), 529–576 (1973)

Nordhaus, W.D., van der Heyden, L.: Modeling Technological Change: Use of Mathematical Programming Models in the Energy Sector (CFP 457), New Haven (1977)

Papineau, M.: An economic perspective on experience curves and dynamic economies in renewable energy technologies. Energy Policy 34(4), 422–432 (2006)

Poponi, D.: Analysis of diffusion paths for photovoltaic technology based on experience curves. Solar Energy 74(4), 331–340 (2003)

Rapping, L.: Learning and World War II Production Functions. The Review of Economics and Statistics 47(1), 81–86 (1965)

Romer, P.M.: Endogenous Technological Change. The Journal of Political Economy 98(5), S71–S102 (1990)

Rosenberg, N.: Inside the black box: Technology and economics. Cambridge Univ. Press, Cambridge (1985)

Rotmans, J., Kemp, R.: Managing Societal Transitions: Dilemmas and Uncertainties: The Dutch energy case-study. In: OECD Workshop on Benefits of Climate Policy: Improving Information for Policy Makers (2003)

Rout, U.K., Fahl, U., Remme, U., Blesl, M., Voß, A.: Endogenous implementation of technology gap in energy optimization models — a systematic analysis within TIMES G5 model. Energy Policy 37(7), 2814–2830 (2009)

Rubin, E.S., Yeh, S., Antes, M., Berkenpas, M., Davison, J.: Use of experience curves to estimate future cost of power plants with CO_2 capture. International Journal of Greenhouse Gas Control 1(2), 188–197 (2007)

Sagar, A.D., van der Zwaan, B.: Technological innovation in the energy sector: R&D, deployment, and learning-by-doing. Energy Policy 34(17), 2601–2608 (2006)

Sano, F., Akimoto, K., Homma, T., Tomoda, T.: Analysis of Technological Portfolios for CO_2 Stabilizations and Effects of Technological Changes. The Energy Journal 27(special issue 1) (2006)

Schaeffer, G.J., Alsema, E., Seebregts, A.J., Beurskens, L., Moor, H.H., van Sark, W., Durstewitz, M., Perrin, M., Boulanger, P., Laukamp, H., Zuccaro, C.: Learning from the Sun: Analysis of the use of experience curves for energy policy purposes: The case of photovoltaic power. Final report of the Photex project, Petten (2004)

Schumpeter, J.A.: The Theory of Economic Development: An Inquiry into Profits, Capital, Credit, Interest, and the Business Cycle. Harvard Economic Studies, vol. 46. Harvard Univ. Press, Cambridge (1934)

Seebregts, A., Kram, T., Schaeffer, G.J., Stoffer, A.: Endogenous Technological Learning: Experiments with MARKAL: Contribution to Task 2.3 of the EU-TEEM Project, Petten (1998)

Seebregts, A.J., Schaeffer, G.J., Barreto, L., Kypreos, S., Messner, S., Schrattenholzer, L.: Technology dynamics in energy systems models with perfect foresight. International Journal of Global Energy Issues 14(1-4), 48–64 (2000)

Söderholm, P., Klaassen, G.: Wind Power in Europe: A Simultaneous Innovation–Diffusion Model. Environmental and Resource Economics 36(2), 163–190 (2007)

Solow, R.: A contribution to the theory of economic growth. The Quarterly Journal of Economics 70(1), 65–94 (1956)

Sue Wing, I.: Representing induced technological change in models for climate policy analysis. Energy Economics 28(5-6), 539–562 (2006)

Tawarmalani, M., Sahinidis, N.V.: A polyhedral branch-and-cut approch to global optimization. Mathematical programming (Series B) 103(2), 225–249 (2005)

van der Zwaan, B., Rabl, A.: The learning potential of photovoltaics: implications for energy policy. Energy Policy 32(13), 1545–1554 (2004)

Viebahn, P., Kronshage, S., Trieb, F., Lechon, Y.: Final Report on technical data, costs, and life cycle inventories of solar thermal power plants: Deliverable 12.2 — RS 1a of the NEEDS project (2008)

Weyant, J., Olavson, T.: Issues in modeling induced technological change in energy, environmental, and climate policy. Environmental Modeling and Assessment 4(2-3), 67–85 (1999)

Williams, R.H., Terzian, G.: A Benefit/Cost Analysis of Accelerated Development of Photovoltaic Technology. Princeton (1993)

Wright, T.P.: Factors Affecting the Cost of Airplanes. Journal of the Aeronautical Sciences 3(4), 122–128 (1936)

Chapter 5
Optimal Investment Strategy for Competing Learning Technologies: An Analytical Approach

A major drawback of employing experience curves to model future technological change is the sensitivity of results to the rate of change, which in turn is typically associated with a high degree of uncertainty. In this chapter a model is developed and analytically solved to examine properties of models with endogenous technological change under uncertainty. It has three main objectives: illustrate specific properties of stochastic programs with experience curves, examine whether technological diversification is an optimal strategy in presence of experience curves and uncertainty, and provide a tool for decision making support. After a review of objectives, approach and current state of research in section 5.1, the basic model is developed in section 5.2 and successively expanded in sections 5.3 and 5.4. In section 5.5 several assumptions required for the preceding sections are critically reviewed. The chapter closes with an illustrative application of the model to two renewable technologies in section 5.6 and a discussion of key findings and implications in section 5.7.

5.1 Objective and Approach

The previous chapters have highlighted the importance of considering cost reductions when assessing the future prospects of renewable power technologies. In case experience curves are used to endogenize this process in energy system models, results are highly sensitive to learning rate assumptions. Future learning rates, however, are highly uncertain. For this reason explicit consideration of uncertainty would be desirable in model-based technology assessment. A small-scale power sector model with endogenized experience curves[1] and uncertainty is presented in this chapter. It has three main objectives:

[1] For the sake of brevity, technologies modeled in such a manner are referred to as *learning* ones in this chapter.

F. Wagner, *Renewables in Future Power Systems*,
Green Energy and Technology,
DOI: 10.1007/978-3-319-05780-4_5,

(1) The model's primary purpose is to illustrate specific properties of stochastic programs with learning that need to be taken into account in the process of model formulation. This serves as preparation for the stochastic enhancement of the large-scale power system model described in the following chapters.

(2) The second objective is to provide industry users with a tool for decision support. Commercial technology assessments are carried out frequently, not only by the companies themselves engaged in research and production of the technologies under consideration but also by other parties, such as analysts or investors. Initial technology assessment often only requires an approximate indication on whether a particular technology might be commercially viable and therefore warrants more detailed investigations. In particular if such assessments concern new, uncompetitive technologies, the use of experience curves to forecast cost reduction prospects is often desired. Unfortunately, the know-how and software needed to build models featuring experience curves, especially in case of changing learning rates over time, is usually not available in-house. An analytical model solution, entered into a spreadsheet, could remedy this shortfall and serve as tool to carry out initial competitiveness assessment. It goes without questions that if a more detailed technology assessment is desired after this initial analysis, more sophisticated approaches are needed.

(3) The third objective is to investigate whether model-endogenous learning in presence of uncertainty results in optimality of a technological diversification strategy, if two competing substitute technologies are available for investment. The background and existing research relevant to this question are covered in the following paragraphs.

In a deterministic setting with strictly increasing returns, where a choice must be made on how much to invest into two substitute technologies, allocation of all investments to the more attractive technology is optimal (cf. Bramoullé and Olson, 2005; Arthur, 1989). Model-endogenous experience curves imply this property of strictly increasing returns, unless other model features do not overcompensate the effect. A sufficiently strict resource constraint applying to a particular technology would be an example of such a compensating model feature.

For model setups considering both experience curves and uncertainty, several numerical studies with large-scale system models have presented results implying a technology diversification strategy. Using a two-stage stochastic programming approach, Barreto (2001) finds that investment into the only technology with uncertain learning rate follows an "intermediate hedging path" until uncertainty is resolved. Mattsson (2002) finds that both learning technologies in his stochastic program are optimally deployed and no lock-out of one or the other occurs. Similarly, sampling-based model approaches (e.g. Gritsevskyi and Nakićenović 2000 or Grübler and Gritsevskyi 2002, pp. 280ff.) have resulted in a large number of different, alternative investment strategies that are characterized by similar overall system costs.

In contrast, mathematical programs characterized by an uncertain objective function coefficient only and no uncertainty in its constraints may, in fact, be deterministic ones given no recourse exists. In such cases, "the only possible action [before the resolution of uncertainty] is to plan in view of the expected cost" (Birge and Louveaux 1997, p. 57). Learning rates are part of the objective function in cost-minimization model but not present in its constraints. In absence of recourse, this would indicate that the optimal investment strategy into two substitute technologies before uncertainty resolution should be based on the expected value of the uncertain parameters.[2] In such a setting, allocation of all investments into one technology would again be optimal, given strictly increasing returns. The expected benefits of higher cost reductions by concentrating on one technology instead of splitting the investment between technologies outweigh the benefit of having the other technology already a bit "further down" on the experience curve.[3] In this chapter such a strategy, where investment is concentrated on one technology in the first stochastic stage, will be referred to as *pure*, whereas strategies, where investment into both technologies in the first stage takes place, will be referred to as *mixed*. In case technologies are not perfect substitutes, mixed strategies may be optimal despite cost reductions from learning, as long as the benefits of diversification exceed the opportunity costs of not installing the technology that has already progressed further on the experience curve. The presence of resource constraints, learning spillovers or different levels of demand may also lead to optimality of such a mixed strategy, depending on model parameterization. For the remainder of the chapter perfect substitutability of the two technologies considered is assumed.

The findings from the numerical models above seem to contradict the discussion for the case of two competing technologies. Yet, in complex numerical models multiple factors and interactions affect technology competitiveness. Mattsson (2002), for example, states that the "number of technologies [...] is not large enough to provoke competition between technologies for the learning

[2] This may be the case in a model, where the point of uncertainty resolution depends on the passage of time and is independent of the investment level. Uncertainty resolution for learning rates may also, potentially to an even higher degree, depend on cumulative investments. An approach considering this is Mattsson (2002), featuring two technologies with uncertain learning. Yet, such an approach with one occurrence of uncertainty resolution per technology will result in a multi-stage stochastic program with (n+1) stages, n denoting the number of technologies with uncertain learning. Unfortunately the computational or manual efforts required to solving a model using this approach and containing several technologies with uncertain learning are very high. Uncertainty resolution in this work occurs at specified points in time.

[3] Assuming positive opportunity cost of capital and absence of lead time between the decision to construct a plant and its commissioning date. The latter assumption implies that a decision maker can postpone the decision to invest until uncertainty has been resolved — otherwise the problem may be characterized by recourse.

opportunities by the global energy system." Moreover, the two technologies with uncertain learning rates in his model behave in a complementary fashion via a spillover mechanism. This example illustrates the difficulty to identify which effects are truly attributable to the presence of experience curve under uncertainty in large-scale numerical models. In a small-scale, analytical model, by contrast, the influence of other factors than learning and uncertainty can either be avoided or at least controlled.

Analytical models featuring endogenized experience curves have previously been employed in the context of climate policy assessment. Goulder and Mathai (2000) investigate optimal levels of CO_2 abatement and carbon taxes, Bramoullé and Olson (2005) optimal investment strategies into abatement technologies. Both studies are characterized by a deterministic setting.[4]

The analytical model in this chapter is a two-technology, two-stage one with endogenous learning via one-factor experience curves. The setup and the solution of all model versions is based on the important hypothesis that *pure* investment strategies in this context are always superior to *mixed* ones, as the two featured technologies are perfect substitutes and the model is characterized by strictly increasing returns to scale. The investor is considered to be risk-neutral and can wait with decisions on second stage investment until uncertainty has been resolved.

The chapter is organized into five main sections. In section 5.2, a deterministic model version is formulated and solved. As the model is non-convex, two local optima with their corresponding optimal investment strategies exist. Such different strategies are referred to as *cases* throughout the chapter. The model is solved following a two-step process: (1) Solutions are derived for both optima. (2) Based on these solutions, a decision rule is formulated to identify the global optimum among the local ones. The basic deterministic model is expanded to include an initial capacity endowment in section 5.3. Uncertainty is then introduced in section 5.4. Specific considerations for the model setup resulting from the presence of experience curves, which are subject to changes in the learning rate at stage transition, are highlighted. The model is then solved in the same manner as the deterministic one — only this time twelve relevant local optima exist. Finally, in section 5.5, the hypothesis of *pure* strategy optimality, which has been used as a basis for the preceding solution processes, is critically tested.

[4] Other analytical models with endogenous learning mechanism and uncertainty in the context of climate change have been presented by Bosetti and Tavoni (2009) and Baker and Shittu (2008). They feature cost reductions via an R&D stock and/or a backstop technology. Ingham et al. (2007), Baker (2005) and Ulph and Ulph (1997) also present analytical models with learning and uncertainty. However, in these studies the term 'learning' refers to the fact that uncertainty is resolved and not to the presence of an endogenous cost reduction mechanism.

5.2 Basic Deterministic Model

5.2.1 Setup and Description

The basic power system model is set up as a two-period model with two competing technologies available for investment and power production. Let set T with $t = \{1, 2\}$ represent these two periods and set I with $i = \{A, B\}$ denote the two competing technologies. Both technologies produce the same good, here electricity, and are considered perfect substitutes. The model is set up as a Nonlinear Program (NLP) with the objective of minimizing total discounted system costs, which are divided into capital, fixed operating and variable operating costs. Capital costs decrease with increasing investment. This renders the model nonlinear and non-convex. An inelastic, uniformly distributed demand has to be fulfilled in each period.[5] Power production from each technology is constrained by its installed capacity stock. Investments increase, end-of-life capacity retirements decrease this capacity stock. The model is described algebraically by Equations (5.1) through (5.4).

Minimize

$$z = \sum_i \sum_t \psi_t \left[\theta_t \alpha_i w_{it} + \beta_i y_{it} + \phi_i \left(\int_{\kappa_{i0}}^{\kappa_{i1}} \kappa_{it}^{-\gamma_i} d\kappa + \frac{\theta_2}{\pi_i} \int_{\kappa_{i1}}^{\kappa_{i2}} \kappa_{it}^{-\gamma_i} d\kappa \right) \right] \tag{5.1}$$

Subject to

$$\sum_i y_{it} \geq \delta_t \theta_t \tag{5.2}$$

$$\sigma_i \theta_t w_{it} \geq y_{it} \tag{5.3}$$

$$w_{i(t-1)} + x_{it} - v_{it} = w_{it} \tag{5.4}$$

The three main summands in the objective function represent fixed operating, variable operating and capital costs, respectively. Fixed operating costs are calculated based on the specific annual fixed operating costs α_i, the installed capacity base w_{it} and the number of years in the particular period θ_t. Variable costs are based on the specific variable costs per unit of production β_i and total production within the period y_{it}. The expression for capital costs is discussed in detail further below. Constraint (5.2) ensures that power demand is fulfilled based on annual demand level δ_t. The capacity restriction is enforced in constraint (5.3); constraint (5.4) specifies how capacity additions x_{it} and retirements v_{it} affect the installed capacity base. The formulation of this last

[5] For not uniformly distributed demand and given different capital versus operating cost ratios, the technologies can no longer be considered perfect substitutes. They would effectively produce different goods, e.g. base versus peak load power.

Table 5.1 Basic deterministic model parameters and variables

Parameters		Variables	
r	Opportunity costs of capital (%/a)	z	Total system costs ($)
α_i	Specific fixed operating costs ($/Wa)	x_{it}	Capacity additions (W)
β_i	Specific variable costs ($/Wh)	y_{it}	Power production (Wh)
γ_i	Learning index (%)	w_{it}	Installed capacity (W)
δ_t	Demand level (W)	κ_{it}	Cumulative capacity (W)
θ_t	Period duration (a)	v_{it}	Capacity retirements (W)
σ_i	Technology availability (%)		
ϕ_i	First unit specific costs ($/W)		
π_i	Plant lifetime (a)		
ψ_t	Present value factor (%)		

constraint as equality instead of inequality is deliberately chosen to reduce the complexity of the solution process. It prevents capacity retirements before the end of a plant's useful life. Table 5.1 provides a comprehensive overview of the variables and parameters used.

Several restrictions on model parameter values are required. Initial cumulative capacity κ_{i0} has to be strictly positive to ensure definiteness of the problem. For the model to be meaningful in an economic context, all parameters require restrictions to nonnegativity. Moreover, the demand level of the second period must not be lower than the one of the first period, i.e. $\delta_2 \geq \delta_1$. Increasing demand is realistic for the power generation sector but must be carefully reviewed for applications of the model to other industries. The purpose of this restriction is again to reduce complexity in the solution process of the stochastic model.[6] Both for the calculation of capital costs and end-of-life divestments, assumptions on the relationship between plant lifetime π_i and period durations θ_t must be made: (1) Plant lifetimes must be greater than any period's duration, i.e. $\pi_i \geq \theta_t$, but (2) plant lifetimes must be shorter than the entire duration covered by the model, i.e. $\pi_i < \sum_t \theta_t$. These assumptions ensure that there is no replacement need for capacity within a single period but also that some replacement is required across the model horizon.[7] These assumptions imply that capacity additions from period one will have reached the end of their useful life sometime during period two.

Endogenous cost reductions via experience curves, development of a capacity base over time and time value of money are dynamic concepts. Yet,

[6] Actually $\delta_2 \geq \delta_1 - v_{it}$ would be sufficiently strict to prevent added complexity. The main benefit of the stricter restriction used is that it exclusively consists of parameters and does not include variable v_{it}, whose value is dependent upon decisions from the first model period.

[7] Although using $\pi_i = \sum_t \theta_t$ would be convenient from a computational perspective, it is not used due to the significant restrictions it would place on model application to real world problems.

to enable application of solution methods associated with comparative-static analysis, approximations have been made to represent these mechanisms in the model at hand. The introduction of experience curves (see chap 4.2) into the model requires allocation of investment costs to the two model periods. Starting point is the generic function of the experience curve, given by Equation (5.5).

$$c_{it} = \phi_i \cdot \kappa_{it}^{-\gamma_i}, \tag{5.5}$$

where c_{it} are the current specific capital costs, ϕ_i the specific capital costs of the "first unit" of capacity,[8] κ_{it} the cumulative capacity to date,[9] and γ_i the learning index. The learning index differs from the learning rate, which indicates the fraction of cost decrease over a doubling of cumulative capacity. Conversion of a learning rate l_i into a learning index is carried out as given by Equation (5.6), conversion of current investment costs c_{i0} into first unit costs ϕ_i as given by Equation (5.7).

$$\gamma_i = -\log_2(1 - l_i) \tag{5.6}$$

$$\phi_i = \frac{c_{i0}}{\kappa_{i0}^{-\gamma_i}} \tag{5.7}$$

The area beneath the experience curve corresponds to cumulative capital costs $\hat{c}_{it}$, which are required for the calculation of capital costs in a particular period. The function describing cumulative costs development is given in Equation (5.8).

$$\hat{c}_{it} = \phi_i \int \kappa_{it}^{-\gamma_i} \, d\kappa \tag{5.8}$$

Capital costs per period are then obtained by identifying the cumulative costs in between the cumulative capacities at the beginning and the end of the period under consideration, as given by Equation (5.9). For the model solution process this expression has been transformed into an integral-free notation.

$$\hat{c}_{it} = \phi_i \int_{\kappa_{i0}}^{\kappa_{i1}} \kappa_{it}^{-\gamma_i} \, d\kappa = \phi_i \left| _{\kappa_{i0}}^{\kappa_{i1}} \frac{1}{(1-\gamma_i)} \kappa_{it}^{1-\gamma_i} = \frac{\phi_i}{(1-\gamma_i)} \left[\kappa_{i1}^{(1-\gamma_i)} - \kappa_{i0}^{(1-\gamma_i)} \right] \right. \tag{5.9}$$

This expression is used to represent capital costs in the objective function (5.1). The corresponding term for the second period, however, is modified to prevent unrealistic results towards the end of the model horizon. The purpose of multiplier $\frac{\theta_2}{\pi_i}$, which is the ratio of period two duration and the useful life of capacity invested in period two, is to match capital costs with

[8] Alternatively one could use a definition in terms of current specific costs. Yet, this will increase the size of formulas in the remainder of the chapter significantly. Therefore, this more compact form has been used.

[9] Defined as $\kappa_{it} = \kappa_{i(t-1)} + x_{it}$.

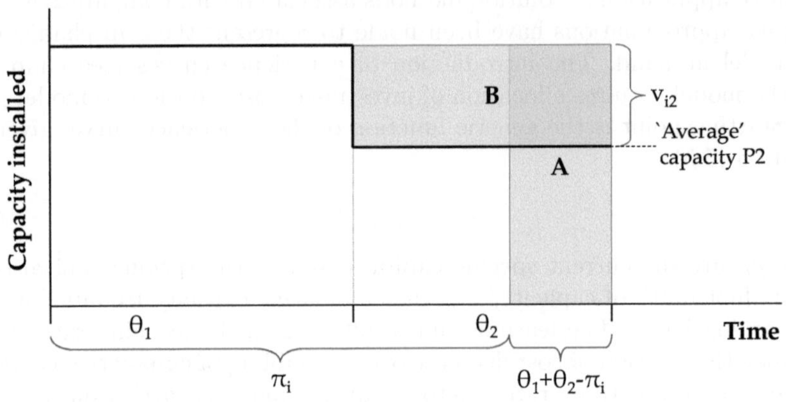

Fig. 5.1 Divestment of capacity installed in period one
Source: Own representation

their associated benefits. If this adjustment were not made, the full capital costs for new investments would be allocated to a period shorter than the actual life of the technology. This would lead to unrealistic advantages for technologies with low capital-intensity. For capacity additions in the first period, this adjustment is not required.

Capacity retirements v_{it} depend upon the useful life of the particular technology π_i and period duration θ_t. For the solution of the model, v_{it} is expressed in terms of corresponding capacity additions x_{it}. The parameter restrictions $\pi_i \geq \theta_t$ and $\pi_i < \sum_t \theta_t$ are the starting point of this transformation, implying that there are no capacity retirements in period one.[10] All capacity installed in period one, x_{i1}, is thus available across the entire first period. At some point during period two, however, the capacity installed in period one reaches the end of its useful life (see Fig. 5.1). Since it is not possible to divide the period further into sub-periods, an "average" remaining capacity of period one investments in period two is used to capture capacity retirement before the end of period two. To obtain the divestment amount required to calculate this average capacity, the area of rectangle A in Figure 5.1, which represents the actual capacity gap in period two, is set equal to the area of rectangle B, which has the duration of the second period θ_2 as one of its sides. The other side of rectangle B, denoted by v_{i2}, then equals the divestment amount required to calculate the "average" installed capacity across period two. In algebraic terms:

$$\begin{aligned} A_{\Box A} &= A_{\Box B} \\ \Leftrightarrow \quad x_{i1} \cdot (\theta_1 + \theta_2 - \pi_i) &= v_{i2} \cdot \theta_2 \\ \Leftrightarrow \quad v_{i2} &= \frac{\theta_1 + \theta_2 - \pi_i}{\theta_2} x_{i1} = \mu_i x_{i1} \end{aligned} \tag{5.10}$$

[10] In absence of an initial capacity endowment — this will be introduced in sec. 5.3.

For reasons of compactness, v_{i2} is expressed in terms of x_{i1} and corresponding coefficient μ_i in the solution process, as defined in eq. (5.10).

Analogously to divestments above, *discounting factors* need to be converted into average factors ψ_t for both periods. With set $n = \{1, 2, \ldots, \sum_t \theta_t\}$ representing all sub-periods within the model horizon, the average discount factors are calculated as given by eq. (5.11). An example for sub-periods would be single years, when period durations span several years.

$$\psi_1 = \frac{\sum_{n=1}^{\theta_1} \frac{1}{(1+r)^n}}{\theta_1} \quad \text{and} \quad \psi_2 = \frac{\sum_{n=\theta_1+1}^{\theta_2} \frac{1}{(1+r)^n}}{\theta_2} \tag{5.11}$$

Calculation of approximated average values across entire model periods and other approximations, such as the method used for preventing game-ending model behavior, come at the expense of accuracy. For example, cost reductions from learning are likely to be overstated to a certain degree under this approach, as the same average discount factor in a period is applied to both higher costs at the beginning of the period and lower ones towards its end.[11] Yet, in a truly dynamic setting, higher discount factors would apply to the higher costs earlier in time and lower ones later on. Although it is important to bear these inaccuracies in mind when interpreting model results, they are unlikely to jeopardize the achievement of the chapter's objectives.

5.2.2 Solution

The model is non-convex, which implies potential existence of multiple local cost minima. It depends on the parameter values of the model, which of these local minima will constitute the global one. To address this problem, the model is solved for all minima that could, depending on the parameter configuration, constitute the global one. Afterwards a decision rule is devised, which enables identification of the global optimum among the local minima. As demonstrated by Arthur (1989) or Bramoullé and Olson (2005), concentration of investment into one or the other substitute technology is always optimal for a model with strictly increasing returns. Therefore, in this perfect-foresight setting, only two local optima exist that are able to constitute the global optimum — dominance of technology A or B, but not a mix including both technologies.

The model is solved for both local minima drawing upon Karush-Kuhn-Tucker (KKT) conditions (see chap. 3.4.1). First, the Lagrangian function is constructed from the extensive form of the model.[12] Next, the Lagrangian function is partially differentiated once with respect to all of the model's variables. The KKT conditions can then be derived from these first-order partial derivatives. Conveniently, only the system made up of equations

[11] Assuming that investment into a particular technology takes place at all.

[12] See Appendix B.1.1 for the extensive form.

relating to derivatives with respect to the Lagrangian multipliers λ_m need to be solved to obtain the cost minima for the deployment of technology A or B. The equations relating to the derivatives with respect to model variables are only required to verify that the identified solutions indeed fulfill all KKT conditions.

The Lagrangian form[13] of the model is then given by

$$\begin{aligned}
Z = \quad & \psi_1\theta_1\left(\alpha_A w_{A1} + \alpha_B w_{B1}\right) + \psi_2\theta_2\left(\alpha_A w_{A2} + \alpha_B w_{B2}\right) \\
& + \psi_1\left(\beta_A y_{A1} + \beta_B y_{B1}\right) + \psi_2\left(\beta_A y_{A2} + \beta_B y_{B2}\right) \\
& + \frac{\phi_A\psi_1}{1-\gamma_A}\left((\kappa_{A0} + x_{A1})^{(1-\gamma_A)} - \kappa_{A0}^{(1-\gamma_A)}\right) \\
& + \frac{\phi_A\psi_2\theta_2}{\pi_A(1-\gamma_A)}\left((\kappa_{A0} + x_{A1} + x_{A2})^{(1-\gamma_A)} - (\kappa_{A0} + x_{A1})^{(1-\gamma_A)}\right) \\
& + \frac{\phi_B\psi_1}{1-\gamma_B}\left((\kappa_{B0} + x_{B1})^{(1-\gamma_B)} - \kappa_{B0}^{(1-\gamma_B)}\right) \\
& + \frac{\phi_B\psi_2\theta_2}{\pi_B(1-\gamma_B)}\left((\kappa_{B0} + x_{B1} + x_{B2})^{(1-\gamma_B)} - (\kappa_{B0} + x_{B1})^{(1-\gamma_B)}\right) \quad (5.12) \\
& + \lambda_1(\theta_1\delta_1 - y_{A1} - y_{B1}) + \lambda_2(\theta_2\delta_2 - y_{A2} - y_{B2}) \\
& + \lambda_3(y_{A1} - \theta_1\sigma_A w_{A1}) + \lambda_4(y_{A2} - \theta_2\sigma_A w_{A2}) \\
& + \lambda_5(y_{B1} - \theta_1\sigma_B w_{B1}) + \lambda_6(y_{B2} - \theta_2\sigma_B w_{B2}) \\
& + \lambda_7(w_{A1} - x_{A1}) + \lambda_8(w_{A2} - w_{A1} - x_{A2} + \mu_A x_{A1}) \\
& + \lambda_9(w_{B1} - x_{B1}) + \lambda_{10}(w_{B2} - w_{B1} - x_{B2} + \mu_B x_{B1})
\end{aligned}$$

From the Lagrangian function, partial, first-order derivatives are obtained.[14] First-order derivatives with respect to λ_m are

$$Z_{\lambda_1} = \theta_1\delta_1 - y_{A1} - y_{B1} \quad (5.13)$$
$$Z_{\lambda_2} = \theta_2\delta_2 - y_{A2} - y_{B2} \quad (5.14)$$
$$Z_{\lambda_3} = y_{A1} - \theta_1\sigma_A w_{A1} \quad (5.15)$$
$$Z_{\lambda_4} = y_{A2} - \theta_2\sigma_A w_{A2} \quad (5.16)$$
$$Z_{\lambda_5} = y_{B1} - \theta_1\sigma_B w_{B1} \quad (5.17)$$
$$Z_{\lambda_6} = y_{B2} - \theta_2\sigma_B w_{B2} \quad (5.18)$$
$$Z_{\lambda_7} = w_{A1} - x_{A1} \quad (5.19)$$
$$Z_{\lambda_8} = w_{A2} - w_{A1} - x_{A2} + \mu_A x_{A1} \quad (5.20)$$
$$Z_{\lambda_9} = w_{B1} - x_{B1} \quad (5.21)$$
$$Z_{\lambda_{10}} = w_{B2} - w_{B1} - x_{B2} + \mu_B x_{B1} \quad (5.22)$$

The KKT conditions for a minimization problem, given by Equations (5.23) through (5.26), then follow from these partial, first-order derivatives, the non-negativity restriction applying to all model variables and the complementary slackness condition.[15]

---

[13] Line breaks inside parentheses or brackets are avoided to improve readability.

[14] Partial derivatives with respect to model variables other than λ_m are shown in appendix B.1.2.

[15] Boundary irregularities may result in invalidity of the KKT conditions. For this problem only linear constraints are used, which circumvents this potential issue.

$$Z_{w_{it}} \geq 0 \qquad w_{it} \geq 0 \qquad Z_{w_{it}} \perp w_{it} \tag{5.23}$$
$$Z_{x_{it}} \geq 0 \qquad x_{it} \geq 0 \qquad Z_{x_{it}} \perp x_{it} \tag{5.24}$$
$$Z_{y_{it}} \geq 0 \qquad y_{it} \geq 0 \qquad Z_{y_{it}} \perp y_{it} \tag{5.25}$$
$$Z_{\lambda_m} \leq 0 \qquad \lambda_m \geq 0 \qquad Z_{\lambda_m} \perp \lambda_m \tag{5.26}$$

For the problem at hand it is known that investment into one technology precludes investment into the other when cost-optimality is desired. This is used as starting point for solving the model for case A, where only investment into technology A takes place. A solution approach using KKT conditions requires up-front "guessing" whether Lagrangian variables λ_m are greater or equal to zero, i.e. whether a constraint is binding. Here only the solution sequence leading to the solution, which fulfills all KKT conditions, is shown.

Solution process for *optimality of technology A*:

Starting point is:

$$x^*_{At} > 0, \quad x^*_{Bt} = 0$$

From Equation (5.21) and $x_{B1} = 0$:

$$w^*_{B1} = x^*_{B1} = 0 \tag{5.27}$$

From Equation (5.22) in combination with $x_{B2} = 0$:

$$w^*_{B2} = x_{B2} + w_{B1} - \mu_B x_{B1} = 0 \tag{5.28}$$

From Equations (5.17) and (5.18):

$$y^*_{B1} = \theta_1 \sigma_B w_{B1} = 0 \tag{5.29}$$
$$y^*_{B2} = \theta_2 \sigma_B w_{B2} = 0 \tag{5.30}$$

From Equations (5.13), (5.14) and $y_{Bt} = 0$:

$$y^*_{A1} = \theta_1 \delta_1 - y_{B1} = \theta_1 \delta_1 \tag{5.31}$$
$$y^*_{A2} = \theta_2 \delta_2 - y_{B2} = \theta_2 \delta_2 \tag{5.32}$$

From Equations (5.15) and (5.16):

$$w^*_{A1} = \frac{y_{A1}}{\theta_1 \sigma_A} = \frac{\delta_1}{\sigma_A} \tag{5.33}$$
$$w^*_{A2} = \frac{y_{A2}}{\theta_2 \sigma_A} = \frac{\delta_2}{\sigma_A} \tag{5.34}$$

Finally, from Equations (5.19) and (5.20):

$$x^*_{A1} = w_{A1} = \frac{\delta_1}{\sigma_A} \tag{5.35}$$

$$x^*_{A2} = \mu_A x_{A1} + w_{A2} - w_{A1} = (\mu_A - 1)\frac{\delta_1}{\sigma_A} + \frac{\delta_2}{\sigma_A} \tag{5.36}$$

The KKT conditions, as given by Equations (5.23) through (5.26), are fulfilled for the values obtained above. The solution, however, only constitutes a local optimum. Only in case it leads to a lower objective function value than the solution for local optimality of technology B, it is also the global optimum. Following, the solution for the opposite case is derived, where technology B is dominating. Analogously to the process described for domination of A, investment into technology B implies zero investment into technology A.

Solution process for *optimality of technology B*:
Starting point is:

$$x_{At}^* = 0, \quad x_{Bt}^* > 0$$

From Equation (5.19) and $x_{A1} = 0$

$$w_{A1}^* = x_{A1}^* = 0 \tag{5.37}$$

From Equation (5.20) in combination with $x_{A2} = 0$

$$w_{A2}^* = x_{A2} + w_{A1} - \mu_A x_{A1} = 0 \tag{5.38}$$

From Equations (5.15) and (5.16)

$$y_{A1}^* = \theta_1 \sigma_A w_{A1} = 0 \tag{5.39}$$
$$y_{A2}^* = \theta_2 \sigma_A w_{A2} = 0 \tag{5.40}$$

From Equations (5.13), (5.14) and $y_{At} = 0$

$$y_{B1}^* = \theta_1 \delta_1 - y_{A1} = \theta_1 \delta_1 \tag{5.41}$$
$$y_{B2}^* = \theta_2 \delta_2 - y_{A2} = \theta_2 \delta_2 \tag{5.42}$$

From Equations (5.17) and (5.18)

$$w_{B1}^* = \frac{y_{B1}}{\theta_1 \sigma_B} = \frac{\delta_1}{\sigma_B} \tag{5.43}$$
$$w_{B2}^* = \frac{y_{B2}}{\theta_2 \sigma_B} = \frac{\delta_2}{\sigma_B} \tag{5.44}$$

Finally, from Equations (5.21) and (5.22)

$$x_{B1}^* = w_{B1} = \frac{\delta_1}{\sigma_B} \tag{5.45}$$
$$x_{B2}^* = \mu_B x_{B1} + w_{B2} - w_{B1} = (\mu_B - 1)\frac{\delta_1}{\sigma_B} + \frac{\delta_2}{\sigma_B} \tag{5.46}$$

Inserting the analytical solution into the KKT conditions again shows that it fulfills all conditions. The solution in this case also only constitutes a local optimum, as long as it is not characterized by a lower objective function value z than the opposite case of technology A dominance.

5.2.3 Investment Decision Rule

The final step in the solution process is the identification of the global optimum among the two local ones. For this purpose the solutions are re-substituted into the objective function, enabling expression of the value of z^* in terms of exogenous parameters. Let z^A denote the objective function value in case technology A dominates and z^B in case B dominates, as given by Eqs. (5.47) and (5.48), respectively.

$$\begin{aligned} z^A = &(\psi_1\theta_1\delta_1 + \psi_2\theta_2\delta_2)\left(\frac{\alpha_A}{\sigma_A} + \beta_A\right) + \frac{\phi_A\psi_1}{1-\gamma_A}\left[\left(\kappa_{A0} + \frac{\delta_1}{\sigma_A}\right)^{(1-\gamma_A)} - \kappa_{A0}^{(1-\gamma_A)}\right] \\ &+ \frac{\phi_A\psi_2\theta_2}{\pi_A(1-\gamma_A)}\left[\left(\kappa_{A0} + \mu_A\frac{\delta_1}{\sigma_A} + \frac{\delta_2}{\sigma_A}\right)^{(1-\gamma_A)} - \left(\kappa_{A0} + \frac{\delta_1}{\sigma_A}\right)^{(1-\gamma_A)}\right] \end{aligned} \tag{5.47}$$

$$\begin{aligned} z^B = &(\psi_1\theta_1\delta_1 + \psi_2\theta_2\delta_2)\left(\frac{\alpha_B}{\sigma_B} + \beta_B\right) + \frac{\phi_B\psi_1}{1-\gamma_B}\left[\left(\kappa_{B0} + \frac{\delta_1}{\sigma_B}\right)^{(1-\gamma_B)} - \kappa_{B0}^{(1-\gamma_B)}\right] \\ &+ \frac{\phi_B\psi_2\theta_2}{\pi_B(1-\gamma_B)}\left[\left(\kappa_{B0} + \mu_B\frac{\delta_1}{\sigma_B} + \frac{\delta_2}{\sigma_B}\right)^{(1-\gamma_B)} - \left(\kappa_{B0} + \frac{\delta_1}{\sigma_B}\right)^{(1-\gamma_B)}\right] \end{aligned} \tag{5.48}$$

The sign of $D = z^A - z^B$ indicates which of the two local constitutes the global optimum.

$$\begin{aligned} D &= z^A - z^B \\ &= (\psi_1\theta_1\delta_1 + \psi_2\theta_2\delta_2)\left(\frac{\alpha_A}{\sigma_A} - \frac{\alpha_B}{\sigma_B} + \beta_A - \beta_B\right) \\ &+ \frac{\phi_A\psi_1}{1-\gamma_A}\left[\left(\kappa_{A0} + \frac{\delta_1}{\sigma_A}\right)^{(1-\gamma_A)} - \kappa_{A0}^{(1-\gamma_A)}\right] \\ &+ \frac{\phi_A\psi_2\theta_2}{\pi_A(1-\gamma_A)}\left[\left(\kappa_{A0} + \mu_A\frac{\delta_1}{\sigma_A} + \frac{\delta_2}{\sigma_A}\right)^{(1-\gamma_A)} - \left(\kappa_{A0} + \frac{\delta_1}{\sigma_A}\right)^{(1-\gamma_A)}\right] \\ &- \frac{\phi_B\psi_1}{1-\gamma_B}\left[\left(\kappa_{B0} + \frac{\delta_1}{\sigma_B}\right)^{(1-\gamma_B)} - \kappa_{B0}^{(1-\gamma_B)}\right] \\ &- \frac{\phi_B\psi_2\theta_2}{\pi_B(1-\gamma_B)}\left[\left(\kappa_{B0} + \mu_B\frac{\delta_1}{\sigma_B} + \frac{\delta_2}{\sigma_B}\right)^{(1-\gamma_B)} - \left(\kappa_{B0} + \frac{\delta_1}{\sigma_B}\right)^{(1-\gamma_B)}\right] \end{aligned} \tag{5.49}$$

The corresponding rule for an optimal decision is then:

$$D > 0 \quad \Rightarrow \quad \text{Invest into technology B only}$$
$$D = 0 \quad \Rightarrow \quad \text{Invest into technology A or B}$$
$$D < 0 \quad \Rightarrow \quad \text{Invest into technology A only}$$

Although fairly straightforward, this decision rule, in combination with decision indicator D, can be used to compare the competitiveness of two learning substitute technologies against each other. Unfortunately, the absence of the possibility to specify an initial capacity endowment requires treating demand within this model version as incremental to the demand satisfied by the capacity, which is already in place at the beginning of the model horizon.[16] Although this may not be a serious issue when two pre-commercial technologies are compared, it will considerably increase the complexity of a comparison when more mature technologies are involved. The practical applicability of this decision rule is thus still quite limited. As a logical next step, the deterministic model is enhanced to incorporate an initial capacity endowment in the following section.

5.3 Deterministic Model with Capacity Endowment

5.3.1 Setup and Description

The basic deterministic model from the previous section has been enhanced to allow for capacity endowments at the beginning of the model horizon. This is especially relevant when one or both of the competing technologies are already in a commercial development stage. The description of the model from the previous section, given by Equations (5.1) through (5.4) remains, with some additions, valid in presence of initial capacity endowments. The introduction of the latter, denoted by parameter w_{i0}, affects demand δ_t as well as capacity retirements v_{it}. The impact on demand is only of conceptual nature and has no impact on the formulation of the model. δ_t now merely denotes the total demand in the market under consideration instead of the incremental one not being served by capacity from before the model horizon. The required changes to capacity retirements v_{it}, however, increase the complexity of the analytical solutions and the corresponding investment decision rule. In the previous section, only retirement of capacity installed in the first period, being retired in the second period, had to be taken into account. In the extended model, retirement of the initial capacity endowment occurs in both periods.

Starting point for the integration of these additional divestments are assumptions $\pi_i \geq \delta_t$ and $\pi_i < \sum_t \delta_t$ from the previous section. The vintage of the endowment is assumed to be uniformly distributed leading to a linear reduction of the initial endowment over time. Taken together, these assumptions imply that there will be retirement of initial endowment capacity in

[16] Additional complexity arises from the fact that this capacity is likely to depreciate across the model horizon.

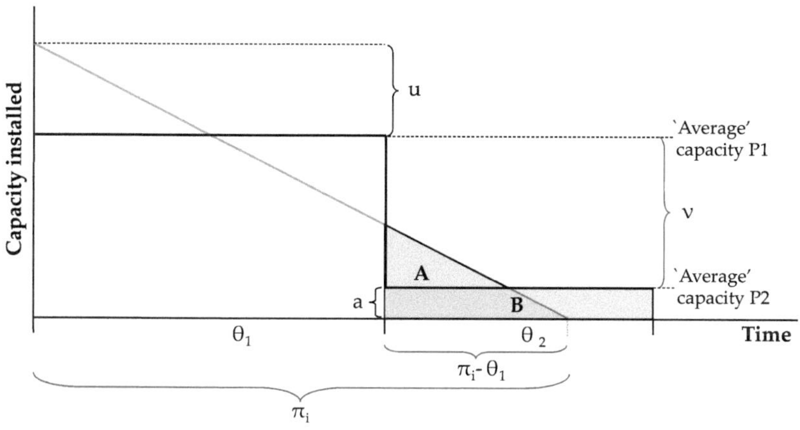

Fig. 5.2 Divestment of initial capacity endowment Source: Own representation

both model stages but sometime during period two all of the initial endowment will have reached the end of its useful life. Similarly to capacity installed within the model horizon, initial endowment capacity is represented as an average value for each period and correspondingly average divestments need to be calculated. Let u denote the divestment of the initial capacity endowment in period one, ν the one in period two and a the average remaining initial endowment capacity in period two (see Fig. 5.2). Then,

$$u = \frac{\theta_1}{2\pi_i} w_{i0} \tag{5.50}$$

For reasons of compactness during the solution process, the shorthand ϵ_i is introduced, denoting average initial capacity across the first period:

$$\epsilon_i = w_{i0} - u = \left(1 - \frac{\theta_1}{2\pi_i}\right) w_{i0} \tag{5.51}$$

Parameter a, representing the average remaining capacity from the initial endowment in the second period, is obtained by equating the area of shaded triangle A in Fig. 5.2 with rectangle B. The latter has the duration of the second period θ_2 as one of its sides.

$$\begin{aligned} A_{\triangle A} = A_{\square B} \quad &\Leftrightarrow \quad \frac{1}{2}\left(1 - 2u_1\right)\left(\pi_i - \theta_1\right) w_{i0} = a \cdot \theta_2 \\ &\Leftrightarrow \quad a = \frac{\left(1 - \frac{\theta_1}{\pi_i}\right)\left(\pi_i - \theta_1\right)}{2\theta_2} w_{i0} \end{aligned} \tag{5.52}$$

With known a, the retirement of initial capacity from the first to the second period, ν_i, is obtained, as given by Equation (5.53).

$$\nu_i = \left[1 - \frac{\theta_1}{2\pi_i} - \frac{\pi_i}{2\theta_2}\left(1 - \frac{\theta_1}{\pi_i}\right)^2\right] w_{i0} \tag{5.53}$$

As period two is not only characterized by divestment of initial endowment capacity but also of capacity installed in the first period, the divestments from both mechanisms need to be combined. Equations (5.54) and (5.55) state the total capacity retirements for the first and the second period, respectively.

$$v_{i1} = \frac{\theta_1}{2\pi_i} w_{i0} \tag{5.54}$$

$$v_{i2} = \left(1 - \frac{\theta_1}{2\pi_i} - \frac{\pi_i - 2\theta_1 + \frac{\theta_1^2}{\pi_i}}{2\theta_2}\right) w_{i0} + \frac{\theta_1 + \theta_2 - \pi_i}{\theta_2} x_{i,1} = \nu_i + \mu_i x_{i1} \tag{5.55}$$

The enhanced model formulation with initial capacity endowment only differs from the basic formulation with respect to the constraints describing capacity development. Those four affected constraints are shown below, additions to the basic model are marked in bold. Capacity retirement for period one in Equations (5.56) and (5.58) are not directly stated — rather they are already considered in the *net* initial endowment, denoted by ϵ_i. In contrast, the total capacity retirements for the second period are explicitly stated as ν_i.

$$x_{A1} + \boldsymbol{\epsilon_A} = w_{A1} \tag{5.56}$$

$$w_{A1} + x_{A2} - \mu_A x_{A1} - \boldsymbol{\nu_A} = w_{A2} \tag{5.57}$$

$$x_{B1} + \boldsymbol{\epsilon_B} = w_{B1} \tag{5.58}$$

$$w_{B1} + x_{B2} - \mu_B x_{B1} - \boldsymbol{\nu_B} = w_{B2} \tag{5.59}$$

5.3.2 Solution

The solution process is identical to the one applied in section 5.2.2. The Lagrangian form of the model is constructed (see Appendix B.2.1) and partial derivatives are calculated, from which the KKT conditions can be formulated. As only four constraints have been changed over the basic model, only the corresponding first-order partial derivatives are shown below in Equations (5.60) through (5.63).

$$Z_{\lambda_7} = w_{A1} - x_{A1} - \boldsymbol{\epsilon_A} \tag{5.60}$$

$$Z_{\lambda_8} = w_{A2} - w_{A1} - x_{A2} + \mu_A x_{A1} + \boldsymbol{\nu_A} \tag{5.61}$$

$$Z_{\lambda_9} = w_{B1} - x_{B1} - \boldsymbol{\epsilon_B} \tag{5.62}$$

$$Z_{\lambda_{10}} = w_{B2} - w_{B1} - x_{B2} + \mu_B x_{B1} + \boldsymbol{\nu_B} \tag{5.63}$$

The process of solving for the two local minima is the same as the one for the basic model. For this reason only the final solutions are stated below. As under the basic setup, the solutions for dominance of technology A and B are "mirror images" of each other with respect to dimension i.

Solution for *dominance of technology A*:

$$
\begin{aligned}
w_{A1}^* &= \frac{\delta_1 - \sigma_B \epsilon_B}{\sigma_A} & w_{A2}^* &= \frac{\delta_2 - \sigma_B(\epsilon_B - \nu_B)}{\sigma_A} \\
w_{B1}^* &= \epsilon_B & w_{B2}^* &= \epsilon_B - \nu_B \\
y_{A1}^* &= \theta_1(\delta_1 - \sigma_B \epsilon_B) & y_{A2}^* &= \theta_2[\delta_2 - \sigma_B(\epsilon_B - \nu_B)] \\
y_{B1}^* &= \theta_1 \sigma_B \epsilon_B & y_{B2}^* &= \theta_2 \sigma_B(\epsilon_B - \nu_B) \\
x_{B1}^* &= 0 & x_{B2}^* &= 0 \\
x_{A1}^* &= \frac{\delta_1 - \sigma_B \epsilon_B}{\sigma_A} - \epsilon_A \\
x_{A2}^* &= (\mu_A - 1)\frac{\delta_1 - \sigma_B \epsilon_B}{\sigma_A} + \frac{\delta_2 - \sigma_B(\epsilon_B - \nu_B)}{\sigma_A} - \mu_A \epsilon_A + \nu_A
\end{aligned}
\tag{5.64}
$$

Solution for *dominance of technology B*:

$$
\begin{aligned}
w_{A1}^* &= \epsilon_A & w_{A2}^* &= \epsilon_A - \nu_A \\
w_{B1}^* &= \frac{\delta_1 - \sigma_A \epsilon_A}{\sigma_B} & w_{B2}^* &= \frac{\delta_2 - \sigma_A(\epsilon_A - \nu_A)}{\sigma_B} \\
y_{A1}^* &= \theta_1 \sigma_A \epsilon_A & y_{A2}^* &= \theta_2 \sigma_A(\epsilon_A - \nu_A) \\
y_{B1}^* &= \theta_1(\delta_1 - \sigma_A \epsilon_A) & y_{B2}^* &= \theta_2[\delta_2 - \sigma_A(\epsilon_A - \nu_A)] \\
x_{A1}^* &= 0 & x_{A2}^* &= 0 \\
x_{B1}^* &= \frac{\delta_1 - \sigma_A \epsilon_A}{\sigma_B} - \epsilon_B \\
x_{B2}^* &= (\mu_B - 1)\frac{\delta_1 - \sigma_A \epsilon_A}{\sigma_B} + \frac{\delta_2 - \sigma_A(\epsilon_A - \nu_A)}{\sigma_B} - \mu_B \epsilon_B + \nu_B
\end{aligned}
\tag{5.65}
$$

5.3.3 Investment Decision Rule

As for the basic model, the solutions for dominance of technologies A and B are substituted back into the objective function. These terms are again denoted by z^A and z^B, respectively.

Objective function value for dominance of technology A:

$$
\begin{aligned}
z^A = {} & \psi_1 \theta_1 \left[(\delta_1 - \sigma_B \epsilon_B) \left(\frac{\alpha_A}{\sigma_A} + \beta_A \right) + \epsilon_B(\alpha_B + \sigma_B \beta_B) \right] \\
& + \psi_2 \theta_2 \left[(\delta_2 - \sigma_B(\epsilon_B - \nu_B)) \left(\frac{\alpha_A}{\sigma_A} + \beta_A \right) + (\epsilon_B - \nu_B)(\alpha_B + \sigma_B \beta_B) \right] \\
& + \frac{\phi_A \psi_1}{1 - \gamma_A} \left[\kappa_{A1}^{(1-\gamma_A)} - \kappa_{A0}^{(1-\gamma_A)} \right] + \frac{\phi_A \psi_2 \theta_2}{\pi_A(1 - \gamma_A)} \left[(\kappa_{A2}^{1-\gamma_A)} - \kappa_{A1}^{(1-\gamma_A)} \right]
\end{aligned}
\tag{5.66}
$$

with cumulative capacities κ_{A1} and κ_{A2} defined as

$$\kappa_{A1} = \kappa_{A0} + \frac{\delta_1 - \sigma_B \epsilon_B}{\sigma_A} - \epsilon_A$$
$$\kappa_{A2} = \kappa_{A0} + \mu_A \frac{\delta_1 - \sigma_B \epsilon_B}{\sigma_A} + \frac{\delta_2 - \sigma_B(\epsilon_B - \nu_B)}{\sigma_A} - \epsilon_A(\mu_A + 1) + \nu_A$$

Objective function value for dominance of technology B:

$$\begin{aligned} z^B = {} & \psi_1 \theta_1 \left[(\delta_1 - \sigma_A \epsilon_A) \left(\frac{\alpha_B}{\sigma_B} + \beta_B \right) + \epsilon_A (\alpha_A + \sigma_A \beta_A) \right] \\ & + \psi_2 \theta_2 \left[(\delta_2 - \sigma_A(\epsilon_A - \nu_A)) \left(\frac{\alpha_B}{\sigma_B} + \beta_B \right) + (\epsilon_A - \nu_A)(\alpha_A + \sigma_A \beta_A) \right] \\ & + \frac{\phi_B \psi_1}{1 - \gamma_B} \left[\kappa_{B1}^{(1-\gamma_B)} - \kappa_{B0}^{(1-\gamma_B)} \right] + \frac{\phi_B \psi_2 \theta_2}{\pi_B (1 - \gamma_B)} \left[(\kappa_{B2}^{1-\gamma_B)} - \kappa_{B1}^{(1-\gamma_B)} \right] \end{aligned} \tag{5.67}$$

with cumulative capacities κ_{B1} and κ_{B2} defined as

$$\kappa_{B1} = \kappa_{B0} + \frac{\delta_1 - \sigma_A \epsilon_A}{\sigma_B} - \epsilon_B$$
$$\kappa_{B2} = \kappa_{B0} + \mu_B \frac{\delta_1 - \sigma_A \epsilon_A}{\sigma_B} + \frac{\delta_2 - \sigma_A(\epsilon_A - \nu_A)}{\sigma_B} - \epsilon_B(\mu_B + 1) + \nu_B$$

The sign of $D = z^A - z^B$ again indicates the optimal investment decision with κ_{i1} and κ_{i2} defined as given above.[17]

$$\begin{aligned} D = {} & z^A - z^B \\ = {} & \psi_1 \theta_1 \left[(\delta_1 - \sigma_B \epsilon_B) \left(\frac{\alpha_A}{\sigma_A} + \beta_A \right) + \epsilon_B (\alpha_B + \sigma_B \beta_B) \right] \\ & - \psi_1 \theta_1 \left[(\delta_1 - \sigma_A \epsilon_A) \left(\frac{\alpha_B}{\sigma_B} + \beta_B \right) + \epsilon_A (\alpha_A + \sigma_A \beta_A) \right] \\ & + \psi_2 \theta_2 \left[(\delta_2 - \sigma_B(\epsilon_B - \nu_B)) \left(\frac{\alpha_A}{\sigma_A} + \beta_A \right) + (\epsilon_B - \nu_B)(\alpha_B + \sigma_B \beta_B) \right] \\ & - \psi_2 \theta_2 \left[(\delta_2 - \sigma_A(\epsilon_A - \nu_A)) \left(\frac{\alpha_B}{\sigma_B} + \beta_B \right) + (\epsilon_A - \nu_A)(\alpha_A + \sigma_A \beta_A) \right] \\ & + \frac{\phi_A \psi_1}{1 - \gamma_A} \left[\kappa_{A1}^{(1-\gamma_A)} - \kappa_{A0}^{(1-\gamma_A)} \right] + \frac{\phi_A \psi_2 \theta_2}{\pi_A (1 - \gamma_A)} \left[(\kappa_{A2}^{1-\gamma_A)} - \kappa_{A1}^{(1-\gamma_A)} \right] \\ & - \frac{\phi_B \psi_1}{1 - \gamma_B} \left[\kappa_{B1}^{(1-\gamma_B)} - \kappa_{B0}^{(1-\gamma_B)} \right] - \frac{\phi_B \psi_2 \theta_2}{\pi_B (1 - \gamma_B)} \left[(\kappa_{B2}^{1-\gamma_B)} - \kappa_{B1}^{(1-\gamma_B)} \right] \end{aligned} \tag{5.68}$$

and the same decision rule as for the basic model applies

[17] Although a more concise description would be possible, the form in Eq. (5.68) has been chosen to enhance readability of the solution by avoiding line breaks within parentheses or brackets.

$D > 0$	$\Rightarrow$	Invest into technology B only
$D = 0$	$\Rightarrow$	Invest into technology A or B
$D < 0$	$\Rightarrow$	Invest into technology A only

Although this model version now allows for specification of an initial capacity endowment, it does not consider potential changes in cost coefficients from one period to the next and still assumes perfect foresight. As decision makers may find these assumptions unsatisfactory, the next section describes a model extension allowing for changes in cost coefficients from the first to the second period as well as explicit representation of uncertainty in the development of several parameters.

5.4 Dynamic Model with Uncertainty

The assumption that exogenous model parameters remain constant over time and are known ex-ante have been convenient in the construction of the power sector model from the previous sections. Yet, they are also quite unrealistic. Both the direction and rate of change in model parameters over time are characterized by uncertainty. Although one still requires knowledge of the probability distributions for uncertain, future parameter values, stochastic programs are a means to explicitly take uncertainty into account. In this section, the deterministic model is extended to reflect changes in learning rates and variable costs over time as well as the uncertainty associated with this development. Analogously to the previous sections, the model is solved to obtain a decision indicator and rule to identify the global among the local optima.

5.4.1 Setup and Description

The hypothetical decision maker in this model is assumed to possess knowledge about period one parameter values. He is also knowledgeable of period two probability distributions describing the uncertain parameters, i.e. learning index and specific variable costs. At the beginning of period two, uncertainty is resolved and investment decisions for period two can be made, based upon the realized values.[18] For this model, period durations equal the duration of the stochastic stages.

Set $s = \{a, b\}$ is used to represent the two possible states of the world in stage two. In state a technology A is more competitive than in state b,

[18] An alternative approach would be to have the investor learn about the actual value of the uncertain parameters at the end of a period. However, this may be a less realistic approach when using period durations of several years, since an investor is able to observe actual cost developments in the market instead of learning about actual costs after a lag of several years.

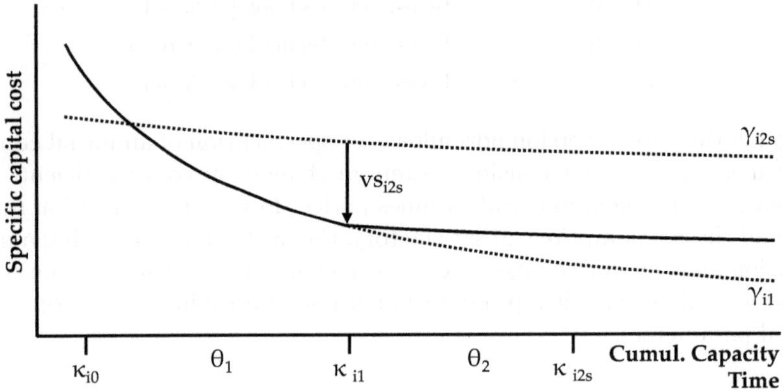

Fig. 5.3 Required vertical shift of experience curve between stages
Source: Own representation

whereas the opposite is true for technology B, which is more competitive in state b than in a. This, however, does not mean that technology A is necessarily more competitive than technology B in state b. Random variables for this basic model are β_{i2s}, representing variable operating costs, and γ_{i2s}, representing the learning index. The probability for occurrence of each of the two states of the world s is given by p_s, with $\sum_s p_s = 1$. The known specific variable cost coefficients and learning indexes for stage one will further be represented by β_{i1} and γ_{i1}, respectively.[19]

The presence of learning in combination with a potential change in the learning index between stages complicate model specification. At the interface between the two stages, which share the same cumulative capacity κ_{i1} specific unit investment costs should be equal. Yet, for the case that $\gamma_{i1} \neq \gamma_{i2s}$ this will unfortunately not be the case if the previously certain learning indexes in stage two are simply replaced by their equivalent notations under uncertainty. This difference is illustrated in Figure 5.3, where $\gamma_{i2s} < \gamma_{i1}$. To correct for this unwanted effect, the experience curve applicable in stage two needs to be shifted downwards by vs_{i2s}. This way, an unrealistic "jump" in specific capital costs at stage transition is avoided but learning still continues at the lower learning rate γ_{i2s}. This required shift will be further referred to as *vertical shift* vs_{i2s}. Its sign and magnitude are given by

$$vs_{i2s} = \kappa_{i1}^{-\gamma_{i1}} - \kappa_{i1}^{-\gamma_{i2s}} \tag{5.69}$$

[19] Implicitly this setup assumes that uncertainty resolution for learning rates is triggered by external events. Alternatively uncertainty resolution could also be triggered by the amount actually invested into a technology — yet this would imply asynchrony of periods and stages for the vast majority of parameter configurations.

The same definition also holds for the opposite case, where $\gamma_{i2s} > \gamma_{i1}$. A result of introducing this vertical shift is a kink in the experience curve at κ_{i1} for any case where $\gamma_{i1} \neq \gamma_{i2s}$ and $x_{i1} > 0$. Using this formulation, the implicit assumption is made that the learning index before the model's horizon equals the one in the first stage, otherwise an additional vertical shift would have to be introduced at capacity κ_{i0}.

The stochastic model reflecting the need for the described vertical shift is given by Equations (5.70) through (5.76). The three lines of the objective function describe total costs in the first stage, variable and fixed operating costs in the second stage, and finally capital costs in the second stage, respectively. Note that the absolute vertical shift in the objective function is given by product $x_{i2s}(\kappa_{i1}^{-\gamma_{i1}} - \kappa_{i1}^{-\gamma_{i2s}})$, whereas the vertical shift vs_{i2s} in Equation (5.69) only refers to specific costs. As the capital costs in the objective function are absolute values across period two, the multiplication of vs_{i2s} by period two capacity installations x_{i2s} becomes necessary. Constraints are similar to the ones from the deterministic model versions — only variables are updated to reflect the presence of two states of the world in stage two but do not contain any uncertain parameters. To limit complexity in the solution process, the stochastic model is set up without initial capacity endowment, thus building upon the basic deterministic model from section 5.2.

Minimize

$$\begin{aligned} z = \;& \psi_1 \sum_i \alpha_i \theta_1 w_{i1} + \beta_{i1} y_{i1} + \frac{\phi_i}{1-\gamma_{i1}} \left(\kappa_{i1}^{1-\gamma_{i1}} - \kappa_{i0}^{1-\gamma_{i1}} \right) \\ & + \psi_2 \sum_i \sum_s p_s \left(\alpha_i \theta_2 w_{i2s} + \beta_{i2s} y_{i2s} \right) \\ & + \psi_2 \sum_i \sum_s \frac{\phi_i p_s \theta_2}{\pi_i} \left[\frac{1}{1-\gamma_{i2s}} \left(\kappa_{i2s}^{1-\gamma_{i2s}} - \kappa_{i1}^{1-\gamma_{i2s}} \right) + x_{i2s} \left(\kappa_{i1}^{-\gamma_{i1}} - \kappa_{i1}^{-\gamma_{i2s}} \right) \right] \end{aligned} \tag{5.70}$$

Subject to

$$\sum_i y_{i1} \geq \delta_1 \theta_1 \tag{5.71}$$

$$\sum_i y_{i2s} \geq \delta_2 \theta_2 \tag{5.72}$$

$$\sigma_i \theta_1 w_{i1} \geq y_{i1} \tag{5.73}$$

$$\sigma_i \theta_2 w_{i2s} \geq y_{i2s} \tag{5.74}$$

$$x_{i1} = w_{i1} \tag{5.75}$$

$$w_{i1} + x_{i2s} - v_{i2s} = w_{i2s} \tag{5.76}$$

5.4.2 Solution

In principle, the solution process for the stochastic model is the same one as used for the deterministic model versions: First, the model is stated in Lagrangian form (see App. B.3.1), then, first-order derivatives are calculated and used to construct the KKT conditions. Based on the latter, the model is solved for all local minima that could constitute the global one. Finally, a decision rule is devised to identify the global among the local optima.

Once again, only the first-order partial derivatives with respect to λ_m are stated here. For the first-order partial derivatives with respect to the model's decision variables the interested reader is referred to App. B.3.2.

$$Z_{\lambda 1} = \delta_1 \theta_1 - y_{A1} - y_{B1} \quad (5.77)$$

$$Z_{\lambda 2} = \delta_2 \theta_2 - y_{A2a} - y_{B2a} \quad (5.78)$$

$$Z_{\lambda 3} = \delta_2 \theta_2 - y_{A2b} - y_{B2b} \quad (5.79)$$

$$Z_{\lambda 4} = y_{A1} - \sigma_A \theta_1 w_{A1} \quad (5.80)$$

$$Z_{\lambda 5} = y_{B1} - \sigma_B \theta_1 w_{B1} \quad (5.81)$$

$$Z_{\lambda 6} = y_{A2a} - \sigma_A \theta_2 w_{A2a} \quad (5.82)$$

$$Z_{\lambda 7} = y_{B2a} - \sigma_B \theta_2 w_{B2a} \quad (5.83)$$

$$Z_{\lambda 8} = y_{A2b} - \sigma_A \theta_2 w_{A2b} \quad (5.84)$$

$$Z_{\lambda 9} = y_{B2b} - \sigma_B \theta_2 w_{B2b} \quad (5.85)$$

$$Z_{\lambda 10} = w_{A1} - x_{A1} \quad (5.86)$$

$$Z_{\lambda 11} = w_{B1} - x_{B1} \quad (5.87)$$

$$Z_{\lambda 12} = w_{A2a} - w_{A1} - x_{A2a} + \mu_A x_{A1} \quad (5.88)$$

$$Z_{\lambda 13} = w_{B2a} - w_{B1} - x_{B2a} + \mu_B x_{B1} \quad (5.89)$$

$$Z_{\lambda 14} = w_{A2b} - w_{A1} - x_{A2b} + \mu_A x_{A1} \quad (5.90)$$

$$Z_{\lambda 15} = w_{B2b} - w_{B1} - x_{B2b} + \mu_B x_{B1} \quad (5.91)$$

For the deterministic model without changes in parameter values over time only two relevant cases exist: either technology A or Technology B is pursued at the global optimum. The number of relevant cases in the stochastic model is considerably larger. Here the hypothesis is used that only *pure* strategies, i.e. concentration of investment on one technology in the first stage, can be globally optimal under the given setup. This is based on the reasoning that, uncertain learning indexes and specific variable costs both being objective function coefficients, the "stochastic" model is de-facto deterministic — optimal decisions in stage one have to be made based on the expected value of the uncertain parameters. In a deterministic setting with strictly increasing returns, concentration on one technology will yield the cost-minimum. The validity of this critical assumption is demonstrated in the next section of this chapter.

For this stochastic model version twelve different cases for local minima and thus optimal investment strategies that could constitute the global optimum exist. This number of relevant configuration cases has been derived in an eneumeration and elimination process. Starting point are the following three assumptions already specified further above.

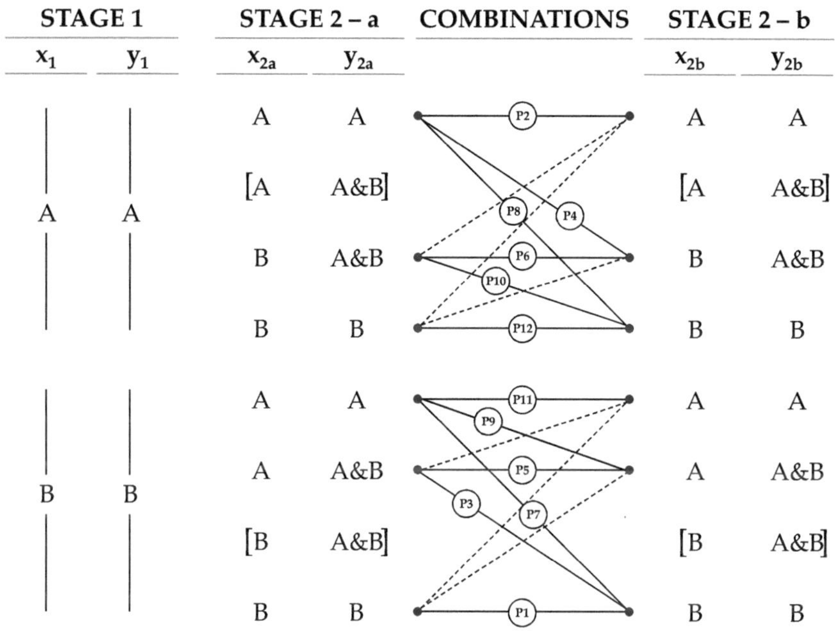

Fig. 5.4 Relevant pure strategy cases
Source: Own calculation

1. *Pure strategy optimality*: In stage one, there can be investment into and therefore also production from either technology A or B but not both.
2. *Stage two investment requirement*: In stage two, an investment must occur, again into either one of the two available technologies. Yet this investment may be different in the states of the world a and b.
3. *No redundant capacity*: Investments made in stage two are always used for production.

Based on those assumptions, the list of possible variable configurations can be enumerated as shown in Fig. 5.4. x_1 and y_1 represent the investment and production in stage one, respectively. Investments and production in stage two are analogously denoted by x_2 and y_2 and are distinguished for both possible scenarios, a and b.

In line one of Fig. 5.4, for example, technology A is installed in the first stage and consequently power is also only produced from technology A. In the second stage, in both states of the world, technology A is installed to close the capacity gap and power is once more only produced from technology A. This variable configuration is further referred to as case P2. Line three of Fig. 5.4 shows a somewhat different case. Again technology A is installed and used for production in stage 1, yet technology B is installed in both states of the world to fill the capacity gap. Both technologies are then used for power production in stage 2 (Case P6). Another case (P10) features the same variable configuration in state of the world A, yet a more extreme shift to B in state b, where technology B replaces the remaining capacity of A from stage one and covers the entire power demand. It can found in lines three and four of Fig. 5.4 for states of the world a and b, respectively.

From the total enumerated combinations in Fig. 5.4 eliminations must be made. Stage two configurations marked in brackets are not permissible, as production from both technologies would require an investment into both technologies beforehand, which does not occur. Also, combinations must be eliminated, where the previous definition of the two scenarios a and b is violated. In state a technology A is more competitive than in state b, whereas the opposite is true for technology B. The combinations, for which this definition does not hold true, are marked as dotted lines in Fig. 5.4. The remaining solid lines then represent the twelve relevant cases.

In the following, the optimal investment strategies corresponding to those twelve cases are described. Due to the increased number of cases, a notation change compared to the deterministic model is required. Instead of the notation z^i, the solution superscript contains the case identification number assigned further below. As each even-numbered case is exactly the opposite of the preceding odd-numbered one with respect to technology i and state of the world s, only odd-numbered cases are discussed in detail.

Case P1 — Dominance of technology B: Only investment into technology B takes place in both stages. Thus, B produces all power across the model horizon. This case corresponds to case B from the deterministic model.

Case P3 — Advantage of technology B: Technology B is installed in stage one. For $s = b$ it is also installed in stage two. For $s = a$, however, technology A is installed in stage two. In the latter scenario, the remaining capacity of B in stage two is further used for power generation. The scenario has been called *advantage*, since the investment of B in the first stage has increased the chance of lock-out for A in the second stage and even for $s = a$ only the capacity gap is filled with technology A.

Case P5 — Investment shift B to A: Technology B is installed in stage one, technology A in stage two for any realized scenarios. In this parameter configuration a significant deterministic drift in the random variables from stage one to two must be present. Existing capacity of technology B is further used for power generation in stage two.

Case P7 — Advantage B, pot. full shift to A: Signs of the investment are identical to the ones for case P3. For $s = b$, technology B, for $s = a$ technology A is invested in stage two. However, in case of $s = a$, investment into A is sufficiently large for A to satisfy the entire power demand in stage two. The remaining capacity of B in stage two consequently remains idle for $s = a$.[20]

[20] Capacity retirement before the end of a plant's useful life is prevented by Eq. (5.76) being specified as equality.

Case P9 — Potential full shift B to A: Whereas investments in stage one go into technology B, the ones in stage two are directed towards technology A, irrespective of the realized scenario. The scenario realized, however, determines the amount of investment into A and the role of the remaining technology B capacity in the second stage. For $s = a$ enough of A is installed to cover the entire power demand and capacity from B remains idle. For $s = b$, however, the remaining capacity of B produces power and only so much of A is installed as needed to bridge the capacity gap.

Case P11 — Full shift B to A: This is the most extreme case, where the remaining capacity B is idle in stage two, no matter of the realized scenario, and a sufficiently large technology A investment takes place to satisfy the entire power demand in stage two.

Table 5.2 lists the case configurations in terms of x_{its} and y_{i2s} for the twelve relevant *pure* investment strategies. Cases P1–P6 are the ones with highest practical relevance, whereas cases P7–P12 have lower practical applicability, since they imply occurrence of such massive parameter value changes across stages that it will be optimal to have some idle capacity in stage two.[21] The higher the demand and the duration of the first stage relative to the second, the higher the lock-in effect for the technology invested in the first stage. This means that technology shifts across stages become less likely the more of a particular technology is added in stage one.

Table 5.2 Sign configurations for pure strategies under uncertainty

Case	*Description*	x_{A1}	x_{B1}	x_{A2a}	x_{A2b}	x_{B2a}	x_{B2b}	y_{A2a}	y_{A2b}	y_{B2a}	y_{B2b}
P1	Dominance B	$=0$	>0	$=0$	$=0$	>0	>0	$=0$	$=0$	>0	>0
P2	Dominance A	>0	$=0$	>0	>0	$=0$	$=0$	>0	>0	$=0$	$=0$
P3	Advantage B	$=0$	>0	>0	$=0$	$=0$	>0	>0	$=0$	>0	>0
P4	Advantage A	>0	$=0$	>0	$=0$	$=0$	>0	>0	>0	$=0$	>0
P5	Invest B to A	$=0$	>0	>0	>0	$=0$	$=0$	>0	>0	>0	>0
P6	Invest A to B	>0	$=0$	$=0$	$=0$	>0	>0	>0	>0	>0	>0
P7	Advantage B FS	$=0$	>0	>0	$=0$	$=0$	>0	>0	$=0$	$=0$	>0
P8	Advantage A FS	>0	$=0$	>0	$=0$	$=0$	>0	>0	$=0$	$=0$	>0
P9	Pot. full shift BA	$=0$	>0	>0	>0	$=0$	$=0$	>0	>0	$=0$	>0
P10	Pot. full shift AB	>0	$=0$	$=0$	$=0$	>0	>0	>0	$=0$	>0	>0
P11	Full shift BA	$=0$	>0	>0	>0	$=0$	$=0$	>0	>0	$=0$	$=0$
P12	Full shift AB	>0	$=0$	$=0$	$=0$	>0	>0	$=0$	$=0$	>0	>0

The following subsections describe the solution process for the local minima corresponding to the six odd-numbered cases in detail. The solutions for even-numbered cases can be found in Appendix B.3.3, as they "mirror" the

[21] The preclusion of premature capacity retirements by assumption aggravates this property.

ones for their respective odd-numbered counterparts with respect to dimensions i and s. Once again, only the solution sequences leading to a fulfillment of the KKT conditions are presented.

Case P1 — Dominance of Technology B

Investment strategy for case P1 consists of investment into and power production through technology B only, regardless of the state of the world realized in stage two. Starting point for the solution process is therefore:

$$x^*_{A1} = 0 \qquad x^*_{A2a} = 0 \qquad x^*_{A2b} = 0$$

From Equations (5.86), (5.88) and (5.90)

$$w^*_{A1} = x_{A1} = 0 \tag{5.92}$$

$$w^*_{A2a} = w_{A1} + x_{A2a} - \mu_A x_{A1} = 0 \tag{5.93}$$

$$w^*_{A2b} = w_{A1} + x_{A2b} - \mu_A x_{A1} = 0 \tag{5.94}$$

From Equations (5.80), (5.82) and (5.84)

$$y^*_{A1} = \sigma_A \theta_1 w_{A1} = 0 \tag{5.95}$$

$$y^*_{A2a} = \sigma_A \theta_2 w_{A2a} = 0 \tag{5.96}$$

$$y^*_{A2b} = \sigma_A \theta_2 w_{A2b} = 0 \tag{5.97}$$

From Equations (5.77) through (5.79)

$$y^*_{B1} = \delta_1 \theta_1 - y_{A1} = \delta_1 \theta_1 \tag{5.98}$$

$$y^*_{B2a} = \delta_2 \theta_2 - y_{A2a} = \delta_2 \theta_2 \tag{5.99}$$

$$y^*_{B2b} = \delta_2 \theta_2 - y_{A2b} = \delta_2 \theta_2 \tag{5.100}$$

From Equations (5.81), (5.83) and (5.85)

$$w^*_{B1} = \frac{y_{B1}}{\sigma_B \theta_1} = \frac{\delta_1}{\sigma_B} \tag{5.101}$$

$$w^*_{B2a} = \frac{y_{B2a}}{\sigma_B \theta_2} = \frac{\delta_2}{\sigma_B} \tag{5.102}$$

$$w^*_{B2b} = \frac{y_{B2b}}{\sigma_B \theta_2} = \frac{\delta_2}{\sigma_B} \tag{5.103}$$

And finally, from Equations (5.87), (5.89) and (5.91)

$$x^*_{B1} = w_{B1} = \frac{\delta_1}{\sigma_B} \tag{5.104}$$

$$x^*_{B2a} = w_{B2a} + \mu_B x_{B1} - w_{B1} = (\mu_B - 1)\frac{\delta_1}{\sigma_B} + \frac{\delta_2}{\sigma_B} \tag{5.105}$$

$$x^*_{B2b} = w_{B2b} + \mu_B x_{B1} - w_{B1} = (\mu_B - 1)\frac{\delta_1}{\sigma_B} + \frac{\delta_2}{\sigma_B} \tag{5.106}$$

Case P3 — Advantage of Technology B

Investment strategy for case P3 consists of technology B investment in stage one and, for $s = b$, also in stage two. For $s = a$, technology A is invested in stage two in order to fill the capacity gap between the remaining capacity of B from stage one and demand in stage two. Starting point is:

$$x^*_{A1} = 0 \qquad x^*_{A2b} = 0 \qquad x^*_{B2a} = 0$$

From Equations (5.86) and (5.90)

$$w^*_{A1} = x_{A1} = 0 \tag{5.107}$$
$$w^*_{A2b} = w_{A1} + x_{A2b} - \mu_A x_{A1} = 0 \tag{5.108}$$

From Equations (5.84), (5.79), (5.80) and (5.77)

$$y^*_{A2b} = \sigma_A \theta_2 w_{A2b} = 0 \tag{5.109}$$
$$y^*_{B2b} = \delta_2\theta_2 - y_{A2b} = \delta_2\theta_2 \tag{5.110}$$
$$y^*_{A1} = \sigma_A \theta_1 w_{A1} = 0 \tag{5.111}$$
$$y^*_{B1} = \delta_1\theta_1 - y_{A1} = \delta_1\theta_1 \tag{5.112}$$

From Equations (5.85), (5.81) and (5.87)

$$w^*_{B2b} = \frac{y_{B2b}}{\sigma_B\theta_2} = \frac{\delta_2}{\sigma_B} \tag{5.113}$$
$$w^*_{B1} = \frac{y_{B1}}{\sigma_B\theta_1} = \frac{\delta_1}{\sigma_B} \tag{5.114}$$
$$x^*_{B1} = w_{B1} = \frac{\delta_1}{\sigma_B} \tag{5.115}$$

From Equation (5.89)

$$w^*_{B2a} = x_{B2a} + w_{B1} - \mu_B x_{B1} = (1-\mu_B)\frac{\delta_1}{\sigma_B} \tag{5.116}$$

From Equation (5.91)

$$x^*_{B2b} = w_{B2b} - w_{B1} + \mu_B x_{B1} = (\mu_B - 1)\frac{\delta_1}{\sigma_B} + \frac{\delta_2}{\sigma_B} \tag{5.117}$$

From Equation (5.83)

$$y^*_{B2a} = \sigma_B\theta_2 w_{B2a} = (1-\mu_B)\delta_1\theta_2 \tag{5.118}$$

From Equation (5.78)

$$y^*_{A2a} = \delta_2\theta_2 - y_{B2a} = \theta_2[\delta_2 - \delta_1(1-\mu_B)] \tag{5.119}$$

From Equation (5.82)

$$w^*_{A2a} = \frac{y_{A2a}}{\sigma_A\theta_2} = \frac{\delta_2 - \delta_1(1-\mu_B)}{\sigma_A} \tag{5.120}$$

Finally, from Equation (5.88)

$$x^*_{A2a} = w_{A2a} + \mu_A x_{A1} - w_{A1} = \frac{\delta_2 - \delta_1(1-\mu_B)}{\sigma_A} \tag{5.121}$$

Case P5 — Investment Shift from Technology B to A

Investment strategy for case P5 consists of technology B investment in stage two and, for both states of the world, investment into A in stage two. The remaining capacity stock of technology B is used for power production in stage two. Starting point is:

$$x_{A1} = 0 \qquad x_{B2b} = 0 \qquad x_{B2a} = 0$$

From Equations (5.86), (5.80) and (5.77)

$$w^*_{A1} = x_{A1} = 0 \tag{5.122}$$
$$y^*_{A1} = \sigma_A\theta_1 w_{A1} = 0 \tag{5.123}$$
$$y^*_{B1} = \delta_1\theta_1 - y_{A1} = \delta_1\theta_1 \tag{5.124}$$

From Equations (5.81) and (5.87)

$$w^*_{B1} = \frac{y_{B1}}{\sigma_B\theta_1} = \frac{\delta_1}{\sigma_B} \tag{5.125}$$
$$x^*_{B1} = w_{B1} = \frac{\delta_1}{\sigma_B} \tag{5.126}$$

From Equations (5.89) and (5.91)

$$w^*_{B2a} = x_{B2a} + w_{B1} - \mu_B x_{B1} = (1-\mu_B)\frac{\delta_1}{\sigma_B} \tag{5.127}$$
$$w^*_{B2b} = x_{B2b} + w_{B1} - \mu_B x_{B1} = (1-\mu_B)\frac{\delta_1}{\sigma_B} \tag{5.128}$$

From Equations (5.83) and (5.85)

$$y^*_{B2a} = \sigma_B\theta_2 w_{B2a} = (1-\mu_B)\delta_1\theta_2 \tag{5.129}$$
$$y^*_{B2b} = \sigma_B\theta_2 w_{B2b} = (1-\mu_B)\delta_1\theta_2 \tag{5.130}$$

From Equations (5.78) and (5.79)

$$y^*_{A2a} = \delta_2\theta_2 - y_{B2a} = \theta_2[\delta_2 - \delta_1(1-\mu_B)] \tag{5.131}$$
$$y^*_{A2b} = \delta_2\theta_2 - y_{B2b} = \theta_2[\delta_2 - \delta_1(1-\mu_B)] \tag{5.132}$$

From Equations (5.82) and (5.84)

$$w^*_{A2a} = \frac{y_{A2a}}{\sigma_A \theta_2} = \frac{\delta_2 - \delta_1(1-\mu_B)}{\sigma_A} \tag{5.133}$$

$$w^*_{A2b} = \frac{y_{A2b}}{\sigma_A \theta_2} = \frac{\delta_2 - \delta_1(1-\mu_B)}{\sigma_A} \tag{5.134}$$

Finally, from Equations (5.88) and (5.90)

$$x^*_{A2a} = w_{A2a} + \mu_A x_{A1} - w_{A1} = \frac{\delta_2 - \delta_1(1-\mu_B)}{\sigma_A} \tag{5.135}$$

$$x^*_{A2b} = w_{A2b} + \mu_A x_{A1} - w_{A1} = \frac{\delta_2 - \delta_1(1-\mu_B)}{\sigma_A} \tag{5.136}$$

Case P7 — Advantage Technology B with Potential Full Shift to A

Similarly to P3, investment strategy for case P7 consists of technology B investment in stage one and, for $s = b$, also for stage two. Yet, for $s = a$, a sufficiently large investment into A occurs in stage two in order to satisfy all demand. Remaining capacity of technology B from stage one is consequently idle given $s = a$.

$$x^*_{A1} = 0 \qquad x^*_{A2b} = 0 \qquad x^*_{B2a} = 0$$

$$y^*_{A2b} = 0 \qquad y^*_{B2a} = 0$$

From Equations (5.86), (5.80) and (5.77)

$$w^*_{A1} = x_{A1} = 0 \tag{5.137}$$

$$y^*_{A1} = \sigma_A \theta_1 w_{A1} = 0 \tag{5.138}$$

$$y^*_{B1} = \delta_1 \theta_1 - y_{A1} = \delta_1 \theta_1 \tag{5.139}$$

From Equations (5.81) and (5.87)

$$w^*_{B1} = \frac{y_{B1}}{\sigma_B \theta_1} = \frac{\delta_1}{\sigma_B} \tag{5.140}$$

$$x^*_{B1} = w_{B1} = \frac{\delta_1}{\sigma_B} \tag{5.141}$$

From Equations (5.78) and (5.79)

$$y^*_{A2a} = \delta_2 \theta_2 - y_{B2a} = \delta_2 \theta_2 \tag{5.142}$$

$$y^*_{B2b} = \delta_2 \theta_2 - y_{A2b} = \delta_2 \theta_2 \tag{5.143}$$

From Equations (5.82) and (5.85)

$$w^*_{A2a} = \frac{y_{A2a}}{\sigma_A \theta_2} = \frac{\delta_2}{\sigma_A} \tag{5.144}$$

$$w^*_{B2b} = \frac{y_{B2b}}{\sigma_B \theta_2} = \frac{\delta_2}{\sigma_B} \tag{5.145}$$

From Equations (5.90) and (5.89)

$$w^*_{A2b} = w_{A1} + x_{A2b} - \mu_A x_{A1} = 0 \tag{5.146}$$

$$w^*_{B2a} = w_{B1} + x_{B2a} - \mu_B x_{B1} = (1 - \mu_B)\frac{\delta_1}{\sigma_B} \tag{5.147}$$

From Equations (5.88) and (5.91)

$$x^*_{A2a} = w_{A2a} + \mu_A x_{A1} - w_{A1} = \frac{\delta_2}{\sigma_A} \tag{5.148}$$

$$x^*_{B2b} = w_{B2b} + \mu_B x_{B1} - w_{B1} = (\mu_B - 1)\frac{\delta_1}{\sigma_B} + \frac{\delta_2}{\sigma_B} \tag{5.149}$$

Case P9 — Potential Full Shift from Technology B to A

In this case, technology B is invested in stage one and, indifferent of the realization of the random variables, technology A in stage two. For $s = b$, the remaining capacity stock of technology B is further used for power generation in stage two. For $s = a$, on the other hand, this capacity remains idle in stage two and it is optimal to install sufficient capacity of technology A to cover the entire stage two power demand. Starting point is:

$$x^*_{A1} = 0 \qquad x_{B2b} = 0$$

$$x^*_{B2a} = 0 \qquad y_{B2a} = 0$$

From Equations (5.86), (5.80) and (5.77)

$$w^*_{A1} = x_{A1} = 0 \tag{5.150}$$

$$y^*_{A1} = \sigma_A \theta_1 w_{A1} = 0 \tag{5.151}$$

$$y^*_{B1} = \delta_1 \theta_1 - y_{A1} = \delta_1 \theta_1 \tag{5.152}$$

From Equations (5.81) and (5.87)

$$w^*_{B1} = \frac{y_{B1}}{\sigma_B \theta_1} = \frac{\delta_1}{\sigma_B} \tag{5.153}$$

$$x^*_{B1} = w_{B1} = \frac{\delta_1}{\sigma_B} \tag{5.154}$$

From Equations (5.78) and (5.82) and (5.88)

$$y^*_{A2a} = \delta_2\theta_2 - y_{B2a} = \delta_2\theta_2 \tag{5.155}$$

$$w^*_{A2a} = \frac{y_{A2a}}{\sigma_A\theta_2} = \frac{\delta_2}{\sigma_A} \tag{5.156}$$

$$x^*_{A2a} = \mu_A x_{A1} + w_{A2a} - w_{A1} = \frac{\delta_2}{\sigma_A} \tag{5.157}$$

From Equations (5.89) and (5.91)

$$w^*_{B2a} = w_{B1} + x_{B2a} - \mu_B x_{B1} = (1-\mu_B)\frac{\delta_1}{\sigma_B} \tag{5.158}$$

$$w^*_{B2b} = w_{B1} + x_{B2b} - \mu_B x_{B1} = (1-\mu_B)\frac{\delta_1}{\sigma_B} \tag{5.159}$$

From Equations (5.85) and (5.79)

$$y^*_{B2b} = \sigma_B\theta_2 w_{B2b} = (1-\mu_B)\delta_1\theta_2 \tag{5.160}$$

$$y^*_{A2b} = \delta_2\theta_2 - y_{B2b} = \theta_2[\delta_2 + (\mu_B - 1)\delta_1] \tag{5.161}$$

From Equations (5.84) and (5.90)

$$w^*_{A2b} = \frac{y_{A2b}}{\sigma_A\theta_2} = \frac{\delta_2 + (\mu_B - 1)\delta_1}{\sigma_A} \tag{5.162}$$

$$x^*_{A2b} = \mu_A x_{A1} + w_{A2b} - w_{A1} = \frac{\delta_2 + (\mu_B - 1)\delta_1}{\sigma_A} \tag{5.163}$$

Case P11 — Full Shift from Technology B to A

In this last case an investment shift from technology B in stage one to technology A in stage two occurs. The investment into technology A is, for both scenarios, sufficiently large to supply all power demand in stage two. Remaining capacity of technology B is idle in stage two. Starting point is:

$$x^*_{A1} = 0 \qquad x^*_{B2a} = 0 \qquad x^*_{B2b} = 0$$

$$y^*_{B2a} = 0 \qquad y^*_{B2b} = 0$$

From Equations (5.86), (5.80) and (5.77)

$$w^*_{A1} = x_{A1} = 0 \tag{5.164}$$

$$y^*_{A1} = \sigma_A\theta_1 w_{A1} = 0 \tag{5.165}$$

$$y^*_{B1} = \delta_1\theta_1 - y_{A1} = \delta_1\theta_1 \tag{5.166}$$

From Equations (5.81) and (5.87)

$$w^*_{B1} = \frac{y_{B1}}{\sigma_B \theta_1} = \frac{\delta_1}{\sigma_B} \tag{5.167}$$

$$x^*_{B1} = w_{B1} = \frac{\delta_1}{\sigma_B} \tag{5.168}$$

From Equations (5.78) and (5.79)

$$y^*_{A2a} = \delta_2\theta_2 - y_{B2a} = \delta_2\theta_2 \tag{5.169}$$

$$y^*_{A2b} = \delta_2\theta_2 - y_{B2b} = \delta_2\theta_2 \tag{5.170}$$

From Equations (5.89) and (5.91)

$$w^*_{B2a} = x_{B2a} + w_{B1} - \mu_B x_{B1} = (1-\mu_B)\frac{\delta_1}{\sigma_B} \tag{5.171}$$

$$w^*_{B2b} = x_{B2b} + w_{B1} - \mu_B x_{B1} = (1-\mu_B)\frac{\delta_1}{\sigma_B} \tag{5.172}$$

From Equations (5.82) and (5.84)

$$w^*_{A2a} = \frac{y_{A2a}}{\sigma_A \theta_2} = \frac{\delta_2}{\sigma_A} \tag{5.173}$$

$$w^*_{A2b} = \frac{y_{A2b}}{\sigma_A \theta_2} = \frac{\delta_2}{\sigma_A} \tag{5.174}$$

And finally, from Equations (5.88) and (5.90)

$$x^*_{A2a} = w_{A2a} + \mu_A x_{A1} - w_{A1} = \frac{\delta_2}{\sigma_A} \tag{5.175}$$

$$x^*_{A2b} = w_{A2b} + \mu_A x_{A1} - w_{A1} = \frac{\delta_2}{\sigma_A} \tag{5.176}$$

5.4.3 Investment Decision Rule

In the same manner as in the previous sections, a decision rule is used to enable identification of the global optimum among the twelve local ones. To this end, the solutions obtained in the previous section have been re-substituted into the objective function. z^n hereby represent the objective function value for the particular case, with $n = \{P1, P2, \ldots, P12\}$. The twelve cases can broadly be grouped into six cases, which would also exist if there were no uncertainty, and six cases, which will only be relevant in presence of uncertain stage two parameter values. This property is visible from the structure of the objective function values and will be discussed for each case separately. In the following, only the rules for odd-numbered cases are shown. Their even-numbered counterparts can be found in Appendix B.3.4.

$$
\begin{aligned}
z^{P1} &= \psi_1\theta_1\delta_1\left(\frac{\alpha_B}{\sigma_B}+\beta_B\right)+\frac{\phi_B\psi_1}{1-\gamma_{B1}}\left(\kappa_{B1}^{1-\gamma_{B1}}-\kappa_{B0}^{1-\gamma_{B1}}\right)\\
&+\psi_2\theta_2\delta_2\left(\frac{\alpha_B}{\sigma_B}+(p_a\beta_{B2a}+p_b\beta_{B2b})\right)\\
&+\frac{\phi_B\psi_2\theta_2}{\pi_B}\left[\frac{p_a}{1-\gamma_{B2a}}\left(\kappa_{B2s}^{1-\gamma_{B2a}}-\kappa_{B1}^{1-\gamma_{B2a}}\right)+\frac{p_b}{1-\gamma_{B2b}}\left(\kappa_{B2s}^{1-\gamma_{B2b}}-\kappa_{B1}^{1-\gamma_{B2b}}\right)\right]\\
&+\frac{\phi_B\psi_2\theta_2}{\pi_B}\left[(\mu_B-1)\frac{\delta_1}{\sigma_B}+\frac{\delta_2}{\sigma_B}\right]\left[\kappa_{B1}^{-\gamma_{B1}}-\left(p_a\kappa_{B1}^{-\gamma_{B2a}}+p_b\kappa_{B1}^{-\gamma_{B2b}}\right)\right]
\end{aligned}
\tag{5.177}
$$

with $\kappa_{B1}=\kappa_{B0}+\frac{\delta_1}{\sigma_B}$ and $\kappa_{B2s}=\kappa_{B0}+\mu_B\frac{\delta_1}{\sigma_B}+\frac{\delta_2}{\sigma_B}$

From Equation (5.177) it can be inferred that the solution is identical to the solution of the deterministic equivalent problem. All random variables have been arranged in a manner, so that they represent their expected values. The term $p_a\beta_{B2a}+p_b\beta_{B2b}$ in line two is nothing else than $E(\beta_{B2s})$. Further, the term $\frac{p_a}{1-\gamma_{B2a}}(\kappa_{B2s}^{1-\gamma_{B2a}}-\kappa_{B1}^{1-\gamma_{B2a}})+\frac{p_b}{1-\gamma_{B2b}}(\kappa_{B2s}^{1-\gamma_{B2b}}-\kappa_{B1}^{1-\gamma_{B2b}})$ in line three is the expected specific capital cost multiplier before consideration of the vertical shift and $\kappa_{B1}^{-\gamma_{B1}}-(p_a\kappa_{B1}^{-\gamma_{B2a}}+p_b\kappa_{B1}^{-\gamma_{B2b}})$ in line four is the expected specific capital cost multiplier for the vertical shift. Therefore, case P1 and the corresponding solution would also exist in a deterministic model with changing parameter values across stages. This observation is altogether not surprising, as case P1 corresponds to the strategy "Dominance of technology B" in the deterministic model.

$$
\begin{aligned}
z^{P3} &= \psi_1\theta_1\delta_1\left(\frac{\alpha_B}{\sigma_B}+\beta_{B1}\right)+\frac{\phi_B\psi_1}{1-\gamma_{B1}}\left(\kappa_{B1}^{1-\gamma_{B1}}-\kappa_{B0}^{1-\gamma_{B1}}\right)\\
&+p_a\psi_2\theta_2\left[(\delta_2-\delta_1(1-\mu_B))\left(\frac{\alpha_A}{\sigma_A}+\beta_{A2a}\right)+\delta_1(1-\mu_B)\left(\frac{\alpha_B}{\sigma_B}+\beta_{B2a}\right)\right]\\
&+p_b\psi_2\theta_2\delta_2\left(\frac{\alpha_B}{\sigma_B}+\beta_{B2b}\right)\\
&+\frac{p_a\phi_A\psi_2\theta_2}{\pi_A(1-\gamma_{A2a})}\left(\kappa_{A2a}^{1-\gamma_{A2a}}-\kappa_{A0}^{1-\gamma_{A2a}}\right)\\
&+\frac{p_b\phi_B\psi_2\theta_2}{\pi_B(1-\gamma_{B2b})}\left(\kappa_{B2b}^{1-\gamma_{B2b}}-\kappa_{B1}^{1-\gamma_{B2b}}\right)\\
&+\frac{p_a\phi_A\psi_2\theta_2[\delta_2-\delta_1(1-\mu_B)]}{\pi_A\sigma_A}\left(\kappa_{A1}^{-\gamma_{A1}}-\kappa_{A1}^{-\gamma_{A2a}}\right)\\
&+\frac{p_b\phi_B\psi_2\theta_2}{\pi_B}\left[(\mu_B-1)\frac{\delta_1}{\sigma_B}+\frac{\delta_2}{\sigma_B}\right]\left(\kappa_{B1}^{-\gamma_{B1}}-\kappa_{B1}^{-\gamma_{B2b}}\right)
\end{aligned}
\tag{5.178}
$$

with $\kappa_{B1} = \kappa_{B0} + \dfrac{\delta_1}{\sigma_B}$ $\qquad \kappa_{A2a} = \kappa_{A0} + \dfrac{\delta_2 - \delta_1(1-\mu_B)}{\sigma_A}$

$\kappa_{B2b} = \kappa_{B0} + \dfrac{\mu_B\delta_1 + \delta_2}{\sigma_B}$

In contrast to case P1, the objective function value for case P3 is clearly not identical to any deterministic equivalent but rather contains specific terms for various realized states of the world in stage two. Probabilities and corresponding random variables cannot be arranged so as to represent the expected value of the random variable — this holds for variable costs in lines two and three, capital costs before a potential vertical shift in lines four and five, and the vertical shift in lines six and seven. Thus, case P3 would not exist in absence of explicit uncertainty representation.

$$\begin{aligned}
z^{P5} = \psi_1 &\left[\theta_1\delta_1\left(\frac{\alpha_B}{\sigma_B} + \beta_{B1}\right) + \frac{\phi_B}{1-\gamma_{B1}}\left(\kappa_{B1}^{1-\gamma_{B1}} - \kappa_{B0}^{1-\gamma_{B1}}\right)\right] \\
&+ \psi_2\theta_2[\delta_2 - \delta_1(1-\mu_B)]\left[\frac{\alpha_A}{\sigma_A} + (p_a\beta_{A2a} + p_b\beta_{A2b})\right] \\
&+ \psi_2\theta_2\delta_1(1-\mu_B)\left[\frac{\alpha_B}{\sigma_B} + (p_a\beta_{B2a} + p_b\beta_{B2b})\right] \\
&+ \frac{\phi_A\psi_2\theta_2}{\pi_A}\left[\frac{p_a}{1-\gamma_{A2a}}\left(\kappa_{A2s}^{1-\gamma_{A2a}} - \kappa_{A0}^{1-\gamma_{A2a}}\right) + \frac{p_b}{1-\gamma_{A2b}}\left(\kappa_{A2s}^{1-\gamma_{A2b}} - \kappa_{A0}^{1-\gamma_{A2b}}\right)\right] \\
&+ \frac{\phi_A\psi_2\theta_2[\delta_2 - \delta_1(1-\mu_B)]}{\pi_A\sigma_A}\left[\kappa_{A0}^{-\gamma_{A1}} - \left(p_a\kappa_{A0}^{-\gamma_{A2a}} + p_b\kappa_{A0}^{-\gamma_{A2b}}\right)\right]
\end{aligned} \tag{5.179}$$

with $\kappa_{B1} = \kappa_{B0} + \dfrac{\delta_1}{\sigma_B}$ and $\kappa_{A2s} = \kappa_{A0} + \dfrac{\delta_2 - \delta_1(1-\mu_B)}{\sigma_A}$

Once again, this case would also exist in the deterministic model with equivalent parameters in stage two. This can be demonstrated again by rearranging terms containing uncertain parameters to their respective expected values. $p_a\beta_{A2a} + p_b\beta_{A2b}$ and $p_a\beta_{B2a} + p_b\beta_{B2b}$ represent the expected value of the specific variable costs for technologies A and B, respectively. The expected value of the specific unit cost multiplier of technology A in stage two is given by $\frac{p_a}{1-\gamma_{A2a}}(\kappa_{A2s}^{1-\gamma_{A2a}} - \kappa_{A1}^{1-\gamma_{A2a}}) + \frac{p_b}{1-\gamma_{A2b}}(\kappa_{A2s}^{1-\gamma_{A2b}} - \kappa_{A1}^{1-\gamma_{A2b}})$ and the expected specific capital cost impact from a potential vertical shift is given by $\kappa_{A1}^{-\gamma_{A1}} - (p_a\kappa_{A1}^{-\gamma_{A2a}} + p_b\kappa_{A1}^{-\gamma_{A2b}})$. For the deterministic model versions in the previous sections this solution had not been relevant, as a deterministic parameter value drift is a necessary condition for this case to constitute the global optimum.

$$
\begin{aligned}
z^{P7} &= \psi_1 \left[\theta_1 \delta_1 \left(\frac{\alpha_B}{\sigma_B} + \beta_{B1}\right) + \frac{\phi_B}{1-\gamma_{B1}} \left(\kappa_{B1}^{1-\gamma_{B1}} - \kappa_{B0}^{1-\gamma_{B1}}\right)\right] \\
&+ p_a \psi_2 \theta_2 \left[\delta_2 \left(\frac{\alpha_A}{\sigma_A} + \beta_{A2a}\right) + \delta_1 (1-\mu_B) \frac{\alpha_B}{\sigma_B}\right] + p_b \psi_2 \theta_2 \delta_2 \left(\frac{\alpha_B}{\sigma_B} + \beta_{B2b}\right) \\
&+ \frac{p_a \phi_A \psi_2 \theta_2}{\pi_A} \left[\frac{1}{1-\gamma_{A2a}} \left(\kappa_{A2a}^{1-\gamma_{A2a}} - \kappa_{A1}^{1-\gamma_{A2a}}\right) + \frac{\delta_2}{\sigma_A} \left(\kappa_{A1}^{-\gamma_{A1}} - \kappa_{A1}^{-\gamma_{A2a}}\right)\right] \\
&+ \frac{p_b \phi_B \psi_2 \theta_2}{\pi_B (1-\gamma_{B2b})} \left(\kappa_{B2b}^{1-\gamma_{B2b}} - \kappa_{B1}^{1-\gamma_{B2b}}\right) \\
&+ \frac{p_b \phi_B \psi_2 \theta_2}{\pi_B} \left((\mu_B - 1)\frac{\delta_1}{\sigma_B} + \frac{\delta_2}{\sigma_B}\right) \left(\kappa_{B1}^{-\gamma_{B1}} - \kappa_{B1}^{-\gamma_{B2b}}\right)
\end{aligned}
\tag{5.180}
$$

with $\kappa_{A1} = \kappa_{A0}$ $\quad \kappa_{B1} = \kappa_{B0} + \frac{\delta_1}{\sigma_B}$

$\kappa_{A2a} = \kappa_{A0} + \frac{\delta_2}{\sigma_A}$ $\quad \kappa_{B2b} = \kappa_{B0} + \mu_B \frac{\delta_1}{\sigma_B} + \frac{\delta_2}{\sigma_B}$

Case P7 only exists in presence of uncertainty. It is obvious that z^{P7} cannot be expressed so that the uncertain parameters combine to their expected values. This is the first of six pure strategy cases implying idle capacity in stage two, as the model has not been formulated to allow for premature divestments. If those divestments were allowed, the term $p_a \psi_2 \theta_2 \delta_1 (1-\mu_B) \frac{\alpha_B}{\sigma_B}$ would drop out of the expression above.

$$
\begin{aligned}
z^{P9} &= \psi_1 \left[\theta_1 \delta_1 \left(\frac{\alpha_B}{\sigma_B} + \beta_{B1}\right) + \frac{\phi_B}{1-\gamma_{B1}} \left(\kappa_{B1}^{1-\gamma_{B1}} - \kappa_{B0}^{1-\gamma_{B1}}\right)\right] \\
&+ \psi_2 \theta_2 \left[p_a \delta_2 \left(\frac{\alpha_A}{\sigma_A} + \beta_{A2a}\right) + (1-\mu_B)\delta_1 \left(\frac{\alpha_B}{\sigma_B} + p_b \beta_{B2b}\right)\right] \\
&+ \psi_2 \theta_2 p_b [\delta_2 + (\mu_B - 1)\delta_1] \left(\frac{\alpha_A}{\sigma_A} + \beta_{A2b}\right) \\
&+ \frac{\phi_A \psi_2 \theta_2}{\pi_A} \left[\frac{p_a}{1-\gamma_{A2a}} \left(\kappa_{A2a}^{1-\gamma_{A2a}} - \kappa_{A1}^{1-\gamma_{A2a}}\right) + \frac{p_b}{1-\gamma_{A2b}} \left(\kappa_{A2b}^{1-\gamma_{A2b}} - \kappa_{A1}^{1-\gamma_{A2b}}\right)\right] \\
&+ \frac{\phi_A \psi_2 \theta_2}{\pi_A \sigma_A} \left[p_a \delta_2 \left(\kappa_{A1}^{-\gamma_{A1}} - \kappa_{A1}^{-\gamma_{A2a}}\right) + p_b [\delta_2 + (\mu_B - 1)\delta_1] \left(\kappa_{A1}^{-\gamma_{A1}} - \kappa_{A1}^{-\gamma_{A2b}}\right)\right]
\end{aligned}
\tag{5.181}
$$

with $\kappa_{B1} = \kappa_{B0} + \frac{\delta_1}{\sigma_B}$ $\quad \kappa_{A1} = \kappa_{A0}$

$\kappa_{A2a} = \kappa_{A0} + \frac{\delta_2}{\sigma_A}$ $\quad \kappa_{A2b} = \kappa_{A0} + \frac{\delta_2 + (\mu_B - 1)\delta_1}{\sigma_A}$

P9 is again a case that only exists in the presence of uncertainty. Yet, no matter what state of the world materializes, there is a shift of investment from technology B to A. This property can be observed in the last two lines of Equation (5.181), which represent the expected value for capital costs and the corresponding vertical shift. In contrast, uncertainty associated with variable costs is the reason behind different optimal levels of investment into technology A in stage two. If this were not the case, either cases P5 or P11 would apply.

$$
\begin{aligned}
z^{P11} = {} & \psi_1 \left[\theta_1\delta_1\left(\frac{\alpha_B}{\sigma_B} + \beta_{B1}\right) + \frac{\phi_B}{1-\gamma_{B1}}\left(\kappa_{B1}^{1-\gamma_{B1}} - \kappa_{B0}^{1-\gamma_{B1}}\right)\right] \\
& + \psi_2\theta_2\left[\delta_2\left(\frac{\alpha_A}{\sigma_A} + p_a\beta_{A2a} + p_b\beta_{A2b}\right) + (1-\mu_B)\delta_1\frac{\alpha_B}{\sigma_B}\right] \\
& + \frac{\phi_A\psi_2\theta_2}{\pi_A}\left[\frac{p_a}{1-\gamma_{A2a}}\left(\kappa_{A2a}^{1-\gamma_{A2a}} - \kappa_{A1}^{1-\gamma_{A2a}}\right) + \frac{p_b}{1-\gamma_{A2b}}\left(\kappa_{A2b}^{1-\gamma_{A2b}} - \kappa_{A1}^{1-\gamma_{A2b}}\right)\right] \\
& + \frac{\phi_A\psi_2\theta_2\delta_2}{\pi_A\sigma_A}\left[\kappa_{A1}^{-\gamma_{A1}} - \left(p_a\kappa_{A1}^{-\gamma_{A2a}} + p_b\kappa_{A1}^{-\gamma_{A2b}}\right)\right]
\end{aligned}
\tag{5.182}
$$

$$
\text{with} \quad \kappa_{B1} = \kappa_{B0} + \frac{\delta_1}{\sigma_B} \qquad \kappa_{A1} = \kappa_{A0}
$$

$$
\kappa_{A2a} = \kappa_{A0} + \frac{\delta_2}{\sigma_A} \qquad \kappa_{A2b} = \kappa_{A0} + \frac{\delta_2}{\sigma_A}
$$

For case z^{P11} the uncertain parameters can, similarly to cases P1 and P5, be expressed as their expected values. Consequently this case would also exist under a deterministic setup, in which parameter values were allowed to vary across stages.

After the objective function values for the different cases P1–P12 have been determined, the decision indicator D and the investment decision rule can be used to identify the global optimum. Due to the higher number of cases than in the previous model versions, the decision indicator D takes a different form. It is now defined as the lowest objective function value among all twelve pure strategy alternatives.

$$
D = \min\left(z^{P1}, z^{P2}, z^{P3}, z^{P4}, z^{P5}, z^{P6}, z^{P7}, z^{P8}, z^{P9}, z^{P10}, z^{P11}, z^{P12}\right)
\tag{5.183}
$$

with corresponding investment decision rule

Condition	*Action stage 1*	*Action stage 2*
$D = z^{P1}$	Install B	Fill B
$D = z^{P2}$	Install A	Fill A
$D = z^{P3}$	Install B	Fill A for $s = a$, fill B for $s = b$
$D = z^{P4}$	Install A	Fill A for $s = a$, fill B for $s = b$
$D = z^{P5}$	Install B	Fill A
$D = z^{P6}$	Install A	Fill B
$D = z^{P7}$	Install B	Install A for $s = a$, fill B for $s = b$,
$D = z^{P8}$	Install A	Fill A for $s = a$, install B for $s = b$
$D = z^{P9}$	Install B	Install A for $s = a$, fill A for $s = b$
$D = z^{P10}$	Install A	Fill B for $s = a$, install B for $s = b$,
$D = z^{P11}$	Install B	Install A
$D = z^{P12}$	Install A	Install B

Install here signifies that investment must be large enough to cover the entire power demand in the respective period, whereas *fill* means that investment should be equal to the difference between the remaining capacity stock from period one and the capacity stock required to satisfy power demand. Equality in the decision indicators for two different cases is possible. In this case, two globally optimal solutions exist and it is up to the decision maker to choose between those alternatives.

If a case only exists in presence of uncertainty, this has important implications for practical model applications. If the model is to be used for decision making support and model-endogenous representation of uncertainty is desired, all twelve objective function values, z^{P1} through z^{P12} must be considered, for example, by entering them in a spreadsheet. In contrast, if only deterministic changes in parameter values are to be considered, implementation of z^{P1}, z^{P2}, z^{P5}, z^{P6}, z^{P11} and z^{P12} is sufficient. When applying this model version to real world problems, it is important to bear in mind that it does not provide for an initial capacity endowment. Demand thus has to be viewed as incremental to the one satisfied by capacity installed before the model horizon. The option to specify an initial capacity endowment is available in the model version from sec. 5.3 — yet at the expense of allowing for changes in variable costs and learning rates over time.

5.5 Pure versus Mixed Investment Strategies

In the previous section it has been presumed that pure investment strategies always outperform mixed ones. The argument for this has been that, in a defacto deterministic setting with strictly increasing returns the benefit of lower costs from concentration on one technology in stage one always outweigh the benefit from being "a bit further down" the experience curve of the other technology at the beginning of stage two.

Ideally the assertion of pure strategy optimality would be demonstrated by proving the existence of strictly increasing returns in the objective function. For this, the sign of the second-order total derivative would have to be positive. Due to the complexity of the model and the further issue that z is not continuously differentiable in presence of the *vertical shift* from stage one to two, this proof is left for further research.

Instead, an approach based on the analytically derived pure strategy optima from the previous section and an exemplary solution for a mixed strategy is followed to demonstrate the superiority of pure over mixed strategies. Starting point for this analysis are the two practically most relevant pure strategy cases, which only exist in the stochastic model, P3 and P4. These pure strategies are then compared to their corresponding mixed strategy, denoted by M7.[22] Prerequisite for this comparison are the analytical solution and corresponding objective function value for case M7, which are derived in section 5.5.1. Then, a numerical approach is used to demonstrate the superiority of pure investment strategies over mixed strategy M7, i.e. there is no parameter configuration for which M7 can ever constitute the global optimum. In this process, the different solution properties arising from different directions of the *vertical shift* are highlighted. Finally, a more general approach is taken to support the findings from the numerical analysis.

5.5.1 Solution for Mixed Strategies

The first step in the solution process for mixed strategy cases is to obtain the total number of local cost minima. From a methodological perspective, this process is the same as for pure strategies (see pp. 96f.). The major difference is that both technologies A and B are used in stage one, which precludes the existence of cases with infeasible power production configurations in stage two. After the elimination of cases, which violate the scenario definition, ten relevant cases remain (see Fig. 5.5). Table 5.3 summarizes the sign configurations for the decision variables in those ten cases.

In contrast to pure strategies, it is not sufficient to solve the equation system corresponding to the partial derivatives with respect to λ_m, as there is investment into both technology A and B in stage one. Instead, the equation system based on all partial derivatives with respect to Lagrangians, as given by Eqs. (5.77) through (5.91), as well as decision variables (see App. B.3.2) must be solved. Although a unique solution can be obtained in this manner, it is unfortunately available only implicitly.

This solution process has been carried out for case M7. In a first step, solutions for the decision variables are obtained by solving the equation system derived from the partial, first-order derivatives with respect to λ_m.

[22] Optimal investment strategies for P3, P4 and M7 are equal in stage two. In stage one, however, either technology B or A is invested in cases P3 and P4, respectively, whereas any combination of A and B can be invested in stage one for M7.

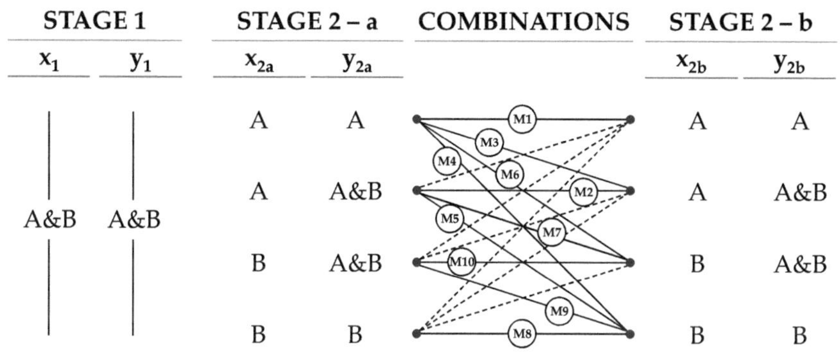

Fig. 5.5 Relevant mixed strategy cases Source: Own calculation

Table 5.3 Sign configurations for mixed strategies under uncertainty

Case	x_{A1}	x_{B1}	x_{A2a}	x_{A2b}	x_{B2a}	x_{B2b}	y_{A2a}	y_{A2b}	y_{B2a}	y_{B2b}
M1	>0	>0	>0	>0	$=0$	$=0$	>0	>0	$=0$	$=0$
M2	>0	>0	>0	>0	$=0$	$=0$	>0	>0	>0	>0
M3	>0	>0	>0	>0	$=0$	$=0$	>0	>0	$=0$	>0
M4	>0	>0	>0	$=0$	$=0$	>0	>0	$=0$	$=0$	>0
M5	>0	>0	>0	$=0$	$=0$	>0	>0	$=0$	>0	>0
M6	>0	>0	>0	$=0$	$=0$	>0	>0	>0	$=0$	>0
M7	>0	>0	>0	$=0$	$=0$	>0	>0	>0	>0	>0
M8	>0	>0	$=0$	$=0$	>0	>0	$=0$	$=0$	>0	>0
M9	>0	>0	$=0$	$=0$	>0	>0	>0	$=0$	>0	>0
M10	>0	>0	$=0$	$=0$	>0	>0	>0	>0	>0	>0

Throughout this process, for which the solutions are stated in the following, the arbitrarily chosen variable x_{B1} is treated as parameter.

$$w^*_{A1} = \frac{\delta_1 - \sigma_B x_{B1}}{\sigma_A} \qquad w^*_{B1} = x_{B1}$$

$$w^*_{A2a} = \frac{\delta_2 + (\mu_B - 1)\sigma_B x_{B1}}{\sigma_A} \qquad w^*_{A2b} = (1-\mu_A)\frac{\delta_1 - \sigma_B x_{B1}}{\sigma_A}$$

$$w^*_{B2a} = (1-\mu_B)x_{B1} \qquad w^*_{B2b} = \frac{\delta_2 + (\mu_A - 1)(\delta_1 - \sigma_B x_{B1})}{\sigma_B}$$

$$x^*_{A1} = \frac{\delta_1 - \sigma_B x_{B1}}{\sigma_A} \qquad x^*_{B1} = x_{B1}$$

$$x^*_{A2a} = \frac{\delta_2 + (\mu_A - 1)\delta_1 + (\mu_B - \mu_A)\sigma_B x_{B1}}{\sigma_A} \qquad x^*_{A2b} = 0$$

$$x^*_{B2a} = 0 \qquad x^*_{B2b} = \frac{\delta_2 + (\mu_A - 1)\delta_1 + (\mu_B - \mu_A)\sigma_B x_{B1}}{\sigma_B}$$

$$y^*_{A1} = \theta_1(\delta_1 - \sigma_B x_{B1}) \qquad y^*_{B1} = \theta_1 \sigma_B x_{B1}$$

$$y^*_{A2a} = \theta_2[\delta_2 + (\mu_B - 1)\sigma_B x_{B1}] \qquad y^*_{A2b} = \theta_2(1-\mu_A)(\delta_1 - \sigma_B x_{B1})$$

$$y^*_{B2a} = \theta_2(1-\mu_B)\sigma_B x_{B1} \qquad y^*_{B2b} = \theta_2[\delta_2 + (\mu_A - 1)(\delta_1 - \sigma_B x_{B1})]$$

The implicit solution for x_{B1} can then be obtained by solving for all λ_m. Due to the considerable manual efforts associated with this process, the model is only solved for one of the mixed strategy cases, M7. The implicit solution for M7 is then given by Equation (5.184).

$$\begin{aligned}
&\psi_1\left(\frac{\alpha_A}{\sigma_A}-\frac{\alpha_B}{\sigma_B}+\beta_{A1}-\beta_{B1}+\frac{\phi_A}{\sigma_A\theta_1}\kappa_{A1}^{-\gamma_{A1}}-\frac{\phi_B}{\sigma_B\theta_1}\kappa_{B1}^{-\gamma_{B1}}\right)\\
&+(\mu_A-\mu_B)\frac{p_A\phi_A\psi_2\theta_2}{\pi_A\sigma_A\theta_1}\left(\kappa_{A2a}^{-\gamma_{A2a}}-\kappa_{A1}^{-\gamma_{A2a}}+\kappa_{A1}^{-\gamma_{A1}}\right)\\
&+(\mu_A-\mu_B)\frac{p_B\phi_B\psi_2\theta_2}{\pi_B\sigma_B\theta_1}\left(\kappa_{B2b}^{-\gamma_{B2b}}-\kappa_{B1}^{-\gamma_{B2b}}+\kappa_{B1}^{-\gamma_{B1}}\right)\\
&+\frac{\psi_2\theta_2}{\theta_1}p_b(\mu_A-1)\left(\frac{\alpha_B}{\sigma_B}-\frac{\alpha_A}{\sigma_A}+\beta_{B2b}-\beta_{A2b}\right)\\
&-\frac{\psi_2\theta_2}{\theta_1}p_a(\mu_B-1)\left(\frac{\alpha_A}{\sigma_A}-\frac{\alpha_B}{\sigma_B}+\beta_{A2a}-\beta_{B2b}\right)\\
&+\frac{p_a\phi_A\psi_2\theta_2}{\pi_A\sigma_A\theta_1}\left[\frac{\delta_2+(\mu_A-1)\delta_1+(\mu_B-\mu_A)\sigma_Bx_{B1}}{\sigma_A\kappa_{A1}}\left(\gamma_{A2a}\kappa_{A1}^{-\gamma_{A2a}}-\gamma_{A1}\kappa_{A1}^{-\gamma_{A1}}\right)\right]\\
&+\frac{p_a\phi_A\psi_2\theta_2}{\pi_A\sigma_A\theta_1}\left[\kappa_{A2a}^{-\gamma_{A2a}}-\kappa_{A1}^{-\gamma_{A2a}}\right]\\
&+\frac{p_b\phi_B\psi_2\theta_2}{\pi_B\sigma_B\theta_1}\left[\frac{\delta_2+(\mu_A-1)\delta_1+(\mu_B-\mu_A)\sigma_Bx_{B1}}{\sigma_B\kappa_{B1}}\left(\gamma_{B2b}\kappa_{B1}^{-\gamma_{B2b}}-\gamma_{B1}\kappa_{B1}^{-\gamma_{B1}}\right)\right]\\
&+\frac{p_b\phi_B\psi_2\theta_2}{\pi_B\sigma_B\theta_1}\left[\kappa_{B2b}^{-\gamma_{B2b}}-\kappa_{A1}^{-\gamma_{B2b}}\right]\\
&=0
\end{aligned} \tag{5.184}$$

with

$$\kappa_{A1}=\kappa_{A0}+\frac{\delta_1-\sigma_Bx_{B1}}{\sigma_A} \qquad \kappa_{A2a}=\kappa_{A0}+\frac{\delta_2+\mu_A\delta_1-(\mu_A-\mu_B+1)\sigma_Bx_{B1}}{\sigma_A}$$

$$\kappa_{B1}=\kappa_{B0}+x_{B1} \qquad \kappa_{B2b}=\kappa_{B0}+x_{B1}+\frac{\delta_2+(\mu_A-1)\delta_1+(\mu_B-\mu_A)\sigma_Bx_{B1}}{\sigma_B}$$

Since the solution is available implicitly only, a trial-and-error process is required to obtain the optimal value for the model's decision variables. Alternatively, numerical methods can be used. Under such an approach, however, it is important to bear in mind that several local minima may exist for different mixes of technology A and B. Therefore, if simple algorithms, such as Goal Seek in Microsoft Excel are to be used, different starting points should be tested. Analogously to the model's decision variables, the generalized objective value of M7, z^{M7}, is also stated in terms of x_{B1}, as given by Equation (5.185).

$$
\begin{aligned}
z^{M7} &= \psi_1\theta_1 \left[(\delta_1 - \sigma_B x_{B1}) \left(\frac{\alpha_A}{\sigma_A} + \beta_{A1} \right) + x_{B1}(\alpha_B + \beta_{B1}\sigma_B) \right] \\
&+ \frac{\phi_A \psi_1}{1-\gamma_{A1}} \left(\kappa_{A1}^{1-\gamma_{A1}} - \kappa_{A0}^{1-\gamma_{A1}} \right) + \frac{\phi_B \psi_1}{1-\gamma_{B1}} \left(\kappa_{B1}^{1-\gamma_{B1}} - \kappa_{B0}^{1-\gamma_{B1}} \right) \\
&+ p_a \psi_2 \theta_2 [\delta_2 + (\mu_B - 1)\sigma_B x_{B1}] \left(\frac{\alpha_A}{\sigma_A} + \beta_{A2a} \right) \\
&+ p_a \psi_2 \theta_2 (1 - \mu_B) \sigma_B x_{B1} \left(\frac{\alpha_B}{\sigma_B} + \beta_{B2a} \right) \\
&+ \frac{p_a \phi_A \psi_2 \theta_2}{\pi_A (1 - \gamma_{A2a})} \left(\kappa_{A2a}^{1-\gamma_{A2a}} - \kappa_{A1}^{1-\gamma_{A2a}} \right) \\
&+ \frac{p_a \phi_A \psi_2 \theta_2}{\pi_A} \left(\frac{\delta_2 + (\mu_A - 1)\delta_1 + (\mu_B - \mu_A)\sigma_B x_{B1}}{\sigma_A} \right) \left(\kappa_{A1}^{-\gamma_{A1}} - \kappa_{A1}^{-\gamma_{A2a}} \right) \\
&+ p_b \psi_2 \theta_2 [\delta_2 - (1 - \mu_A)(\delta_1 - \sigma_B x_{B1})] \left(\frac{\alpha_B}{\sigma_B} + \beta_{B2b} \right) \\
&+ p_b \psi_2 \theta_2 (1 - \mu_A)(\delta_1 - \sigma_B x_{B1}) \left(\frac{\alpha_A}{\sigma_A} + \beta_{A2b} \right) \\
&+ \frac{p_b \phi_B \psi_2 \theta_2}{\pi_B (1 - \gamma_{B2b})} \left(\kappa_{B2b}^{1-\gamma_{B2b}} - \kappa_{B1}^{1-\gamma_{B2b}} \right) \\
&+ \frac{p_b \phi_B \psi_2 \theta_2}{\pi_B} \left(\frac{\delta_2 + (\mu_A - 1)\delta_1 + (\mu_B - \mu_A)\sigma_B x_{B1}}{\sigma_B} \right) \left(\kappa_{B1}^{-\gamma_{B1}} - \kappa_{B1}^{-\gamma_{B2b}} \right)
\end{aligned} \tag{5.185}
$$

with

$$
\kappa_{A1} = \kappa_{A0} + \frac{\delta_1 - \sigma_B x_{B1}}{\sigma_A} \qquad \kappa_{A2a} = \kappa_{A0} + \frac{\delta_2 + \mu_A\delta_1 - (\mu_A - \mu_B + 1)\sigma_B x_{B1}}{\sigma_A}
$$

$$
\kappa_{B1} = \kappa_{B0} + x_{B1} \qquad \kappa_{B2b} = \kappa_{B0} + x_{B1} + \frac{\delta_2 + (\mu_A - 1)\delta_1 + (\mu_B - \mu_A)\sigma_B x_{B1}}{\sigma_B}
$$

This expression is the starting point for demonstrating global pure strategy superiority.

5.5.2 Global Optimality of Pure Strategies

The main focus in the following demonstration of global pure strategy optimality lies on cases P3, P4 and M7. All three cases exist only in presence of uncertainty and in all three of them investment in the second stage depends entirely on the realization of scenario s. Strictly speaking, as M7 covers the full range of possible stage one investment mixes between technology A and B, it also includes P3 and P4. The latter are corner solutions of M7, where all investment is directed towards technology B or A, respectively. Additionally, cases P1, P2, P5 and P6 are compared to M7 in situations, where a mixed investment in stage one under M7 outperforms P3 and P4.

The assertion to confirm is, that either objective function values z^{P3} and/or z^{P4} will be lower than z^{M7} or, if this is not the case, that another pure strategy will result in lower system costs than a mixed strategy under M7. This problem is primarily addressed in a numerical way, although some of the properties demonstrated numerically are also shown in a more general manner afterwards. Yet, undoubtedly both of these approaches can only support the assertion made but they do not constitute a generally valid proof thereof.

A convenient model property is that specific variable costs under uncertainty are not relevant for the question of mixed versus pure strategy optimality. This is due to the fact that specific variable operating costs in stage two are independent of endogenous model variable levels. This property can be verified by comparing the objective function terms for variable costs in cases P3 and M7, z^{P3}_{Oper} and z^{M7}_{Oper}, respectively. When relative specific variable costs are assumed equal for both technologies, i.e. $\beta_{A1} = \beta_{B1} = \beta_{i1}$, $\beta_{A2a} = \beta_{B2b} = \beta_{i2H}$, $\beta_{A2b} = \beta_{B2a} = \beta_{i2L}$ and $\sigma_A = \sigma_B = \sigma_i$, any technology mix in stage one and two will yield the same overall variable costs, as shown in Eq. (5.186).[23] Obviously, different specific cost or availability assumptions will lead to proportionally lower overall variable costs for increases in the share of technologies with the lower relative cost position.

$$
\begin{aligned}
& z^{M7}_{Oper} - z^{P3}_{Oper} \\
= \; & \psi_1\theta_1\delta_1\left(\frac{\alpha_i}{\sigma_i} + \beta_{i1}\right) - \psi_1\theta_1\delta_1\left(\frac{\alpha_i}{\sigma_i} + \beta_{i1}\right) \\
& + \frac{\psi_2\theta_2}{2}\left(1-\mu_i\right)\left(\delta_1 - \sigma_i x_{B1}\right)\left(\frac{\alpha_i}{\sigma_i} + \beta_{i2H}\right) \\
& + \frac{\psi_2\theta_2}{2}\left(1-\mu_i\right)\left(\sigma_i x_{B1} - \delta_1\right)\left(\frac{\alpha_i}{\sigma_i} + \beta_{i2L}\right) \\
& - \frac{\psi_2\theta_2}{2}\left(1-\mu_i\right)\left(\sigma_i x_{B1} - \delta_1\right)\left(\frac{\alpha_i}{\sigma_i} + \beta_{i2H}\right) \\
& - \frac{\psi_2\theta_2}{2}\left(1-\mu_i\right)\left(\delta_1 - \sigma_i x_{B1}\right)\left(\frac{\alpha_i}{\sigma_i} + \beta_{i2L}\right) \\
= \; & 0
\end{aligned}
\tag{5.186}
$$

In the remainder of the section, operating cost uncertainty will therefore be ignored to focus on specific investment costs, which are a function of capacity additions and thus not independent of endogenous variable levels.

Numerical Evidence

In-depth numerical analyses have been carried out to test the robustness of the assertion that pure strategies perform better than mixed strategies. The attempt was made to identify at least one set of parameter values, for

[23] The same property will also hold for other combinations of cases.

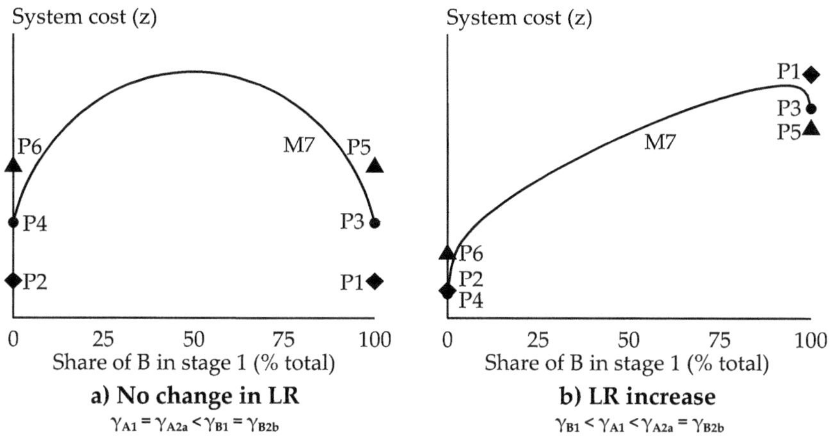

Fig. 5.6 Parameter configurations with general superiority of pure strategies
Source: Own calculations

which a mixed investment strategy under M7[24] yielded a lower objective function value than all of the pure strategies P1–P6. This would have refuted the hypothesis of general pure strategy optimality. The numerical tests have been carried out for the different possible relationships among the values of learning indexes γ_{i1} and γ_{i2s} across stages.[25]

Results are shown in graphical form: four representative sets of objective function values to be discussed in more depth in Figures 5.6 and 5.7, the full set of results in Appendix B.4. The graphs display the objective function value z on the y-axis and the share of technology B installed in stage one on the x-axis. As pure strategies are characterized by investment concentration on either technology A or B in stage one, they appear as points: diamonds represent the "dominance" strategies P1 and P2, triangles the "investment shift" strategies P5 and P6, and bullets the corner solutions of M7, P3 and P4. As z^{M7} is a function of the share of technology B in stage one, it appears as a line.

Two main types of solution sets can be distinguished: (1) Solution sets, where the pure strategies P3 and/or P4 as corner solutions of M7 always perform better than other mixed strategies described by M7 and (2) solution sets, where a mixed strategy M7 performs better than both of the pure strategies P3 and P4. In the latter case, however, another pure strategy has always been found globally superior to the locally optimal mixed one described by M7.

[24] Apart from its pure strategy corner solutions P3 and P4.

[25] Only γ_{A2a} and γ_{B2b} are relevant for the objective function value z^{M7} and therefore also for z^{P3} and z^{P3}. For γ_{A2b} and γ_{B2a}, which impact z^{P1}, z^{P2}, z^{P5} and z^{P6}, the corresponding value from stage one has been used.

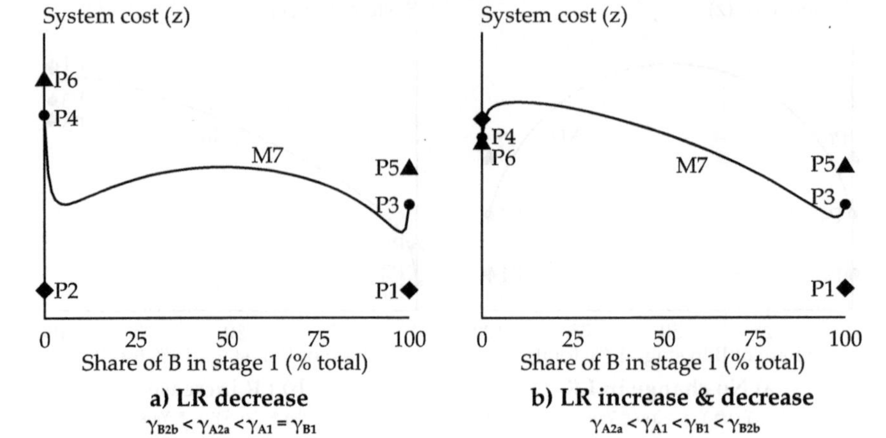

Fig. 5.7 Parameter configurations with global superiority of pure strategies
Source: Own calculations

(1) The first group of solutions, where P3 and/or P4 always perform better than other technology combinations under M7, is associated with no or a positive change in learning indexes across stages. The solution shown in Fig. 5.6a is based on a parameter configuration, where no change in learning indexes occurs.[26] The objective function of M7 takes the shape of a *reversed U* and it is obvious that investment concentration on one technology is optimal — P3 *and* P4 outperform all mixed strategies of M7. This finding is not surprising, as the case of no change in learning indexes corresponds to the deterministic model setup from previous sections, where dominance of one technology has always been found globally optimal. The same observation can be made for an increase in the learning index across stages.[27] Figure 5.6b illustrates this property for a sample parameter configuration. The full set of tested parameter configurations from this group can be found on pp. 267ff. For an increase in learning index, the objective value of M7 is also characterized by a reversed U shape and P3 *or* P4 yield a lower objective function value than any other mix of technologies described by M7. Intuitively, this observation can be explained by the fact, that there is no regret from foregoing investment into a technology in stage one, as the learning index in stage two will not be lower than the one in the first.

The situation is more difficult in presence of a decreasing learning index. Figure 5.7 shows two numerical examples for such a parameter configuration.

[26] In this case the learning index is equal for both stages and technologies; the case, where one technology is characterized by higher learning indexes than the other is given in App. B.4.1.

[27] This includes parameter configurations with a learning rate increase for both technologies as well as configurations, where only the learning rate of one technology increases and the other remains constant.

The solutions in Fig. 5.7a correspond to a parameter configuration, where there is a learning index decrease for both technologies. Figure 5.7b shows a configuration, where the learning index decreases for technology A but increases for B. It is apparent, that strategies P3 or P4 no longer necessarily perform better than mixed technology strategies. However, despite substantial efforts, no parameter configuration could be produced, where a mixed M7 strategy also constituted the global optimum (see pp. 269ff). In fact, given a decreasing learning index,[28] no set of parameter values could be produced that made P3 or P4 globally optimal. In contrast, given opposed developments of learning indexes, P3 or P4 may well constitute the global optimum (see, for example, Figures j or l on p. 271). An explanation for the potential superiority of a mixed M7 strategy over its pure corner solutions P3 or P4 could be optimality of "saving" a small share of the steeper initial section of the experience curve associated with the technology characterized by an increasing learning index. In stage two, the higher learning index would then allow for an even faster reduction of specific capital costs.

In summary, no set of parameter values that ever rendered a mixed investment strategy M7 *globally* optimal could be produced.[29] For any parameter configuration, at least one pure strategy yielded an equal or lower objective function value than given by z^{M7}.[30] The numerical analyses carried out therefore support the assertion that at least one pure stage one investment strategy outperforms mixed ones in the presented model. Yet it is clear that these findings are not sufficient to generally prove the property of pure over mixed strategy optimality. On the one hand, it is impossible to numerically test for all possible parameter configurations. On the other hand, only strategy M7 has been investigated in lieu of all ten possible mixed strategies.

Analytical Evidence

In this subsection, an analytical approach is pursued to support the findings from the numerical approach. To limit complexity, parameter values for technologies A and B are generally assumed to be equal, the development of the learning index in stage two occurs in a diametrically opposed manner.[31] From this assumption follows that $z^{P1} = z^{P2}$ and $z^{P3} = z^{P4}$. Pure strategy superiority is shown for the two developments of γ_{it}, where the corner solutions of M7, P3 and P4, outperform all mixed strategies:

[28] For either both technologies or only one technology in combination with a constant learning index for the other technology.

[29] Further parameter configurations in addition to the ones listed in App. B.4 have been tested.

[30] An equal value is obtained, if P3 or P4 are optimal, which are the corner solutions of M7.

[31] More formally: $\sigma_A = \sigma_B = \sigma_i$, $\phi_A = \phi_B = \phi_i$, $\kappa_{A0} = \kappa_{B0} = \kappa_{i0}$, $p_a = p_b = p_s = 0.5$, $\gamma_{A1} = \gamma_{B1} = \gamma_{i1}$, $\gamma_{A2a} = \gamma_{B2b} = \gamma_{i2H}$ and $\gamma_{A2b} = \gamma_{B2a} = \gamma_{i2L}$.

1. Constant learning index $\gamma_{i1} = \gamma_{i2H}$
2. Increasing learning index $\gamma_{i1} < \gamma_{i2H}$

Unfortunately, demonstration of global pure strategy optimality for the case of $\gamma_{i1} > \gamma_{i2H}$ entails vastly higher efforts than demonstration for the two configurations listed above and is left for future research.[32]

(1) The numerical analyses in the previous section indicate that for $\gamma_{i1} = \gamma_{i2H}$ one of the corner solutions of M7, P3 or P4 will yield a lower objective function value than any mixed strategy. As $z^{P3} = z^{P4}$ here, both strategies P3 and P4 should lead to a lower objective function value than a mixed strategy under M7. It will thus be shown that $z^{M7}_{Inv} - z^{P3}_{Inv}$ is positive,[33] where subscript *Inv* denotes that only the objective function terms associated with investment costs are considered. Conveniently, the expressions representing the vertical shift of the experience curve at stage transition cancel out, as learning indexes remain constant across stages.

$$\begin{aligned}
& z^{M7}_{Inv} - z^{P3}_{Inv} > 0 \\
\Leftrightarrow\ & \frac{\phi_i \psi_1}{1-\gamma_{i1}} \left[\left(\kappa_{A1(M7)}^{1-\gamma_{i1}} - \kappa_{i0}^{1-\gamma_{i1}} \right) + \left(\kappa_{B1(M7)}^{1-\gamma_{i1}} - \kappa_{i0}^{1-\gamma_{i1}} \right) - \left(\kappa_{B1(P3)}^{1-\gamma_{i1}} - \kappa_{i0}^{1-\gamma_{i1}} \right) \right] \\
& + \frac{\phi_i \psi_2 \theta_2}{2\pi_i (1-\gamma_{i2H})} \left[\left(\kappa_{A2a(M7)}^{1-\gamma_{i2H}} - \kappa_{A1(M7)}^{1-\gamma_{i2H}} \right) - \left(\kappa_{A2a(P3)}^{1-\gamma_{i2H}} - \kappa_{i0}^{1-\gamma_{i2H}} \right) \right] \\
& + \frac{\phi_i \psi_2 \theta_2}{2\pi_i (1-\gamma_{i2H})} \left[\left(\kappa_{B2b(M7)}^{1-\gamma_{i2H}} - \kappa_{B1(M7)}^{1-\gamma_{i2H}} \right) - \left(\kappa_{B2b(P3)}^{1-\gamma_{i2H}} - \kappa_{B1(P3)}^{1-\gamma_{i2H}} \right) \right] > 0
\end{aligned} \tag{5.187}$$

Let x_{Tt} represent the total investment required in a particular stage t, with $x_{At} + x_{Bt} = x_{Tt}$.[34] Further, let $\frac{\phi_i \psi_1}{1-\gamma_{i1}} = u$ and $\frac{\phi_i \psi_2 \theta_2}{2\pi_i (1-\gamma_{i2H})} = v$, then

$$\begin{aligned}
& u(x_{T1} - x_{B1} + \kappa_{i0})^{1-\gamma_{i1}} - u\kappa_{i0}^{1-\gamma_{i1}} + u(x_{B1} + \kappa_{i0})^{1-\gamma_{i1}} - u\kappa_{i0}^{1-\gamma_{i1}} \\
& - u(x_{T1} + \kappa_{i0})^{1-\gamma_{i1}} + u\kappa_{i0}^{1-\gamma_{i1}} + v(x_{T2} + x_{T1} - x_{B1} + \kappa_{i0})^{1-\gamma_{i2H}} \\
& - v(x_{T1} - x_{B1} + \kappa_{i0})^{1-\gamma_{i2H}} - v(x_{T2} + \kappa_{i0})^{1-\gamma_{i2H}} + v\kappa_{i0}^{1-\gamma_{i2H}} \\
& + v(x_{T2} + x_{B1} + \kappa_{i0})^{1-\gamma_{i2H}} - v(x_{B1} + \kappa_{i0})^{1-\gamma_{i2H}} \\
& - v(x_{T2} + x_{T1} + \kappa_{i0})^{1-\gamma_{i2H}} + v(x_{T1} + \kappa_{i0})^{1-\gamma_{i2H}} \\
> \ & 0
\end{aligned} \tag{5.188}$$

[32] It would be necessary to compare M7 to P1, P3 and P5 simultaneously.

[33] Alternatively z^{P4}_{Inv} instead of z^{P3}_{Inv} could have been used, since $z^{P3} = z^{P4}$.

[34] This is permissible due to equal availability σ_i of technologies A and B, implying that the same nameplate capacity is required to achieve a particular power production level.

Collecting terms and rearranging yields

$$\begin{aligned}
&\big[u(x_{T1} - x_{B1} + \kappa_{i0})^{1-\gamma_{i1}} - v(x_{T1} - x_{B1} + \kappa_{i0})^{1-\gamma_{i2H}} \\
&\quad + u(x_{B1} + \kappa_{i0})^{1-\gamma_{i1}} - v(x_{B1} + \kappa_{i0})^{1-\gamma_{i2H}}\big] \\
&- \big[u(x_{T1} + \kappa_{i0})^{1-\gamma_{i1}} - v(x_{T1} + \kappa_{i0})^{1-\gamma_{i2H}} + u\kappa_{i0}^{1-\gamma_{i1}} - v\kappa_{i0}^{1-\gamma_{i2H}}\big] \\
&+ v\big[(x_{T2} + x_{T1} - x_{B1} + \kappa_{i0})^{1-\gamma_{i2H}} + (x_{T2} + x_{B1} + \kappa_{i0})^{1-\gamma_{i2H}}\big] \\
&- v\big[(x_{T2} + x_{T1} + \kappa_{i0})^{1-\gamma_{i2H}} + (x_{T2} + \kappa_{i0})^{1-\gamma_{i2H}}\big] \\
&> 0
\end{aligned} \tag{5.189}$$

Equation (5.189) can now be used to show that the asserted inequality holds within the range $0 < x_{B1} < \frac{\delta_1}{\sigma_i}$.[35] The expressions in lines four and five best illustrate the principle employed. The total cumulative capacity across all terms in line four is exactly the same as the one in line five. Yet line four, i.e. $v[(x_{T2} + x_{T1} - x_{B1} + \kappa_{i0})^{1-\gamma_{i2H}} + (x_{T2} + x_{B1} + \kappa_{i0})^{1-\gamma_{i2H}}]$, will take on a higher value than line five, i.e. $v[(x_{T2} + x_{T1} + \kappa_{i0})^{1-\gamma_{i2H}} + (x_{T2} + \kappa_{i0})^{1-\gamma_{i2H}}]$, rendering the difference between those two lines positive. This can be attributed to the following property: $x^{1-\gamma_{i2H}}$ is a concave function of x. Then, $x^{1-\gamma_{i2H}} + y^{1-\gamma_{i2H}} < (x+z)^{1-\gamma_{i2H}} + (y-z)^{1-\gamma_{i2H}}$ for $x < y$ and $z < y - z$. The same property can also be applied to lines one/two and three of Eq. (5.189). Due to differences in the coefficients of the summands, the property is not as obvious as for lines four and five. Coefficient u is always associated with a positive term, v with a negative one.[36] For all parameter configurations, u will take on higher values than v. Since the case $\gamma_{i2H} = \gamma_{i1}$ is considered, terms may be grouped and the concavity of the power function implies that the difference of lines one/two and three is indeed positive. As both the differences between lines one/two and three on the one hand, as well as between lines four and five on the other hand, are positive, the corner solutions of M7 indeed yield a lower objective function value than a mixed strategy for $\gamma_{i1} = \gamma_{i2H}$.

(2) For the parameter configurations with an increase in the learning rate across stages, i.e. $\gamma_{i2H} > \gamma_{i1}$, the numerical results also indicate that pursuing either P3 or P4 will result in lower system costs than any mixed strategy under M7. Therefore, once again, z^{M7} should be larger than z^{P3}.[37] To demonstrate this property, a two-step process is followed. First, Equation (5.189) is used to show that investment costs under z^{M7} are indeed higher than the ones for z^{P3}, given $\gamma_{i2H} > \gamma_{i1}$. Then, it is shown that the same finding also holds for the terms representing the vertical shift, which cannot be ignored as done previously, since the learning index increases from stage one to two.

[35] For $x_{B1} = 0$ or $x_{B1} = \frac{\delta_1}{\sigma_i}$, z^{M7} would equal z^{P4} or z^{P3}, respectively.

[36] After line three has been factored out.

[37] Again, for $0 < x_{B1} < \frac{\delta_1}{\sigma_i}$. Otherwise, the strict inequality would become a weak one.

General positiveness of the vertical shift in isolation can be shown for values of γ_{i2H} below a certain threshold.

When interpreting Equation (5.189) in light of $\gamma_{i2H} > \gamma_{i1}$, the same line of argument as above applies why the difference between lines four and five must be positive. This is due to the fact that only γ_{i2H} appears in the respective terms. Positiveness of the difference between lines one/two and three is not as obvious, not only due to the difference in coefficients already mentioned above but especially due to the differences among power values. In terms of the general form $a \cdot b^p$ for the terms in lines one through three, this means that both coefficients a and power values p differ. The terms with positive sign, however, are characterized by coefficient u and power $1 - \gamma_{i1}$, whereas the negative ones are characterized by coefficient v and power $1 - \gamma_{i2H}$. Since $\gamma_{i2H} > \gamma_{i1}$, the power of the term with positive sign will always be greater than the power of the term with negative sign, which supports the assertion of a positive difference between lines one/two and three. The matter of coefficient a is slightly more difficult. In almost any realistically imaginable parameter configuration, u will take on higher values than v. Only in case γ_{i1} very far below γ_{i2H} and $\gamma_{i2H} > 0.5$ could there be $u < v$. However, in this case the high value of γ_{i2H} will drive down the value of terms with negative sign, overcompensating the effect of the reversed relationship between u and v. After establishing that existence of different coefficients and power values still supports the assertion that $z_{Inv}^{M7} - z_{Inv}^{P3}$, the same argument as for $\gamma_{i2H} = \gamma_{i1}$ holds, demonstrating that the difference between lines one/two and three is indeed positive. Next, it is shown that the difference in the terms representing the vertical shift for M7 and P3 is also positive. This is formally expressed as

$$\begin{aligned} & z_{V.Shift}^{M7} - z_{V.Shift}^{P3} > 0 \\ \Leftrightarrow \quad & \frac{\phi_i \psi_2 \theta_2 [\delta_2 + (\mu_i - 1)\delta_1]}{2\pi_i \sigma_i} \left[\left(\kappa_{A1(M7)}^{-\gamma_{i1}} - \kappa_{A1(M7)}^{-\gamma_{i2H}} \right) + \left(\kappa_{B1(M7)}^{-\gamma_{i1}} - \kappa_{B1(M7)}^{-\gamma_{i2H}} \right) \right] \\ & - \frac{\phi_i \psi_2 \theta_2 [\delta_2 + (\mu_i - 1)\delta_1]}{2\pi_i \sigma_i} \left[\left(\kappa_{i0}^{-\gamma_{i1}} - \kappa_{i0}^{-\gamma_{i2H}} \right) + \left(\kappa_{B1(P3)}^{-\gamma_{i1}} - \kappa_{B1(P3)}^{-\gamma_{i2H}} \right) \right] \\ & > 0 \end{aligned} \tag{5.190}$$

Let x_{Tt} again represent the total investment required at a particular stage t, with $x_{At} + x_{Bt} = x_{Tt}$. Further, let $\frac{\phi_i \psi_2 \theta_2 [\delta_2 + (\mu_i - 1)\delta_1]}{2\pi_i \sigma_i} = g$, then

$$\begin{aligned} & g\left[(x_{T1} - x_{B1} + \kappa_{i0})^{-\gamma_{i1}} + (x_{B1} + \kappa_{i0})^{-\gamma_{i1}} - (x_{T1} + \kappa_{i0})^{-\gamma_{i1}} - \kappa_{i0}^{-\gamma_{i1}} \right] \\ & + g\left[(x_{T1} + \kappa_{i0})^{-\gamma_{i2H}} + \kappa_{i0}^{-\gamma_{i2H}} - (x_{T1} - x_{B1} + \kappa_{i0})^{-\gamma_{i2H}} - (x_{B1} + \kappa_{i0})^{-\gamma_{i2H}} \right] \\ & > 0 \end{aligned} \tag{5.191}$$

As previously, the bases of the power functions in lines one and two, as well as their respective sums, are equal. However, the power functions in Eq. 5.191 are convex instead of concave, as $x^{-\gamma_i 2H}$ is a convex function of x. Therefore, $x^{-\gamma_{i2H}} + y^{-\gamma_{i2H}} > (x+z)^{-\gamma_{i2H}} + (y-z)^{-\gamma_{i2H}}$ for $x < y$ and $z < y - z$. Consequently line one in Eq. 5.191 takes on a negative sign, whereas line two is positive. For the entire expression to be positive, the absolute value of line two must therefore be greater than the absolute value of line one. As starting point Eq. 5.191 is transformed into the expression in Eq. 5.192.

$$\begin{aligned} & g\left[(x_{T1} + \kappa_{i0})^{-\gamma_{i2H}} + \kappa_{i0}^{-\gamma_{i2H}} - (x_{T1} - x_{B1} + \kappa_{i0})^{-\gamma_{i2H}} - (x_{B1} + \kappa_{i0})^{-\gamma_{i2H}} \right] \\ & - g\left[(x_{T1} + \kappa_{i0})^{-\gamma_{i1}} + \kappa_{i0}^{-\gamma_{i1}} - (x_{T1} - x_{B1} + \kappa_{i0})^{-\gamma_{i1}} - (x_{B1} + \kappa_{i0})^{-\gamma_{i1}} \right] \\ & > 0 \end{aligned} \tag{5.192}$$

The terms in lines one and two of Eq. 5.192 differ only with respect to the exponent and sign. Since $\gamma_{i2H} > \gamma_{i1}$ in here, the value of $h = (x_{T1} + \kappa_{i0})^{-\gamma} + \kappa_{i0}^{-\gamma} - (x_{T1} - x_{B1} + \kappa_{i0})^{-\gamma} - (x_{B1} + \kappa_{i0})^{-\gamma}$ must be larger for higher absolute values of exponent γ to render the entire expression in Eq. 5.192 positive. The first-order derivate of h with respect to exponent γ is employed to test this property.

$$h'(\gamma) = \frac{\ln(x_{T1} - x_{B1} + \kappa_{i0})}{(x_{T1} - x_{B1} + \kappa_{i0})^{\gamma}} + \frac{\ln(x_{B1} + \kappa_{i0})}{(x_{B1} + \kappa_{i0})^{\gamma}} - \frac{\ln(x_{T1} + \kappa_{i0})}{(x_{T1} + \kappa_{i0})^{\gamma}} - \frac{\ln(\kappa_{i0})}{(\kappa_{i0})^{\gamma}} \tag{5.193}$$

$h'(\gamma)$, as shown in Eq. 5.193, is positive for values of γ below a certain threshold.[38] The numerators in Eq. 5.193 are again convex functions of x, where $ln(x) + ln(y) < ln(x+z) + ln(y-z)$ for $x < y$ and $z < y - z$. The denominators, however, are made up of concave functions of x. For positive, small values of γ up until a threshold value the impact of the numerator, i.e. the convex functions, prevails for any parameter configuration and the entire expression takes on a positive sign. Beyond this threshold value, however, the expression in Eq. 5.192 may become negative for specific configurations of x_{T1}, x_{B1}, κ_{i0}, γ_{i2H} and γ_{i2H} . Yet higher values of γ will also increase the positive value of the expression describing investment cost differences before a vertical shift in Eq. 5.187, which acts in a compensating manner if the expression in Eq. 5.192 takes on a negative value. The threshold value of γ corresponds to the maximum of $h(\gamma)$, which is available implicitly.[39]

---

[38] For a step by step process to derive $h'(\gamma)$, the interested reader is referred to Appendix B.3.5.

[39] Obtained based on $h'(\gamma) = 0$ and using trial-and-error or numerical techniques to solve for γ.

5.6 Model Application: Offshore Wind and PV

In this section the model is applied to compare the relative competitiveness of two renewable technologies: Offshore wind and PV. Both technologies have reached the phase of large-scale deployment and currently enjoy high growth rates.[40] The total cumulative capacity of PV systems manufactured is with approximately 30 GW in 2011 (EPIA, 2013) higher than the one of offshore wind with 4 GW by 2011 (GWEC, 2013). It is therefore likely that PV has progressed further on its experience curve than Offshore wind — yet Offshore wind benefits from spillovers from Onshore wind technology, for example regarding blade, gearbox and generator technology.

In Germany PV and Offshore wind technologies are key elements of the aspired shift towards renewable power generation. Especially in case the ambitious aspiration of 70%+ power generation from renewables by 2040 is to be reached, the contribution of Offshore wind and PV would have to ramp up from a joint power production of 4 TWh in 2008[41] to more than 150 TWh in 2040 (cf. Nitsch and Wenzel, 2009, p. 11).

For both technologies there have recently been changes to cost (estimates). PV has experienced large deployments over the years from 2008 to 2011, raising global cumulative capacity from roughly 13 to 70 GW. This deployment, alongside shifts of manufacturing capacity towards China, have enabled substantial cost reductions. Whereas specific capital costs for PV generators were more than 4,000 $/kW around 2008 (see App. A.3.2), they have reached around 3,000 $/kW in 2011, with some sources even quoting lower cost points (cf. REN21, 2012, p. 28).[42] The deployment of Offshore wind, on the other hand, has progressed much slower than PV. New implementations have shown that the cost realized in the mid-2000s (see App. A.2.2), largely based on shallow-water implementations near the UK and Denmark, are too optimistic for the deeper water sites being developed today. REN21 (2012, p. 28) states capital offshore capital costs of roughly 3,800 to 5,900 $/kW.

Due to those recent developments, this application considers two different points in time: a decision point in 2008 and one in 2011. For the latter the competitiveness of PV should have increased substantially. In terms of regional scope, the analysis covers Germany, as future planned demand/ production data for PV and Offshore is available (cf. Nitsch and Wenzel, 2009, p. 11). Unfortunately the focus on a particular country has the drawback that experience spillovers from other countries are not considered. Yet this restriction can, from another perspective, also be considered realistic in the

[40] Another interesting comparison would be PV versus CSP. Yet one of the great advantages of CSP, the option of energy storage, cannot be reflected in this small-scale model. It is therefore investigated in the large-scale model presented in the following chapter.

[41] Power production from PV only.

[42] Investment cost include all investments required to enable power feed-in into the grid, not only module cost.

Table 5.4 Assumptions offshore wind versus PV

	Unit	*A: Offshore wind*	*B: Photovoltaics*
Useful life	years	25	25
Full load hours	h/a	3,300	1,000
Cumulative capacity 2008	GW	2.1	13.6
Cumulative capacity 2011	GW	4.1	70.0
Capital cost 2008	\$/kW	2,300	4,200
Capital cost 2011	\$/kW	4,000	3,100
Operating cost	\$/kWa	80	40
Learning rate – Stage 1	%	10%	20%
Learning rate – Stage 2a	%	10%	15%
Learning rate – Stage 2b	%	5%	25%

sense that support schemes continue to be decided on the national level and decisionmakers do not control deployments in other countries. When interpreting results, additional limitations of this simple model should be taken into account. Implications of output fluctuation, renewable potential or interactions with other technologies and sectors, for example, are not reflected. The full set of assumptions for both decision points, 2008 and 2011, can be found in Tab. 5.4. The model horizon is from 2008 or 2011 to 2040, with stage two starting 2020.[43]

For 2008 as decision year, the cost situation, even before detailed investigation speaks for offshore wind. The latter is characterized by a more than three times higher capacity factor than PV at lower specific capital cost. Only its learning rate lies below the one of PV. Model results are expectedly quite clear — "Dominance of technology A" (P2), i.e. offshore wind, is identified as cost-optimal strategy. Table 5.5 illustrates the high robustness of those results and transferability to regions with a different potential structure for the two technologies.[44] Only in regions with extraordinary favorably PV potential[45] and extraordinarily unfavorable offshore potential could the optimal strategy include PV deployment. The latter either as P1,"Dominance of technology B", or as P6, where offshore investment would be optimal for stage one but PV would be installed in stage two to fill the capacity gap.

For 2011 as decision year, specific capital cost for offshore wind now lie above the ones for PV due to the cost estimate revision for the first and deployment-enabled cost reductions for the latter. Yet despite those changes, the cost-optimal strategy clearly remains "Dominance of technology A" and very favorable PV potential in combination with unfavorable offshore wind

[43] Other assumptions include average power demand/ required production across both technologies of 27 TWh in stage 1 and 103 TWh in stage two (based on Nitsch and Wenzel, 2009, p. 11) as well as opportunity cost of capital of 7%.

[44] Assuming the same demand structure and excluding experience spillover benefits.

[45] Yielding more than 2,000 FLH/a.

Table 5.5 Optimal strategy offshore wind vs. PV — Decision 2008

B: PV	*A: Offshore wind FLH (h/a)*							
FLH	*1,600*	*2,000*	*2,400*	*2,800*	*3,200*	*3,300*	*3,600*	*4,000*
1,000	P2	P2	P2	P2	P2	P2	P2	P2
1,200	P2	P2	P2	P2	P2	P2	P2	P2
1,400	P2	P2	P2	P2	P2	P2	P2	P2
1,600	P2	P2	P2	P2	P2	P2	P2	P2
1,800	P2	P2	P2	P2	P2	P2	P2	P2
2,000	P6	P2	P2	P2	P2	P2	P2	P2
2,200	P1	P2	P2	P2	P2	P2	P2	P2
2,400	P1	P6	P2	P2	P2	P2	P2	P2

Table 5.6 Optimal strategy offshore wind vs. PV — Decision 2011

B: PV	*A: Offshore wind FLH (h/a)*							
FLH	*1,600*	*2,000*	*2,400*	*2,800*	*3,200*	*3,300*	*3,600*	*4,000*
1,000	P2	P2	P2	P2	P2	P2	P2	P2
1,200	P2	P2	P2	P2	P2	P2	P2	P2
1,400	P1	P2	P2	P2	P2	P2	P2	P2
1,600	P1	P1	P2	P2	P2	P2	P2	P2
1,800	P1	P1	P1	P2	P2	P2	P2	P2
2,000	P1	P1	P1	P2	P2	P2	P2	P2
2,200	P1	P1	P1	P1	P2	P2	P2	P2
2,400	P1	P1	P1	P1	P1	P2	P2	P2

potential would be required to tip the cost-optimal strategy towards implementation of PV, as shown in Tab. 5.6.

In the previous sections it has been demonstrated that six of the twelve optimal strategies only exist in presence of uncertainty. The added complexity from uncertainty representation raises the question on what the added value of the stochastic model over its deterministic counterpart is for practical applications.

From a practical point of view, the added value from the stochastic over the deterministic model only appears to be an insight on what investment strategies will be optimal in the second stage, depending on the materialization of the uncertain parameters. Yet, this information can also be obtained using scenario analysis and the comparatively small benefit from explicit uncertainty representation in this manner needs to be weighed against the additional effort needed when implementing the corresponding solutions[46] in a spreadsheet. Furthermore, this and other applications of the model with

[46] Strictly speaking, the corresponding objective function values for the different cases, expressed exclusively in terms of parameters.

practically relevant parameterization have also (in the vast majority of cases) produced optimal strategies that correspond to the deterministic strategies P1, P2, P5 and P6. In light of those observations a simplified version of the model excluding cases only present under uncertainty is suggested for practical applications.

5.7 Key Findings and Implications

In this chapter, a small-scale power sector model with two competing, learning technologies under uncertainty has been set up and solved. The issue of non-convexity has been addressed by solving for all local minima, which could constitute the global one, depending on parameter configuration. To identify the global minimum among those local ones, a decision rule based on their objective function values has been constructed.

The primary objective of this chapter has been to highlight special considerations for the setup of stochastic programs featuring endogenous learning. The main finding in this context is the need to adjust for unrealistic jumps of specific capital costs at stage transition if there is a change in learning rate across stochastic stages.[47] This adjustment — referred to as *vertical shift* of the experience curve in this chapter — is not only required in a stochastic model but also in a deterministic one, where learning rates are subject to change over time. A direct consequence of this finding is the reflection of this vertical shift in the formulation of the large-scale power system model SCORE, which features uncertain learning rates in chapter 7. Although formulating or neglecting this adjustment in model formulation has a considerable impact on results, the author is not aware of other publications describing it to date.

The second objective of this chapter has been to provide a framework for decision support to industry participants. For this purpose, the use of a deterministic model version is recommended: for analyses, where parameters are constant over time, the model version from section 5.3; for analyses, where variable costs or learning rates are expected to change, a simplified version of the model presented in section 5.4.[48]

The third objective of this chapter has been to investigate whether presence of uncertainty in combination with learning via experience curves is a sufficient condition for optimality of a technology diversification strategy before uncertainty has been resolved. The investigations in this chapter indicate that this is not the case. On the contrary, in the model with substituting

[47] Provided that learning rates change at a predetermined point in time.

[48] Expected values for variable costs and learning rates in the second period should be used instead of two different states of the world. This way, only calculation of cases P1, P2, P5, P6, P11 and P12 — for example, by entering the corresponding analytical expressions into a spreadsheet — is required to identify the globally optimal strategy.

technologies used in this chapter, technology diversification before uncertainty resolution has never been found optimal. This finding, however, has to be qualified in light of the observation that the model appears not to be characterized by recourse. The optimal investment strategy in the first stage is therefore based on the expected values of the uncertain parameters. Besides a more general proof of these properties than undertaken in this chapter, it would therefore also be interesting to analyze optimal investment strategies in presence of learning within a stochastic framework characterized by recourse. Recourse could, for example, be established by introducing plant construction lead times, although those are comparatively short for most renewable power technologies.

Although the analytical model in this section has revealed important methodological aspects for the setup of models with endogenous learning under uncertainty and can be put to use to compare competitiveness between two substitute technologies, it is not suitable to analyze the prospects for renewable power technologies as a group. The latter requires consideration of numerous other factors apart from learning as well as interactions among technologies within an energy system. In the next chapter, a numerically solved power system model is proposed, which takes such factors into consideration. The chapter thereafter finally introduces uncertainty into the large-scale model, drawing upon the methodological findings from this analytical model.

References

Arthur, W.B.: Competing Technologies, Increasing Returns, and Lock-In by Historical Events. The Economic Journal 99(394), 116–131 (1989)

Baker, E.: Uncertainty and learning in a strategic environment: global climate change. Resource and Energy Economics 27(1), 19–40 (2005)

Baker, E., Shittu, E.: Uncertainty and endogenous technical change in climate policy models. Energy Economics 30(6), 2817–2828 (2008)

Barreto, L.: Technological Learnin. In: Energy Optimisation Models And Deployment of Emerging Technologies: Dissertation. Dissertation, Swiss Federal Institute of Technology, Zurich (2001)

Birge, J.R., Louveaux, F.: Introduction to stochastic programming. Springer, New York (1997)

Bosetti, V., Tavoni, M.: Uncertain R&D, backstop technology and GHGs stabilization. Energy Economics 31(suppl. 1), S18–S26 (2009)

Bramoullé, Y., Olson, L.J.: Allocation of pollution abatement under learning by doing. Journal of Public Economics 89(9-10), 1935–1960 (2005)

EPIA. Global Market Outlook for Photovoltaics 2013-2017, Brussels (2013)

Goulder, L.H., Mathai, K.: Optimal CO_2 Abatement in the Presence of Induced Technological Change. Journal of Environmental Economics and Management 39(1), 1–38 (2000)

Gritsevskyi, A., Nakićenović, N.: Modeling uncertainty of induced technological change. Energy Policy 28(13), 907–921 (2000)

Grübler, A., Gritsevskyi, A.: A Model of Endogenous Technological Change through Uncertain Return on Innovation. In: Grübler, A., Nakićenović, N., Nordhaus, W.D. (eds.) Technological Change and the Environment, pp. 280–319. Resources for the Future and IIASA, Washington D.C. (2002)

GWEC. Global Wind Report — Annual Market Update 2012, Brussels (2013)

Ingham, A., Ma, J., Ulph, A.: Climate change, mitigation and adaptation with uncertainty and learning. Energy Policy 35(11), 5354–5369 (2007)

Mattsson, N.: Introducing uncertain learning in an energy system model: a pilot study using GENIE. International Journal of Global Energy Issues 18(2-4), 253–265 (2002)

Nitsch, J., Wenzel, B.: Langfristszenarien und Strategien für den Ausbau erneuerbarer Energien in Deutschland: Leitszenario 2009, Berlin (2009)

REN21. Renewables 2012 — Global Status Report, Paris (2012)

Ulph, A., Ulph, D.: Global Warming, Irreversibility and Learning. The Economic Journal 107(442), 636–650 (1997)

Chapter 6
Optimal Future Deployment of Renewable Power Technologies: A System Model Approach

In this chapter the large-scale power sector model CORE is presented, which is required to address the research question raised in chapter 2. The chapter begins with a discussion of objectives, general approach and the current state of research. The CORE model formulation is then presented in section 6.2, followed by the key assumptions made in section 6.3. In section 6.4 model results are discussed and the chapter ends with a short summary of findings and their implications in section 6.5.

6.1 Objective and Approach

The research question from chapter 2 asks for identification of the optimal investment strategy into commercially available renewable power generation technologies across the 21st century. A popular approach used in this context is based on the levelized costs of power generation.[1] Future generation costs for particular technologies are forecasted to identify the point in time, when cost-competitiveness with a corresponding incumbent technology is reached. These forecasts are frequently based on experience curves for capital costs in combination with assumptions on future investment activity (see, for example, van der Zwaan and Rabl 2004, Poponi 2003 or Neij 1999).

Apart from other shortfalls, this approach does not provide an answer to the question *whether* the technology should be pursued at all. This is an important question, since diffusion of new technologies is not self-sufficient in most instances. In presence of increasing returns and given long-term beneficial deployment from a societal perspective, an initially uncompetitive cost position may prevent self-sufficient deployment of the technology under consideration. Governments can then step in to overcome this type of market failure. One may be tempted to suggest that the questions of *how much* to invest and *when* investments should occur only arise after the *whether* question has been answered. Yet, as numerous factors impose constraints on the

[1] See p. 7 for an introduction of the concept or App. A for additional information.

F. Wagner, *Renewables in Future Power Systems*,
Green Energy and Technology,
DOI: 10.1007/978-3-319-05780-4_6,

how much and *when*, the answer to each of the three questions is conditional upon the two others. In complex systems, such as the power generation sector, numerous interdependent factors determine cost-optimal investment strategy.

Clearly neither the generation cost approach applied to technologies in isolation nor the two-technology model from the previous chapter are capable of capturing a large set of those key factors determining optimal investment strategy into renewables. A more sophisticated model is therefore required, capturing as many of those key factors as possible. Although it covers the entire power sector, the model presented in this chapter has a clear focus on renewable technologies. It explicitly represents system elements of particular importance to renewable technology competitiveness, such as detailed renewable potentials, endogenous technical change, intermittency or energy storage.[2]

To allow for detailed technology representation and associated interdependencies within the electricity sector, a bottom-up, partial equilibrium approach is chosen.[3] In line with the normative nature of the research question, the model is formulated as optimization, minimizing total discounted system costs. In this neo-classical setting, technologies compete on the basis of costs, which are determined by their fundamental properties. For simplicity, the model will further be referred to as CORE, standing for "Competitiveness of Renewable Electricity." CORE, as presented in this chapter, is characterized by perfect foresight. Uncertainty is introduced into the framework in the next chapter. The global model features 14 power technologies, eight thereof renewable, and is divided into eight regions. As power demand levels vary across the year, CORE also falls into the class of peak-load-pricing models (cf. Weber, 2005; Boiteux, 1960).

From a methodological point of view, the impact of simultaneously considering detailed renewable potentials as well as endogenized experienced curves is analyzed. Explicit consideration of renewable potential imposes a resource bound upon the respective technology. *Bounding* in a setting of increasing returns[4] may prevent dominance of a technology (cf. Arthur, 1989).[5] Detailed representation of renewable potential differentiated by productivity, i.e. output that can be generated per unit of capacity installed instead of merely using average output should have a similar effect as the presence of niche markets.[6] The latter reduce the amount of up-front investment

[2] Hydro and CSP energy storage are considered.

[3] The drawback of this approach is that interactions with other sectors, especially within the energy sector, are not represented endogenously and additional assumptions must be made.

[4] Endogenization of experience curves in CORE implies presence of increasing returns.

[5] Presence of bounding in CORE is not only attributable to potential limitations but also to other sources, e.g. power demand in a particular load segment.

[6] Assuming that rational investors will first exploit the renewable potential characterized by highest output.

required to bring a particular technology to competitive cost levels. Therefore, it is expected that differentiation of potentials by productivity in presence of bounded increasing returns will either increase or at least not decrease the number of technologies pursued as part of the optimal future investment strategy.

CORE aims to bridge the gap between two strands of literature: *renewable potential analyses* (see chapter 2.2.2) on the one hand and *integrated energy system models with endogenous technical change* on the other. Most renewable potential analyses confine themselves to a single technical and economic potential figure per technology and region (e.g. Krewitt et al., 2009; Goldemberg, 2000). Some studies take a step further and differentiate technical potentials based on energy carrier availability (e.g. Hoogwijk et al., 2004; Trieb et al., 2009). On the basis of this data and assumptions on future costs, cost-supply curves can be derived for a particular, future point in time (e.g. for Wind in Hoogwijk et al., 2004). An extension of this approach to more than one technology is de Vries et al. (2007), who consider Wind, Biomass and PV with their respective land requirements simultaneously when deriving their respective potentials. Yet, even these more sophisticated analyses can only provide an answer on how much of a particular technology *could* be deployed in an economically viable manner for an assumed cost position. Optimization-based energy system models, on the other hand, can provide an answer on how much of a technology *should* be employed, especially if they include technical change endogenously. Initial applications (cf. Messner, 1997; Mattsson and Wene, 1997) ignore renewable potential constraints, as have most bottom-up focused models since. An exception is the TIMES model framework, where Hydro, Wind, Biomass and Geothermal potentials can be considered (cf. Loulou and Labriet, 2008). Yet, only for wind the maximum technical potential is further divided into four segments characterized by differences in energy carrier productivity. A recent application of TIMES with endogenous technical change can be found in Rout et al. (2009). Unfortunately, the regional breakdown used in this latter study does not allow for a practically relevant interpretation of effects from incorporating potentials.[7]

The CORE model incorporates potential data, differentiated by region and primary energy productivity, for six renewable technologies in an integrated power sector setting. The remainder of the chapter is divided into four parts. Section 6.2 offers a detailed description of CORE including its algebraic formulation. Critical assumptions follow in section 6.3. Results for the base case and the different sensitivity analyses are then presented and discussed in section 6.4. Finally, section 6.5 summarizes key findings and highlights implications.

[7] Regions include Non-EU OECD and Non-OECD excluding China and India. This implies renewable potential aggregation across oceans. Large-scale power transmission across oceans, though, would be associated with high costs not captured within the model.

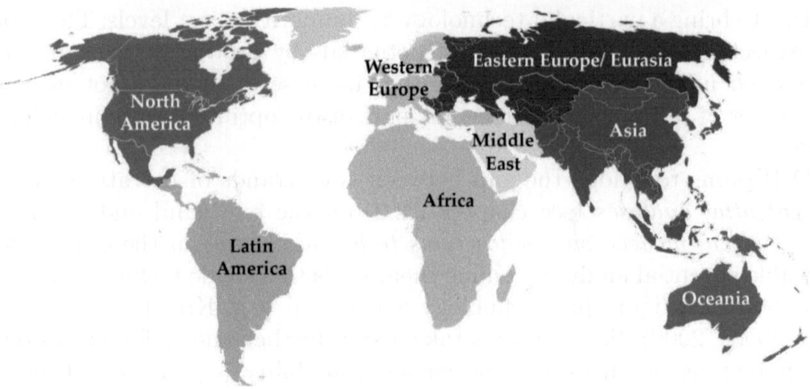

Fig. 6.1 CORE model regions
Source: Modified from IEA (2009)

6.2 Model Description

CORE is a system model confined to the power sector. To more accurately represent key drivers of renewable competitiveness, a higher degree of detail in the power sector is chosen over a more comprehensive coverage of the energy sector.[8] Other primary energy demand sectors, such as transport or residential, as well as fossil fuel extraction are consequently outside the model. Its geographic scope is global with division into eight world regions, which are almost congruent with the ones used by IEA (2009) and shown in Fig. 6.1. There are, however, deviations from this definition for Asia and Oceania.[9]

The period covered reaches from 2008 to 2100. To optimize computational performance, seven representative model years with differing intervals are used instead of calculating each of the more than 90 years.[10] Hence, cost types initially not on a per-period basis[11] must be converted into annuities. Costs are quoted in US$$_{2008}$ across the model horizon; opportunity costs of capital of 7% are assumed. Investment into 14 fossil, nuclear and renewable power generation technologies is possible, as given in the following list. An asterisk marks the technologies, for which cost reductions via endogenized experience curves can take place.

---

[8] Computational limitations, especially when considering the stochastic version of CORE, mandate this trade-off.

[9] An aggregated region including both Rest of Asia and OECD Pacific is not meaningful. In such a region China and India with their comparatively high power demand would have access to the vast solar potential of Australia, which is characterized by comparatively high irradiance. Therefore, a split into Asia and Oceania has been chosen.

[10] 2008, 2010, 2020, 2030, 2050, 2075 and 2100.

[11] Capital costs, decommissioning costs.

- Coal power plants (hard coal, lignite)
- Gas turbines (open cycle*, combined cycle*)
- Oil power plants
- Fissile nuclear power plants
- Biomass power plants*
- Wind turbines (onshore*, offshore*)
- Photovoltaic generators*
- Concentrating solar power plants (base plant*, storage extension*)
- Hydro power stations (new, replacement)

Carbon capture and storage enhanced coal plants (CCS) are excluded as the primary focus of the model is to assess the competitiveness of renewable power and little empirical information on expected cost reduction trajectories, e.g. in form of empirical learning rates, are available.[12] The latter is also the reason for omitting fuel cells[13], nuclear fusion[14], enhanced geothermal power systems and ocean power. Traditional geothermal power has been excluded since global installations have been stable at approximately 200MW per year over the last three decades (IEA, 2008) and a significant step-up in capacity additions appears unlikely. Worldwide geothermal technical potential is also lower than for any of the other renewable technologies covered in the model (cf. Krewitt et al. 2009).

Specific relationships exist among some of the modeled technologies. Offshore wind investment is represented as investment into onshore technology plus addition of an offshore upgrade. This allows for learning spillovers between the two technologies. Concentrating solar power plants are modeled as two separate technology components. The *base plant* without storage consists of the power block and a collector field large enough to produce enough heat to drive the power block during times of sufficient irradiance. The second component is a *storage extension* to the basic plant. It consists of a six-hour thermal storage and an addition to the collector field sufficiently large to charge the storage.[15] Finally, two different types of hydro technologies exist: new and replacement stations. Replacement stations are characterized by lower capital costs than new ones and can only be constructed at sites, where a hydro station had already existed previously, drawing upon the civil components of this former station.

[12] Nuclear power is available as carbon-free substitute to renewable technologies in the model.

[13] Fuel cells would have to be assessed in a framework covering additional energy demand sectors to obtain meaningful results.

[14] Technical feasibility of power generation from fusion has to date not been demonstrated.

[15] This component-based representation is adopted from Trieb et al. (2009).

Table 6.1 CORE sets

Index	Description	Number of elements
i	Conversion technologies	14
f	Fuel types	5
r	World regions	8
t	Model reference years	7
λ	Load duration curve segments	8
ϕ	Fuel cost-supply curve segments	5
μ	Renewable potential curve segments	6
χ	Experience curve segments	5

The model is set up as Mixed-Integer program, which is implemented in GAMS and solved using the branch & cut algorithm of CPLEX.[16] With base case parameter configuration the reduced[17] deterministic program consists of approximately 20,000 rows, 21,000 columns, 132,000 non-zero variables and 200 binaries.[18] On a computer with an Intel Core i5 processor and 16 GB of memory, solution time for the base case after solver tuning is approximately 30 minutes.[19] However, solution times can vary substantially depending on model parameterization. Sensitivity analyses[20] take up to twelve hours to complete.

In the following, the model's algebraic formulation is presented. Equations are grouped into six main categories. An overview of the different sets used can be found in Table 6.1. Variables and parameters are introduced together with the equation, in which they first appear but are also summarized in Tables 6.2 and 6.3. Generally, variables are represented by uppercase and parameters by lowercase Latin or Greek letters.[21] Several model variables, parameters and thus equations exist more than once in slightly modified form, e.g. if a particular set applies only to a subset of technologies or fuel types. To avoid redundancies, only the equation version for technologies, fuels, etc. with the largest number of dimensions is presented.

[16] Binary variables are required for linearly approximating experience curves. See chapter 4.4 for more details on methods to endogenize experience curves in bottom-up models.

[17] After pre-solving.

[18] Before pre-solving the model consists of 90,000 rows, 89,000 columns, 331,000 non-zero variables and 280 discrete rows.

[19] For a maximum difference of 0.01% between best node and best integer solution.

[20] Generally based on 20–25 scenarios.

[21] Accents are used to further distinguish sets, variables and parameters. $\tilde{x}$ indicates a parameter, which contains information on segment sizes or endpoints described by a set x. $\hat{x}$ is the cumulative value to date of x. $\overline{x}$ indicates an average value.

Table 6.2 CORE variables

Variable	Description	Unit
$K^{iins}_{i,r,t,\mu}$	Installed capacity (investment within model horizon)	TW
$K^{bins}_{i,r,t,\mu}$	Installed capacity (investment before model horizon)	TW
$K^{op}_{i,r,t,\mu}$	Operational capacity (installed capacity less early divestments)	TW
$\hat{\kappa}_{i,t}$	Total cumulative capacity manufactured to date	TW
$\hat{\kappa}^{seg}_{i,t,\chi}$	Cumulative capacity by active learning curve segment	TW
$X_{i,r,t,\mu}$	Investment	TW
$\hat{X}_{i,r,t,\mu}$	Cumulative investment to date (within model horizon)	TW
$\dot{X}_{i,r,t,t',\mu}$	Investment tracking variable	TW
$V^{dec}_{i,r,t,\mu}$	Decay divestments at end of useful life	TW
$\hat{V}^{dec}_{i,r,t,\mu}$	Cumulative decay divestments to date	TW
$V^{idec}_{i,r,t,\mu}$	Decay divestments (investment in model horizon)	TW
$V^{bdec}_{i,r,t,\mu}$	Decay divestments (investment before model horizon)	TW
$V^{act}_{i,r,t,\mu}$	Actual divestments (decay and premature)	TW
$\hat{V}^{act}_{i,r,t,\mu}$	Cumulative actual divestment to date	TW
$Y_{i,r,t,\lambda}$	Electricity generated	TWh_{el}
$E_{r,r',t,\lambda}$	Electricity exported	TWh_{el}
$M_{r,r',t,\lambda}$	Electricity imported	TWh_{el}
$W_{i,r,t,\lambda}$	Energy stored	TWh_{el}
$U_{f,r,t,\lambda,\phi}$	Fuel consumed (by fuel type)	$unit^*$
$\hat{U}_{f,t,\phi}$	Cumulative nuclear fuel depleted to date	Gt
$O_{f,r,t,\lambda}$	Carbon dioxide emissions	t^{co_2}
$\hat{C}^{tinv}_{i,t}$	Cumulative investment costs to date	$\$B_{2008}$
$C^{ainv}_{i,t}$	Investment cost annuity	$\$B_{2008}$
$\dot{C}^{ainv}_{i,t,t'}$	Investment cost annuity tracking variable	$\$B_{2008}$
$C^{dec}_{i,r,t}$	Decommissioning cost annuity	$\$B_{2008}$
$C^{op}_{i,r,t}$	Non-fuel operating costs	$\$B_{2008}$
$C^{fue}_{f,r,t,\lambda,\phi}$	Fuel costs	$\$B_{2008}$
$C^{tra}_{r,r',t,\lambda}$	Trans-regional transmission costs	$\$B_{2008}$
C^{sys}	Total discounted system costs	$\$B_{2008}$
$A_{i,t,\chi}$	Active learning curve segment indicator	*binary*

* Metric tons for all fuels but natural gas, which is quoted in 1,000m^3.

6.2.1 Objective Function

The model's objective is the minimization of total discounted system costs C^{sys} over the reference years from 2008 to 2100. Cost types considered are annualized investment costs $C^{ainv}_{i,t}$, fuel costs $C^{fue}_{f,r,t,\lambda,\phi}$, non-fuel operating costs $C^{op}_{i,r,t}$, annualized decommissioning costs $C^{dec}_{i,r,t}$ and trans-regional transmission costs $C^{tra}_{r,r',t,\lambda}$. The objective function is given by Equation (6.1).

Table 6.3 CORE parameters

Parameter	Description	Unit
π_i	Useful economic life	$Years$
d_i^{cst}	Construction duration	$Years$
τ_t	Duration since preceding model period	$Years$
$\hat{\tau}_t$	Cumulative duration since first model period	$Years$
θ_λ	Load duration curve segment duration	$Hours$
$\delta_{r,t}$	Annual power demand	TWh_{el}
$\zeta_{r,r'}$	Trans-regional transmission capacity	TW
$\hat{\kappa}_i^{max}$	Maximum cumulative capacity	TW
$\hat{\kappa}_i^{flo}$	Cumulative capacity at which floor costs are reached	TW
$\tilde{\phi}_{f,r,\phi,t}$	Fuel price segment size	TW
$\tilde{\mu}_{i,r,\mu}$	Renewable potential segment size	TW
$\tilde{\chi}_{i,\chi}$	Experience curve segment upper bound	TW
$x_t^{Max,abs}$	Maximum investment (absolute portion)	TW
$x_t^{Max,rel}$	Maximum investment (relative portion)	$Pct.$
$x^{Max,rel}$	Maximum investment p.a. (relative portion)	$Pct.$
w_i^{Max}	Storage capacity	$Pct.$
y_i^{Min}	Minimum production	$Pct.$
$\overline{\nu}_{i,r,\mu}$	Average annual availability	$Pct.$
$\nu_{i,r,\lambda,\mu}$	Fluctuation-adjusted availability	$Pct.$
$\omega_{i,\lambda}$	Fluctuation adjustment coefficient	$Pct.$
ρ_r	Capacity reserve margin	$Pct.$
$\tilde{\lambda}_\lambda$	Load duration curve segmentation coefficient	$Pct.$
η_i^{con}	Conversion efficiency (thermal to electric)	$Pct.$
η^{ttr}	Trans-regional transmission efficiency	$Pct.$
η_r^{tir}	Intra-regional transmission efficiency	$Pct.$
l_i	Learning rate	$Pct.$
γ_i	Learning index	$Pct.$
r	Opportunity costs of capital	$Pct.$
ψ_t	Present value coefficient	$Pct.$
o_t^{max}	CO_2 emission cap	t_{co_2}
h_f	Fuel energy content	$TWh_{th}/unit^*$
q_f	Specific carbon dioxide emissions from combustion	$t_{co_2}/unit^*$
$c_{f,r,t}^{fue}$	Fossil fuel price	$\$_{2008}/unit^*$
$\tilde{c}_{f,\phi}^{fue}$	Fuel price per segment	$\$_{2008}/t_{fue}$
c_i^{ovn}	Overnight investment costs	$\$_{2008}/kW$
c_i^{inv}	Investment costs including construction financing	$\$_{2008}/kW$
c_i^{1st}	First unit investment costs	$\$_{2008}/kW$
c_i^{dec}	Decommissioning costs at time of commissioning	$\$_{2008}/kW$
c_i^{op}	Annual non-fuel operating costs	$\$_{2008}/kW$
$\alpha_{i,\chi}$	Cumulative cost curve segment slope	$\$_{2008}/kW$
$\beta_{i,\chi}$	Cumulative cost curve segment y-intercept	$\$_{2008}/kW$
c^{tra}	Trans-regional transmission costs	$\$_{2008}/kWh$

* Metric tons for all fuels but natural gas, which is quoted in 1,000m^3.

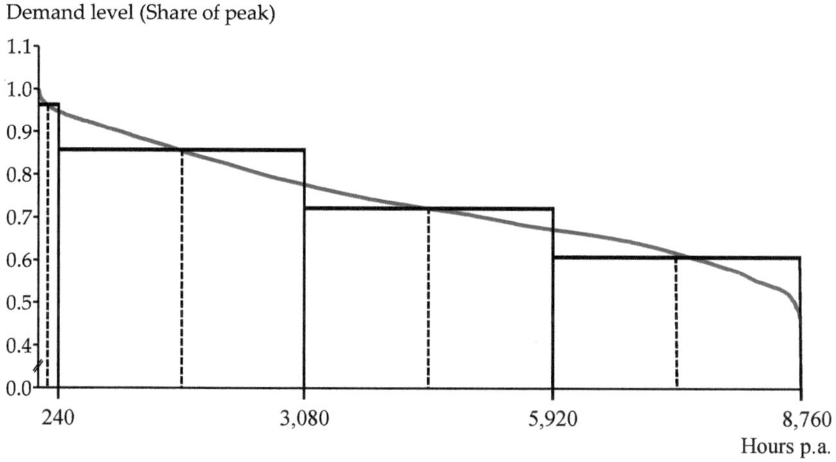

Fig. 6.2 Approximated load duration curve
Source: RTE (2010), own calculations

$$
\begin{aligned}
C^{sys} = & \sum_f \sum_r \sum_t \sum_\lambda \sum_\phi \left(\psi_t \cdot C^{fue}_{f,r,t,\lambda,\phi} \right) \\
& + \sum_r \sum_{r'} \sum_t \sum_\lambda \left(\psi_t \cdot C^{tra}_{r,r',t,\lambda} \right) \\
& + \sum_i \sum_r \sum_t \left(\psi_t \cdot C^{op}_{i,r,t} \right) \\
& + \sum_i \sum_r \sum_t \left(\psi_t \cdot C^{dec}_{i,r,t} \right) \\
& + \sum_i \sum_t \left(\psi_t \cdot C^{ainv}_{i,t} \right)
\end{aligned}
\tag{6.1}
$$

Discounting factor ψ_t is a function of the opportunity costs of capital r and the cumulative years passed since 2008, $\hat{\tau}_t$.

$$\psi_t = \frac{1}{(1+r)^{\hat{\tau}_t}} \tag{6.2}$$

Set I contains the 14 conversion technologies, set T the seven model reference years, set R the eight world regions, F the five fuel types, Λ the eight load duration curve segments and Φ the different segments of the approximated fuel cost-supply curve. The objective function is linear but not continuous, since the calculation of $C^{ainv}_{i,t}$ involves binary variables.

6.2.2 Power Generation, Consumption and Trade

Power demand needs to be satisfied in all model years. Across a model year, the distribution of demand follows a linearly approximated annual load duration curve with four main segments. Each of these four segments is again divided in half for a total of eight (see Fig. 6.2). The second division is required to represent stochastic fluctuations in the supply of intermittent primary energy carriers in an approximated manner. The duration curve is based on historical demand in the French power system[22] (RTE, 2010). To avoid excessive peak-shaving,[23] the first main segment is considerably shorter than the subsequent three.

Power production from wind and solar plants relies upon the supply of the respective primary energy carrier, which is fluctuating. To reflect the resulting fluctuations in plant availability, *average* annual plant availability $\overline{\nu}_{i,r,\mu}$ has been adjusted by means of *fluctuation factor* $\omega_{i,\lambda}$, as shown in Equation (6.3).

$$\nu_{i,r,\lambda,\mu} = \overline{\nu}_{i,r,\mu} \cdot \omega_{i,\lambda} \tag{6.3}$$

The fluctuation factor $\omega_{i,\lambda}$ is derived in a two-step process. First, cyclical fluctuations are reflected by matching hourly demand data with the corresponding average hourly production at reference plant locations.[24] This process then yields plant availability for each of the model's four main load segments, adjusted for cyclical fluctuations of power output over the diurnal and annual cycle (see Fig. 6.3). The cyclical plant availability difference across the main load duration curve segments is more pronounced for solar than for wind. This is mainly driven by generator downtime at night. Despite downtime at night, the adjustment factor for solar does not drop close to zero in the base load segment since this segment still includes a considerable number of hours with positive solar radiation. Early morning hours in spring and summer, when there is considerable radiation but comparatively low demand, are examples of such times.

Second, the fluctuation factor $\omega_{i,\lambda}$ is extended to reflect stochastic fluctuations of the energy carrier. These fluctuations are mainly attributable to short-term changes in meteorological conditions. This type of fluctuations is reflected in an approximated, deterministic manner to avoid the

[22] Based on data from 1996 to 2005.

[23] Reduction of the peak demand level due to averaging demand across a segment.

[24] Beresford (USA) for wind (South Dakota State University, 2010); Albuquerque (USA) for global and direct solar radiance (NREL, 2010). A duration curve based on American demand would have certainly been preferable to match the American production reference locations. Production from wind turbines is based on hourly wind speeds and the power curve of the General Electric 1.5xle turbine (GE Energy, 2009). For PV generators and CSP plants, as approximation, a linear relationship of global and direct radiance, respectively, to power output has been used.

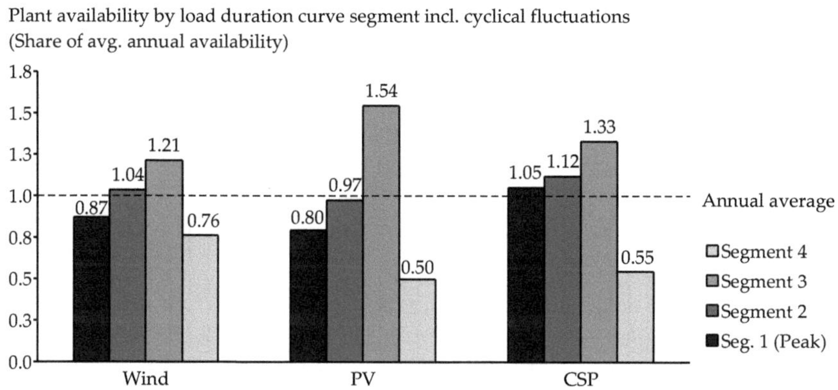

Fig. 6.3 Plant availability reflecting cyclical energy carrier fluctuations
Source: Own calculations based on South Dakota State University (2010), NREL (2010) and RTE (2010)

computational burden associated with stochastic modeling techniques.[25] Each of the four main load duration curve segments is divided into a segment with *low* and *high* energy carrier availability. *Low* energy carrier availability is set to one standard deviation below average, cyclical-fluctuation-adjusted plant availability. Correspondingly, *high* availability is one standard deviation above the corresponding average value. The value of one standard deviation is somewhat arbitrary. However, a very high value of two or even more standard deviations would likely overstate stochastic output variability since production of a single reference site has been used in the process of assumption generation. If a group of interconnected sites, whose meteorological regimes are not perfectly correlated, had been used for reference, this group of sites would have shown lower output variability than the single one (see sec. 2.2.3). Wind turbines and CSP plants are characterized by a high degree of stochastic output fluctuations (see Fig. 6.4). A key reason for CSP availability being more volatile than the one of PV is CSP's dependence on direct irradiance, which is strongly influenced by the presence of clouds.

Satisfaction of demand in each of the load duration curve segments, denoted by $\delta_{r,t}\tilde{\lambda}_{\lambda}$, is enforced as given by Equation (6.4).[26] It is to be satisfied

[25] An additional benefit of representing fluctuations in this manner is that *capacity credit* constraints (also see 2.2.3) can be avoided. Since the capacity credit of a new installation is dependent on the penetration of the particular and other intermittent technologies within a power system, they can only realistically be represented in a non-linear fashion or a corresponding piece-wise linear approximation.

[26] $\delta_{r,t}$ = Annual demand, $\tilde{\lambda}_{\lambda}$ = Share of annual demand attributable to load segment λ.

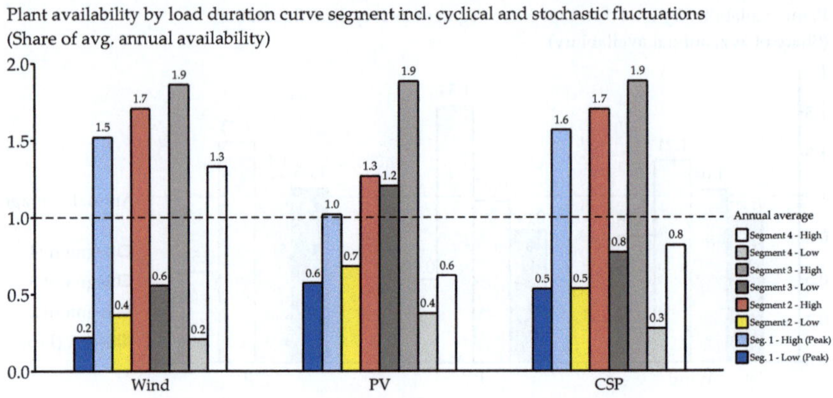

Fig. 6.4 Plant availability reflecting energy carrier fluctuations
Source: Own calculations based on South Dakota State University (2010), NREL (2010) and RTE (2010)

by the sum of power generated, $Y_{i,r,t,\lambda}$, and net imports[27] under consideration of transmission losses η_r^{tir} within the region under consideration.

$$\eta_r^{tir}\left(\sum_i Y_{i,r,t,\lambda} + \sum_{r'} M_{r,r',t,\lambda} - \sum_{r'} E_{r,r',t,\lambda}\right) \geq \delta_{r,t}\tilde{\lambda}_\lambda \tag{6.4}$$

A capacity reserve above the expected peak demand level needs to be maintained. For technologies with storage capability (subset $IS(I)$), i.e. hydro stations and CSP thermal storage, the formulation of this requirement, stated in Eq. (6.5), implies the assumption that sufficient stored energy is available during peak demand to operate the plant at nameplate capacity.[28]

$$\sum_{i \notin is(i)} \sum_{\mu} K^{op}_{i,r,t,\mu}\nu_{i,r,\lambda,\mu} + \sum_{\mu} K^{op}_{hyd,r,t,\mu} + \sum_{\mu} K^{op}_{csp\_sto,r,t,\mu}(1 - \nu_{csp\_bas,r,\lambda,\mu})$$
$$\geq (1+\rho_r)\,\frac{\delta_{r,t}\tilde{\lambda}_\lambda}{\eta_r^{tir}\theta_\lambda} \tag{6.5}$$

$K^{op}_{i,r,t,\mu}$ indicates the nameplate capacity of the particular technology that can be used for power production, $\nu_{i,r,\lambda,\mu}$ the fluctuation-adjusted portion of nameplate capacity available for power generation, ρ_r the reserve margin as share of total required capacity and θ_λ the duration of a particular load segment. Maintenance and repair downtime of plants with storage during peak-load hours are not reflected in this formulation.

[27] Imports $M_{r,r',t,\lambda}$ less exports $E_{r,r',t,\lambda}$.

[28] Model results confirm that this assumption is reasonable. Storage modeling is described in section 6.2.3.

Operational capacity $K^{op}_{i,r,t,\mu}$ and its corresponding availability determine the maximal amount of power generation. Equation (6.6) describes this constraint for all technologies apart from hydro stations and the CSP storage extension. Equation (6.7) limits production from hydro plants to the station's nameplate capacity. Otherwise, given sufficiently high reservoir fill levels, power generation in excess of rated capacity would be possible. A similar rationale holds for CSP storage extensions. Since they draw upon the power block of the CSP base plant, they are only capable of using the share of the power block not simultaneously "occupied" by the base plant, enforced by Eq. (6.8).

$$Y_{i,r,t,\lambda} \leq \theta_\lambda \sum_{\mu} \left(K^{op}_{i,r,t,\mu} \nu_{i,r,\lambda,\mu}\right) \quad \forall\, i \neg\, hyd\,,\; csp\_sto \tag{6.6}$$

$$Y_{hyd,r,t,\lambda} \leq \theta_\lambda \sum_{\mu} \left(K^{op}_{hyd,r,t,\mu}\right) \tag{6.7}$$

$$Y_{csp\_sto,r,t,\lambda} \leq \theta_\lambda \sum_{\mu} \left(K^{op}_{csp\_sto,r,t,\mu}\right) - Y_{csp\_bas,r,t,\lambda} \tag{6.8}$$

Power can be traded trans-regionally. Yet this is only the case for few regional combinations, such as between Western Europe (ENTSO-E grid) and Eastern Europe/Eurasia (IPS/UPS grid). Transmission capacities are kept constant and trans-regional power transmission overall only has a minor impact on model results. Equation (6.9) enforces the capacity constraint with $\zeta_{r,r'}$ representing trans-regional transmission capacity. Equation (6.10) controls the relationship between exports and imports. Transmission losses are represented by means of the trans-regional transmission efficiency coefficient η^{ttr}.

$$E_{r,r',t,\lambda} \leq \zeta_{r,r'} \theta_\lambda \tag{6.9}$$

$$M_{r',r,t,\lambda} = \eta^{ttr} E_{r,r',t,\lambda} \tag{6.10}$$

The amount of fuel required for power generation, $U_{f,r,t,\lambda,\phi}$, is a function of the amount of power produced, specific energy content h_f of the primary energy carrier and specific technology conversion efficiency η^{con}_i, as given by Eq. (6.11). Matrix $g_{i,f}$ matches generation technologies i with their corresponding fuel types f. Hard coal, lignite, oil, nuclear fuel and biomass are quoted in metric Gt, natural gas in Tm^3.[29]

$$\sum_{\phi} U_{f,r,t,\lambda,\phi} = \sum_{i} \left(\frac{g_{i,f} Y_{i,r,t,\lambda}}{\eta^{con}_i h_f 10^9} \right) \tag{6.11}$$

[29] The comparatively large units are required to avoid large variable values, which can result in problems during the model solution process.

Carbon dioxide emissions depend upon the type and amount of fossil fuel consumed. Biomass combustion is considered carbon-neutral as the carbon dioxide emitted has previously been removed from the atmosphere. Emissions per fuel type, $O_{f,r,t,\lambda}$, are then calculated as given by Eq. (6.12), with q_f denoting specific CO_2 emissions per reference fuel unit.

$$O_{f,r,t,\lambda} = q_f \sum_{\phi} U_{f,r,t,\lambda,\phi} \tag{6.12}$$

In order to limit global warming to 2°C, the international community has recognized that "deep cuts in global greenhouse gas emissions are required" (UNFCCC, 2010). Consequently, CORE incorporates a global carbon cap, corresponding to a 450ppm CO_2 concentration, given by Eq. (6.13). As the electricity sector is not the sole emitter of CO_2, the carbon cap o_t^{max} represents the exogenously given share of total emission allowance available to the electricity sector. Carbon cap assumptions are discussed in section 6.3.

$$\sum_{f} \sum_{r} \sum_{\lambda} O_{f,r,t,\lambda} \leq o_t^{max} \tag{6.13}$$

6.2.3 Energy Storage

Hydro stations and CSP plants offer the possibility to store energy at reasonable cost. All new as well as replacement hydro station capacity in CORE is characterized by a reservoir of size w_i^{Max}. To acknowledge the existence of run-of-river stations, the reservoir size w_i^{Max}, expressed as share of nameplate capacity, is only approximately half of the one typically found for storage hydro stations.[30] Storage capacity for CSP plants is modeled as separate technology in CORE. It incorporates both the thermal storage addition to the base plant as well as the enlargement to the collector field required to charge the storage. Stored energy takes the form of potential (hydro) and thermal (CSP) energy. However, for the sake of simplicity, stored energy is quoted in terms of TWh_{el} within this model. The rules governing the development of stored energy $W_{i,r,t,\lambda}$ across time and load duration curve segments are given by Equations (6.14) and (6.15). Equation (6.14) ensures that hydro and CSP storages contain the same amount of energy at the beginning and the end of a reference year. Equation (6.15) defines the stored energy level across load duration curve segments.[31]

[30] Modeling run-of-river and storage hydro stations as different technologies has not proven meaningful in this context. Data on renewable potential, split into storage and run-of-river, is not available.

[31] As load duration curves do not have a chronological order, this explicit representation of energy storage can only be considered an approximation. Despite this shortfall storage is still explicitly represented to account for one of the key advantages of CSP, the option to store energy at competitive cost levels, over PV and other "new" renewable technologies within the model.

$$W_{i,r,t,\lambda} = W_{i,r,t,\lambda_\lambda^{las}} + K_{i,r,t,\mu}^{op} \nu_{i,r,\lambda,\mu} \theta_\lambda - Y_{i,r,t,\lambda} \qquad \forall \lambda = \lambda_\lambda^{1st} \quad (6.14)$$

$$W_{i,r,t,\lambda} = W_{i,r,t,\lambda-1} + K_{i,r,t,\mu}^{op} \nu_{i,r,\lambda,\mu} \theta_\lambda - Y_{i,r,t,\lambda} \qquad \forall \lambda \neq \lambda_\lambda^{1st} \quad (6.15)$$

Parameter λ_λ^{1st} indicates the first load duration curve segment, i.e. peak demand with low intermittent energy carrier availability. Correspondingly, λ_λ^{las} identifies the last load curve segment, i.e. base load with high intermittent energy carrier availability. Storage capacity determines the maximum possible amount of stored energy. It is expressed as share of nameplate capacity, as given by Eq. (6.16).

$$W_{i,r,t,\lambda} \leq w_i^{Max} \sum_\mu K_{i,r,t,\mu}^{op} \cdot 8,760\text{h} \qquad \forall i \in is_i \qquad (6.16)$$

To ensure that the existence of run-of-river plants without or with only limited pondage capability is properly reflected, a minimum hydro production constraint is introduced, as given by Eq. (6.17). y_{hyd}^{Min} expresses this minimum production requirement as share of average available capacity.[32]

$$Y_{hyd,r,t,\lambda} \geq y_{hyd}^{Min} \sum_\mu \left(K_{hyd,r,t,\mu}^{op} \nu_{hyd,r,\lambda,\mu} \right) \cdot \frac{\theta_\lambda}{8,760\text{h}} \qquad (6.17)$$

6.2.4 Capacity Additions and Divestments

Three notions of capacity are used within CORE: *operational* capacity K^{op}, *installed* capacity K^{bins} or K^{iins}, and *cumulative manufactured* capacity $\hat{\kappa}$. *Operational* capacity describes the capacity actually in place and available for power generation. *Installed* capacity covers all capacity installed at a certain point in time and still within its useful economic life. The only difference between the definitions of operational and installed capacity is the inclusion of premature divestments. Whereas operational capacity excludes capacity retired before the end of its useful life, installed capacity still includes it. Installed capacity is thus greater than or at least equal to operational capacity. Installed capacity is further divided into two subcategories. K^{iins} represents capacity that has been added within the model horizon, whereas K^{bins} has been added before the model horizon and is therefore part of the initial capacity endowment. Finally, *cumulative manufactured* capacity, in literature also often referred to as *cumulative capacity*, is the sum of all capacity ever manufactured and installed of a particular technology. It is required for the

[32] Strictly speaking, this is the capacity corresponding to regular water inflow over the considered period. Actually available capacity at a certain point in time depends on storage decisions.

endogenous representation of cost reductions via experience curves (see section 6.2.6).

The development rules for these different capacity definitions are straightforward. The value for installed and operational capacity is dependent on the previous period's value, capacity additions $X_{i,r,t,\mu}$ and capacity retirements, as given by Eqs. (6.18) and (6.20). The key difference here is that only end-of-lifetime retirements $V^{idec}_{i,r,t,\mu}$ are considered for capacity installed, whereas all capacity retirements $V^{act}_{i,r,t,\mu}$, including premature ones, are considered for operational capacity. Installed capacity associated with the initial endowment $K^{bins}_{i,r,t,\mu}$ does not increase, as shown in Eq. (6.19).

$$K^{iins}_{i,r,t,\mu} = K^{iins}_{i,r,t-1,\mu} + X_{i,r,t,\mu} - V^{idec}_{i,r,t,\mu} \tag{6.18}$$
$$K^{bins}_{i,r,t,\mu} = K^{bins}_{i,r,t-1,\mu} - V^{bdec}_{i,r,t,\mu} \tag{6.19}$$
$$K^{op}_{i,r,t,\mu} = K^{op}_{i,r,t-1,\mu} + X_{i,r,t,\mu} - V^{act}_{i,r,t,\mu} \tag{6.20}$$

Capacity addition history is stored in variable $\dot{X}_{i,r,t,t',\mu}$ by means of double indexation, where t' indicates the year of investment. $\dot{X}_{i,r,t,t',\mu}$ includes information on the amount and timing of all capacity additions that are either still within their useful life, or are up for end-of-life retirement within period t. Equation (6.21) describes this information saving procedure with π_t indicating the plant's useful life, τ_t the duration between the current and previous model year[33] and $\hat{\tau}_t$ the cumulative time passed since the beginning of the model. End-of-life divestments for capacity added during the model horizon are then calculated as given by Eq. (6.22).

$$\dot{X}_{i,r,t,t',\mu} = X_{i,r,t',\mu} \qquad \forall\, \hat{\tau}_t - \hat{\tau}_{t'} < \pi_i + \tau_t \tag{6.21}$$
$$V^{idec}_{i,r,t,\mu} = \sum_{t'=t_0}^{\hat{\tau}_t - \pi_i} \dot{X}_{i,r,t,t',\mu} \tag{6.22}$$

End-of-life divestment for capacity, which has been part of the initial endowment, occurs linearly and is described in Eq. (6.23). This assumes that the remaining life of capacity in the initial endowment is, on average, half of total useful economic life. Equation (6.24) ensures that installed capacity from the initial endowment reaches exactly zero at the end of its useful life. This is otherwise not granted in presence of unequal intervals between model years. Equation (6.25) summarizes end-of-life divestments of capacity installed within and before the model horizon. This is also the first step required to establishing the correct relationship between end-of-life and premature divestments.

[33] Time intervals between model years vary.

$$V^{bdec}_{i,r,t,\mu} = (\hat{\tau}_t - \hat{\tau}_{t-1}) \frac{K^{bins}_{i,r,t_0,\mu}}{\pi_i} \qquad \forall \hat{\tau}_t < \pi_i \tag{6.23}$$

$$V^{bdec}_{i,r,t,\mu} = K^{bins}_{i,r,t-1,\mu} \qquad \forall \hat{\tau}_t \geq \pi_i \tag{6.24}$$

$$V^{dec}_{i,r,t,\mu} = V^{bdec}_{i,r,t,\mu} + V^{idec}_{i,r,t,\mu} \tag{6.25}$$

The second step of this process involves calculation of cumulative end-of-life divestments $\hat{V}^{dec}_{i,r,t,\mu}$ and cumulative actual divestments $\hat{V}^{act}_{i,r,t,\mu}$, as given by Equations (6.26) and (6.27).

$$\hat{V}^{dec}_{i,r,t,\mu} = \hat{V}^{dec}_{i,r,t-1,\mu} + V^{dec}_{i,r,t,\mu} \tag{6.26}$$

$$\hat{V}^{act}_{i,r,t,\mu} = \hat{V}^{act}_{i,r,t-1,\mu} + V^{act}_{i,r,t,\mu} \tag{6.27}$$

The final step to establishing the correct relationship between actual and end-of-life retirements is to force cumulative actual divestments to at least include all end-of-lifetime ones, as given by Equation (6.28).

$$\hat{V}^{act}_{i,r,t,\mu} \geq \hat{V}^{dec}_{i,r,t,\mu} \tag{6.28}$$

A problem arising from the presence of endogenized experience curves is *bang-bang* behavior.[34] In an optimization model this may lead to capacity additions, which are unrealistic from a manufacturing standpoint as manufacturing capacity for generation units cannot be established overnight. For this reason Equation (6.29) restricts the possible growth in capacity additions. The constraint does not only contain a relative component, which depends on the previous period's capacity, but also an absolute one to avoid disproportionate disadvantages for technologies with small installed base. Those growth constraints, however, only have a limited impact on model results, as will be shown in section 6.4.7.

$$\sum_r \sum_\mu X_{i,r,t,\mu} \leq \tau_t x^{Max,abs}_t + \sum_r \sum_\mu K^{op}_{i,r,t-1,\mu} x^{Max,rel}_t \tag{6.29}$$

$x^{Max,abs}_t$ controls the absolute component of the growth constraint, $x^{Max,rel}_t$ the relative one. The latter relates operational capacity of the previous period to maximal allowed growth. $x^{Max,rel}_t$ is derived from the year-on-year growth rate limit $x^{Max,rel}$ as described by Eq. (6.30).

$$x^{Max,rel}_t = \left(1 + x^{Max,rel}\right)^{\tau_t} - 1 \tag{6.30}$$

[34] If investment into a technology with initially uncompetitive costs and increasing returns is deemed optimal, this investment should occur as fast as possible to obtain benefits earlier rather than later.

The following Equations, (6.31) through (6.37), are used to represent relationships, which are specific to particular technologies. Equation (6.31) ensures that replacement hydro stations with their lower capital costs can only be added in case there previously existed another hydro station.

$$\hat{X}_{hyd\_rep,r,t,\mu} \leq \hat{V}^{act}_{hyd\_new,r,t,\mu} + \hat{V}^{act}_{hyd\_rep,r,t,\mu} \tag{6.31}$$

where $\hat{X}_{hyd\_rep,r,t,\mu}$ represents cumulative investments, as defined in Eq. (6.32).

$$\hat{X}_{i,r,t,\mu} = \hat{X}_{i,r,t-1,\mu} + X_{i,r,t,\mu} \tag{6.32}$$

The relationship of CSP base plant technology to the storage expansion modules is governed by Eq. (6.33). Following the general approach presented by Trieb et al. (2009), the base plant has a collector field large enough to load the power block[35] in times of insolation. The storage modules each add six hours' worth of storage capacity and a collector field extension to the base plant. During times of sufficient solar radiation the storage is charged. The stored energy can then be used to generate power in times, when the energy from the base plant's collector field is insufficient to drive the power block at rated capacity. Equation (6.33) ensures that storage extensions cannot be built without sufficient CSP base capacity, since the latter incorporates the power block. A maximum of 18 hours of storage is allowed per base plant. This, in combination with the production from the base plant, would in principle allow for continuous power generation.

$$K^{op}_{csp\_sto,r,t,\mu} \leq 3 \cdot K^{op}_{csp\_bas,r,t,\mu} \tag{6.33}$$

Offshore wind can draw upon technological advances made for onshore wind turbines and, to a certain degree, vice versa. To represent this spillover effect (see sec. 4.2.2), the investment into offshore technology is modeled as combination of an investment into onshore capacity and an offshore upgrade. The capital costs of the upgrade are the cost difference between a full offshore and an onshore unit of capacity.[36] The capacity addition process for *offshore* wind in CORE is represented in form of three simultaneous transactions: (1) Addition of *onshore* wind capacity, (2) Instant, premature retirement of this *onshore* wind capacity and (3) Addition of *offshore* capacity. Both onshore wind turbine and offshore wind upgrade are subject to cost reductions

[35] The power block converts thermal energy into electricity. Similar technology to the one in conventional thermal plants is used, i.e. a combination of a turbine and a generator.

[36] The representation of offshore as upgrade to onshore is only applicable to the process of allocating capital costs. In operation, offshore technology has the properties of a regular offshore wind farm in the model, e.g. for operating and decommissioning cost calculation.

via experience curves. Offshore capacity addition leads to cost reductions for both onshore and offshore upgrade. Capacity additions of genuine onshore capacity only lead to cost reductions in onshore technology. Yet, as costs for offshore capacity additions are partly based on onshore technology, specific total offshore capital costs also decrease despite unchanged costs for the offshore upgrade in this case.

Equations (6.34) through (6.36) describe the implementation of this process in CORE. Equation (6.34) ensures that, for every unit of added offshore capacity, the same amount of onshore capacity is installed. Equations (6.35) and (6.36) force the immediate retirement of the added onshore capacity without compromising the model's capability to prematurely retire "true" onshore capacity. For onshore wind, Equation (6.36) effectively replaces (6.28), since it is stricter.

$$X_{win\_on,r,t,\mu} \geq X_{win\_of,r,t,\mu} \tag{6.34}$$
$$V^{act}_{win\_on,r,t,\mu} \geq V^{dec}_{win\_on,r,t,\mu} + X_{win\_of,r,t,\mu} - V^{idec}_{win\_of,r,t,\mu} \tag{6.35}$$
$$\hat{V}^{act}_{win\_on,r,t,\mu} \geq \hat{V}^{dec}_{win\_on,r,t,\mu} + K^{iins}_{win\_of,r,t,\mu} - \hat{V}^{idec}_{win\_of,r,t,\mu} \tag{6.36}$$

Due to the environmental issues associated with lignite mining, maximal regional lignite capacity has been limited to the level existing in 2008, as given by Eq. (6.37).

$$K^{op}_{lig,r,t} \leq K^{op}_{lig,r,t_0} \tag{6.37}$$

6.2.5 Renewable and Nuclear Resources

Renewable power resources are scarce, in particular resources of high productivity. Although it may be tempting to cap renewable capacity at its technical potential (see sec. 2.2.2) and apply an average availability factor across all plants of the same technology, this approach will greatly understate the economic competitiveness of the so-treated technologies. As a rational investor will, ceteris paribus, install renewable plants at sites with most abundant primary energy availability first, these plants are characterized by higher production and thus lower generation costs than a model using the average output across the entire technical potential would imply. It would therefore be ideal to implement marginal output curves for each technology and region. Unfortunately, data availability and the restriction of CORE to linear Equations require an approach somewhere in between this ideal and the other extreme of using a single average power output figure. In CORE, marginal cost curves are approximated by up to six linear potential segments for each region and renewable technology ir_i with $ir_i = \{win\_on, win\_of, pv, csp, hyd, bio\}$. Figure 6.5 illustrates two sample potential curves. Sites characterized by different average wind speeds translate into different levels of average plant

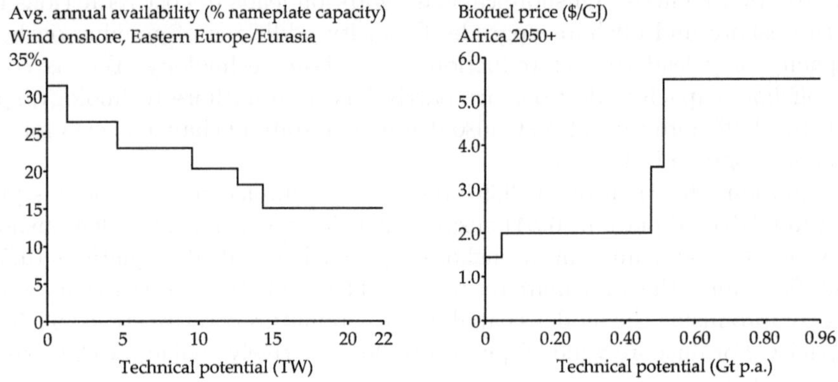

Fig. 6.5 Sample potential curves for onshore wind sites and biofuel
Source: Own calculation based on Hoogwijk et al. (2004) and Hoogwijk (2004)

availability.[37] Fuel prices describe differences in the potential for biomass. Onshore wind, offshore wind and PV technologies are each constrained by an individual potential curve, as given by Eq. (6.38), with $\tilde{\mu}_{i,r,\mu}$ denoting the size of the particular potential segment. The two available CSP technologies, base plant and storage extension, draw upon the same respective potential curve, as given by Eq. (6.39). Equation (6.40) then implements the same mechanism for new and replacement hydro stations. Ideally the model would also consider the correlation of CSP and PV potential curve segments, since many high productivity sites of the one technology will also be high productivity sites of the other. Yet, this correlation has been neglected for several reasons. First, the data for PV and CSP potentials used are from different sources and are thus not congruent in terms of segment sizes. Moreover, even if segmentation were congruent, the degree of correlation is unknown. Finally, the impact of this omission on model results ought to be minor, since regions, where favorable solar potential exists at all, are generally also characterized by comparatively vast amounts thereof.

$$K^{op}_{i,r,t,\mu} \leq \tilde{\mu}_{i,r,\mu} \qquad \forall\, i \in ir_i \neg\, csp,\ hyd,\ bio \tag{6.38}$$

$$K^{op}_{csp\_bas,r,t,\mu} + K^{op}_{csp\_sto,r,t,\mu} \leq \tilde{\mu}_{csp,r,\mu} \tag{6.39}$$

$$K^{op}_{hyd\_new,r,t,\mu} + K^{op}_{hyd\_rep,r,t,\mu} \leq \tilde{\mu}_{hyd,r,\mu} \tag{6.40}$$

Analogously, the restriction for the amount of biofuel available annually at a certain price is given by Eq. (6.41).

[37] Expressed as share of nameplate capacity.

$$\sum_{\lambda} U_{f,r,t,\lambda,\phi} \leq \tilde{\phi}_{f,r,\phi,t} \tag{6.41}$$

In light of nuclear's role as carbon-free alternative to renewable technologies, uranium resource depletion is represented within the model. The available amount of nuclear fuel[38] at different price levels is given by $\tilde{\phi}_{f,\phi}$. The latter includes all uranium reserves and resources according to BGR (2009). Moreover, nuclear fuel from reprocessing[39] is also considered as option for extending the total available amount of nuclear fuel.[40] It is likely that Uranium resources beyond the ones considered in $\tilde{\phi}_{f,\phi}$, at costs above the customary cutoff value of 130 $/kg, exist. However, since an assumption on these additional resources would be arbitrary, only the published figures and additional resources from reprocessing have been considered.

In contrast to biofuel, nuclear fuel resources in price segments ϕ are depleted over time, as given by Eq. (6.42). $\hat{U}_{f,r,t,\phi}$ denotes the amount of fuel depleted in a particular segment ϕ. Equation (6.43) ensures that only nuclear fuel from segments that have not yet been depleted can be used. Equation (6.44) prevents full depletion of all uranium resources exactly at the end of the model horizon. This is unrealistic, as it would imply that existing nuclear plants in 2100 could no longer be used for power generation due to a lack of fuel. Instead, uranium resources have to be depleted in a manner so that the remaining resources in 2100 will suffice to operate existing plants for a specified duration d_i^{res} beyond the model horizon.[41]

$$\hat{U}_{uo2,t,\phi} = \sum_{r}\sum_{\lambda} U_{uo2,r,t,\lambda,\phi} \frac{\tau_t + \tau_{t+1}}{2} \tag{6.42}$$

$$\sum_{r}\sum_{\lambda} U_{uo2,r,t,\lambda,\phi} \leq \tilde{\phi}_{uo2,\phi} - \hat{U}_{uo2,t,\phi} \tag{6.43}$$

$$\sum_{r} \frac{g_{nuc,uo2}\, d_{nuc}^{res}\, K_{nuc,r,2100}^{iins}\, \bar{\nu}_{nuc} \cdot 8,760h}{\eta_{nuc}^{con}\, h_f \cdot 10^9} \leq \sum_{\phi} \left(\tilde{\phi}_{uo2,\phi} - \hat{U}_{uo2,t,\phi}\right) \tag{6.44}$$

[38] Only light water reactor technology with fuel based on U-235 is considered.
[39] Production of mixed oxide fuel (MOX) from nuclear waste.
[40] The maximum possible production of MOX is restricted to 30% of total reserves and resources.
[41] $K_{i,r,t}^{iins}$ is used here instead of $K_{i,r,t}^{op}$ to avoid "game-ending" nuclear capacity retirement, which would defy the intended purpose of the constraint.

6.2.6 Cost Components

This section covers the calculation of costs represented in the objective function, given in Eq. (6.1). The first subsection describes calculation of capital costs for technologies without endogenous cost reductions, the second one capital costs for technologies with endogenous cost reductions. Technologies with endogenous cost reductions are described by set $IL(I)$ with elements $il(i) = \{gas\_oc,\ gas\_cc,\ win\_on,\ win\_of,\ pv,\ csp\_bas,\ csp\_sto,\ bio\}$. The last subsection then covers fuel, non-fuel operating, decommissioning and trans-regional transmission costs.

Capital Costs for Non-learning Technologies

Instead of fully being accounted for in the year of investment, capital costs are represented in form of annuities. This approach has two important advantages. First, it allows for explicit representation of only few years with unequal intervals from one model year to the next instead of having to model all years across the model horizon. Second, it avoids unrealistic model behavior towards the end of the model horizon, also referred to as "game-ending behavior."[42]

For technologies without endogenous cost reduction, capital cost annuities are allocated according to Eqs. (6.45) through (6.47). At the time of capacity addition, the applicable annuity is allocated to all model periods within the plant's useful life by means of variable $\dot{C}^{ainv}_{i,t,t'}$, as shown in Eq. (6.45).

$$\dot{C}^{ainv}_{i,t,t'} = \frac{c^{inv}_i r\,(1+r)^{\pi_i}}{(1+r)^{\pi_i} - 1} \sum_r \sum_\mu X_{i,r,t',\mu} \qquad \forall\, \hat{\tau}_t - \hat{\tau}_{t'} < \pi_i,\ i \notin IL(I) \quad (6.45)$$

Capital cost annuities depend on the specific capital costs including construction financing charges c^{inv}_i, which in turn are a function of overnight capital costs c^{ovn}_i, construction duration d^{cst}_i, and the opportunity costs of capital r. For the calculation of construction financing costs a linear buildup of value over the construction period is assumed. Those costs are accounted for at their future value at the time of plant commissioning. The calculation of specific investment costs c^{inv}_i, including construction financing, is given by Eq. (6.46). Set $J = \{1, 2, \ldots, 10\}$ contains the number of years required for construction.[43] The factor required to discount the construction financing

[42] In case all capital costs apply in the year of capacity addition, the model may lean towards low-capital, high operating cost technologies at the end of the model horizon. This occurs, since investment into higher capital-cost technologies, although optimal from a going-concern perspective, would increase total system costs by more than low-capital cost technologies, whose comparatively high operating costs partially lie outside the model horizon.

[43] Annual compounding of interest is assumed.

costs back to the beginning of the construction period[44] is ψ_j^{con}. Parameter j_j^{cur} indicates the construction year in the summation procedure.[45]

$$c_i^{inv} = c_i^{ovn} \left[1 + r(1+r)^{d_i^{cst}} \cdot \sum_j \left(\frac{2j_j^{cur} - 1}{2d_i^{cst}} \psi_j^{con} \right) \right] \qquad \forall\, j_j^{cur} \leq d_i^{cst} \quad (6.46)$$

The capital costs entering the objective function $C_{i,t}^{ainv}$ are then given by Eq. (6.47).

$$C_{i,t}^{ainv} = \frac{c_i^{inv} r\,(1+r)^{\pi_i}}{(1+r)^{\pi_i} - 1} \sum_r \sum_\mu X_{i,r,t,\mu} + \sum_{t'=t_0}^{t-1} \dot{C}_{i,t,t'}^{ainv} \qquad \forall i \notin IL(I) \quad (6.47)$$

Capital Costs for Learning Technologies

To endogenize technical change in CORE, the experience curve approach[46] is used (see chap. 4.4.1). Specific cost reductions are consequently a function of the endogenous variable cumulated manufactured capacity, $\hat{\kappa}$. The endogenous cost reduction mechanism is implemented for all renewable technologies, apart from hydro, and for open- and combined cycle gas turbines.

The issue of non-convexity is addressed by replacing the concave with a piecewise linear objective function (see chap. 4.4.2), implying a Mixed-Integer program setup. Although retaining non-convexity would be desirable from a computational burden perspective, CORE's complexity would make identification of the global optimum under such a setup highly unlikely. As customary since the first implementations within large-scale energy system models (Mattsson and Wene, 1997; Messner, 1997), experience curves are endogenized by using a linear approximation of the cumulative capital cost function in conjunction with binary variables indicating the active linear segment. The approach followed for CORE is largely based on the one described in Barreto (2001) and Barreto and Kypreos (2002), which in turn are based on the one presented in Mattsson (1997). In contrast to those earlier implementations, the approach in CORE uses annualized capital costs to avoid game-ending behavior. Moreover, *floor costs* are introduced to prevent unrealistically high cost reductions for technologies with comparatively high learning rates, here in particular PV.

First, implementation of the linearized cumulative cost curves within CORE is described in Eqs. (6.48) through (6.56) — these are genuine model

[44] This value is subsequently increased by factor $(1+r)^{d_i^{cst}}$ to obtain the (future) value of construction costs at the time of plant commissioning.

[45] Its numerical values correspond to the ones in set J.

[46] Often also referred to as *learning-by-doing* in literature, although cost reductions are, strictly speaking, not only attributable to learning effects.

equations. Then, beginning with Eq. (6.57), the procedure employed to obtain the linear approximation of the nonlinear cumulative cost functions is discussed. As this process is used to derive parameters, the equations describing it are not model equations in a narrower sense. The approximated cost curves are divided into five segments, contained in set χ. A higher number of segments would be desirable[47] — yet an additional segment would increase computational times in a highly disproportionate manner and test runs with more experience curve segments have shown only minor differences in results.[48]

Cumulative manufactured capacity to date $\hat{\kappa}$ is the independent variable of the cumulative cost function. Cumulative capacity in a particular segment of the approximated cumulative cost curve, $\hat{\kappa}^{seg}_{i,t,\chi}$, describes the entire cumulative capacity to date — not only the cumulative capacity added within the currently active cost curve segment.[49] Equation (6.48) relates $\hat{\kappa}^{seg}_{i,t,\chi}$ to the investments taking place within the model.

$$\sum_{\chi} \hat{\kappa}^{seg}_{i,t,\chi} = \sum_{\chi} \hat{\kappa}^{seg}_{i,t-1,\chi} + \sum_{r} \sum_{\mu} X_{i,r,t,\mu} \qquad \forall i \in IL(I) \qquad (6.48)$$

Total cumulative costs $\hat{C}^{tinv}_{i,t}$, which is the dependent variable of the cumulative cost curve, are calculated based on the marginal specific capital costs of the active cumulative cost curve segment, $\alpha_{i,\chi}$. Furthermore, parameter $\beta_{i,\chi}$ is used to reflect costs from segments preceding the active one, which are characterized by higher specific capital costs. This adjustment term $\beta_{i,\chi}$ is controlled by means of binary variable $A_{i,t,\chi}$, which indicates the active segment of the cumulative cost curve.

$$\hat{C}^{tinv}_{i,t} = \sum_{\chi} \left(\alpha_{i,\chi} \cdot \hat{\kappa}^{seg}_{i,t,\chi} + \beta_{i,\chi} \cdot A_{i,t,\chi} \right) \qquad \forall i \in IL(I) \qquad (6.49)$$

Investment costs for a particular model year $C^{tinv}_{i,t}$ are then calculated as difference between cumulative costs of the current and the previous period, as shown in Eq. (6.50).

$$C^{tinv}_{i,t} = \hat{C}^{tinv}_{i,t} - \hat{C}^{tinv}_{i,t-1} \qquad \forall i \in IL(I) \qquad (6.50)$$

[47] Several applications use six segments.

[48] Applications with more segments generally apply experience curves to fewer technologies and/or are characterized by smaller linear programs, which have to be solved at each node of the branch & cut solution algorithm.

[49] Yet only for the currently active segment. For non-active segments, the value is zero.

In Eqs. (6.51) and (6.52) transformation into annuities takes place, following the same approach as for technologies without endogenous technological change, where costs are allocated to all periods within the useful life of the plant by means of double indexation.

$$\dot{C}^{ainv}_{i,t,t'} = C^{tinv}_{i,t'} \frac{r\,(1+r)^{\pi_i}}{(1+r)^{\pi_i} - 1} \qquad \forall\, \hat{\tau}_t - \hat{\tau}_{t'} < \pi_i,\; i \in IL(I) \quad (6.51)$$

$$C^{ainv}_{i,t} = C^{tinv}_{i,t} \frac{r\,(1+r)^{\pi_i}}{(1+r)^{\pi_i} - 1} + \sum_{t'=t_0}^{t-1} \dot{C}^{ainv}_{i,t,t'} \qquad \forall\, i \in IL(I) \quad (6.52)$$

$A_{i,t,\chi}$ indicates the currently active segment of the cumulative cost curve with a value of one. Equations (6.53) through (6.55) are used to implement this property, where the parameter $\tilde{\chi}_{i,\chi}$ represents the cumulative capacity at the upper border of a particular segment of cumulative cost curve.

$$\hat{\kappa}^{seg}_{i,t,\chi} \geq \tilde{\chi}_{i,\chi-1} \cdot A_{i,t,\chi} \qquad \forall\, i \in IL(I) \quad (6.53)$$

$$\hat{\kappa}^{seg}_{i,t,\chi} \leq \tilde{\chi}_{i,\chi} \cdot A_{i,t,\chi} \qquad \forall\, i \in IL(I) \quad (6.54)$$

$$\sum_{\chi} A_{i,t,\chi} = 1 \qquad \forall\, i \in IL(I) \quad (6.55)$$

Due to the presence of the binary variable $A_{i,t,\chi}$, CORE falls into the class of mixed integer programs. Solving this type of problem requires considerably more computational resources than a linear or continuous nonlinear program would. To mitigate the computational burden, the constraint in Eq. (6.56) is used. It prevents the solver algorithm from trying nonzero values for $A_{i,t,\chi}$ in segments χ preceding the active one — this is meaningful since cumulative manufactured capacity cannot ever decrease. Parameter χ^{act} denotes the numerical value of the currently active cost curve segment.

$$A_{i,t,\chi} = 0 \qquad \forall\, i \in IL(I),\; \chi < \chi^{act} \quad (6.56)$$

The nonlinear and concave cumulative cost function is approximated by a limited number of linear segments, as shown in Fig. 6.6. Since these segments lie beneath the original curve, the approximation will yield lower[50] costs than the nonlinear function.[51] The following equations describe the procedure, which ultimately produces parameters $\tilde{\chi}_{i,\chi}$, $\alpha_{i,\chi}$ and $\beta_{i,\chi}$ starting from the initial, nonlinear function.

[50] or, at most, equal costs.

[51] Defining the linear segments in a way so that they would intersect the nonlinear function at segment midpoint is not meaningful. In this case, there would be a discontinuity in cumulative costs at segment endpoints and cumulative costs of the next higher segment would initially be lower than cumulative costs at the end of the previous one.

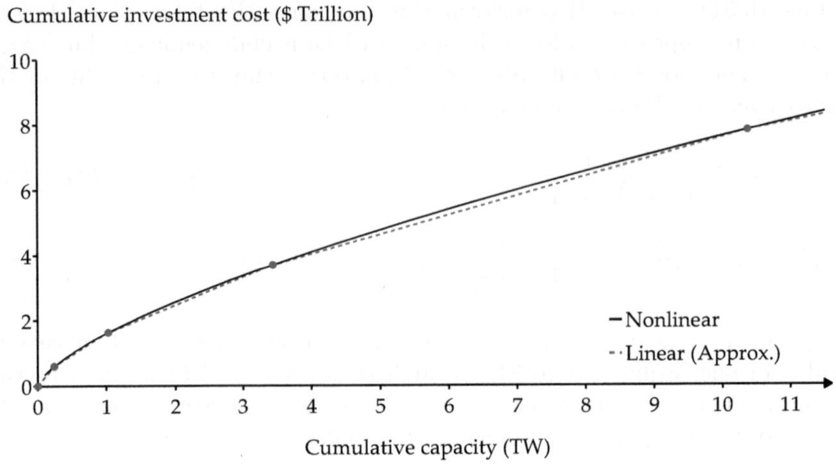

Fig. 6.6 Illustrative cumulative capital costs for a learning technology
Source: Own calculations

(1) The learning rate l_i needs to be transformed into the learning index γ_i, as shown in Eq. (6.57). Moreover, the costs of the *first unit of capacity* c^{1st} is calculated based on specific investment costs and cumulative capacity at model start, as given by Eq. (6.58).[52]

$$\gamma_i = -\log_2(1 - l_i) \tag{6.57}$$

$$c_i^{1st} = \frac{c_i^{inv}}{(\hat{\kappa}_{i,t_0})^{-\gamma_i}} \tag{6.58}$$

(2) Based on those values, the cumulative investment costs at the beginning of the model, $\hat{c}_{i,t_0}^{tinv}$, and at the assumed maximal cumulative capacity, $\hat{c}_i^{tinv,max}$, can be obtained, as shown in Eqs. (6.59) and (6.60). The assumption for maximum cumulative capacity $\hat{\kappa}_i^{max}$ is required as endpoint for the last segment of the approximation. In general, a lower $\hat{\kappa}_i^{max}$ is preferable to a higher one in order to minimize the approximation error. However, too low values for $\hat{\kappa}_i^{max}$ may severely distort model results. Capacity additions beyond $\hat{\kappa}_i^{max}$, although potentially optimal under the nonlinear learning curve or given a higher $\hat{\kappa}_i^{max}$, would be prevented for a too low $\hat{\kappa}_i^{max}$. Taken to the extreme, this could prevent investment into an initially uncompetitive technology altogether.

52 The following equations could also be stated in terms of the current investment costs at model start c_i^{inv}.

$$\hat{c}_{i,t_0}^{tinv} = \frac{c_i^{1st}}{1-\gamma_i} \left(\hat{\kappa}_{i,t_0}\right)^{(1-\gamma_i)} \tag{6.59}$$

$$\hat{c}_i^{tinv,max} = \frac{c_i^{1st}}{1-\gamma_i} \left(\hat{\kappa}_i^{max}\right)^{(1-\gamma_i)} \tag{6.60}$$

(3) Different segmentation procedures are applied, depending on whether a technology is capable of reaching floor costs before the maximum capacity level is reached. In other words, a technology is capable of reaching floor costs if $\hat{\kappa}_i^{flo}$ is smaller than the assumed maximum capacity $\hat{\kappa}_i^{max}$. The values of $\hat{\kappa}_i^{flo}$ and the cumulative investment costs up until the point of reaching floor costs $\hat{c}_i^{tinv,ulf}$ are required for this case differentiation and given by Eqs. (6.61) and (6.62).

$$\hat{\kappa}_i^{flo} = \sqrt[\gamma_i]{\frac{c_i^{1st}}{c_i^{flo}}}, \quad \hat{\kappa}_i^{flo} \geq 0 \tag{6.61}$$

$$\hat{c}_i^{tinv,ulf} = \frac{c_i^{1st}}{1-\gamma_i} \left(\hat{\kappa}_i^{flo}\right)^{(1-\gamma_i)} \tag{6.62}$$

The cumulative investment costs at model start and at the maximum cumulative capacity indicate the endpoints of the experience curve approximation. Generally, segments are divided in such a manner that the next segment contains double the investment costs of the preceding segment.[53] (3a) If floor costs are not reached before $\hat{\kappa}_i^{max}$, the costs contained in each segment $c_{i,\chi}^{tinv,seg}$ are calculated as given by Eqs. (6.63) and (6.64). Parameters χ^{1st}, χ^{las} and χ denote the numerical value of the first, i.e. 1, the last, i.e. 5, and the reference cumulative cost curve segment, respectively. χ should not be confused with χ^{act} from further above. The latter indicates the currently active segment during a solution process, the first, used in the following, merely the reference segment during a process of transforming parameters.

$$c_{i,\chi}^{tinv,seg} = \frac{\frac{1}{2^{(\chi^{las}-\chi)}}\left(\hat{c}_i^{tinv,max} - \hat{c}_{i,t_0}^{tinv}\right)}{\sum\limits_{\chi'} \frac{1}{2^{(\chi^{las}-\chi')}}} + \hat{c}_{i,t_0}^{tinv} \forall i \,|\, \hat{\kappa}_i^{flo} \geq \hat{\kappa}_i^{max} \wedge \chi = \chi^{1st} \tag{6.63}$$

$$c_{i,\chi}^{tinv,seg} = \frac{\frac{1}{2^{(\chi^{las}-\chi)}}\left(\hat{c}_i^{tinv,max} - \hat{c}_{i,t_0}^{tinv}\right)}{\sum\limits_{\chi'} \frac{1}{2^{(\chi^{las}-\chi')}}} \quad \forall i \,|\, \hat{\kappa}_i^{flo} \geq \hat{\kappa}_i^{max} \wedge \chi > \chi^{1st} \tag{6.64}$$

[53] As in Barreto (2001) but other segmentation rules would also be possible.

(3b) In case floor costs are reached within the approximation range, the last segment covers the entire range from the cumulative capacity, at which floor costs are reached, $\hat{\kappa}_i^{flo}$, to the maximum capacity $\hat{\kappa}_i^{max}$. This property is implemented as given by Eq. (6.65).

$$c_{i,\chi}^{tinv,seg} = c_i^{flo}\left(\hat{\kappa}^{max} - \hat{\kappa}^{flo}\right) \qquad \forall\, i \,|\, \hat{\kappa}_i^{flo} < \hat{\kappa}_i^{max} \wedge \chi = \chi^{las} \tag{6.65}$$

The segments before the last one, are divided according to the rule that the next segment contains double the investment costs of the previous one, as given by Eqs. (6.66) and (6.67). The main difference of the latter Equations to (6.63) and (6.64) is the use of $\hat{c}_i^{tinv,ulf}$ instead of $\hat{c}_i^{tinv,max}$.

$$c_{i,\chi}^{tinv,seg} = \frac{\frac{1}{2(\chi^{las}-\chi)}\left(\hat{c}_i^{tinv,ulf} - \hat{c}_{i,to}^{tinv}\right)}{\sum\limits_{\chi'=\chi^{1st}}^{\chi^{las}-1} \frac{1}{2(\chi^{las}-\chi')}} + \hat{c}_{i,to}^{tinv}$$
$$\forall\, i \,|\, \hat{\kappa}_i^{flo} < \hat{\kappa}_i^{max} \wedge \chi = \chi^{1st} \tag{6.66}$$

$$c_{i,\chi}^{tinv,seg} = \frac{\frac{1}{2(\chi^{las}-\chi)}\left(\hat{c}_i^{tinv,ulf} - \hat{c}_{i,to}^{tinv}\right)}{\sum\limits_{\chi'=\chi^{1st}}^{\chi^{las}-1} \frac{1}{2(\chi^{las}-\chi')}}$$
$$\forall\, i \,|\, \hat{\kappa}_i^{flo} < \hat{\kappa}_i^{max} \wedge \chi^{1st} < \chi < \chi^{las} \tag{6.67}$$

(5) After the segmentation has been carried out, the costs contained in the particular segments are cumulated, yielding $\hat{c}_{i,\chi}^{tinv,seg}$, as shown in Eq. (6.68).

$$\hat{c}_{i,\chi}^{tinv,seg} = \sum_{\chi=\chi^{1st}}^{\chi} \hat{c}_{i,\chi-1}^{tinv,seg} + c_{i,\chi}^{tinv,seg} \tag{6.68}$$

(6) This information is then used to derive the cumulative capacity values corresponding to each segment's upper border, as shown in Eqs. (6.69) and (6.70).

$$\tilde{\chi}_{i,\chi} = \left(\frac{1-\gamma_i}{c_i^{1st}\cdot \hat{c}_{i,\chi}^{tinv,seg}}\right)^{\frac{1}{(1-\gamma_i)}} \qquad \forall\, \chi < \chi^{las} \tag{6.69}$$

$$\tilde{\chi}_{i,\chi} = \hat{\kappa}^{max} \qquad \forall\, \chi = \chi^{las} \tag{6.70}$$

(7) Finally, slope $\alpha_{i,\chi}$, i.e. marginal investment costs, and the y-intercept $\beta_{i,\chi}$, describing cumulative investment costs as function of cumulative capacity, can be obtained, as shown in Eqs. (6.71) and (6.72). These operations conclude the approximation procedure.

$$\alpha_{i,\chi} = \frac{\hat{c}^{tinv,seg}_{i,\chi} - \hat{c}^{tinv,seg}_{i,\chi-1}}{\tilde{\chi}_{i,\chi} - \tilde{\chi}_{i,\chi-1}} \tag{6.71}$$
$$\beta_{i,\chi} = \hat{c}^{tinv,seg}_{i,\chi-1} - \alpha_{i,\chi} \cdot \tilde{\chi}_{i,\chi-1} \tag{6.72}$$

Operating and Decommissioning Costs

Calculation of costs in this category is generally straightforward: an activity level variable is multiplied by a specific cost coefficient.[54] Fossil fuel prices $c^{fue}_{f,r,t}$ are exogenous to the model, whereas consumption of nuclear and biofuel has model-endogenous implications on fuel costs $\tilde{c}^{fue}_{f,\phi}$.[55] Equation (6.73) describes the calculation of costs for fossil fuels, (6.74) the calculation for non-fossil ones, i.e. nuclear and biomass. Subset $FF(F)$, containing all fossil fuels, is used for case differentiation here.[56]

$$C^{fue}_{f,r,t,\lambda} = U_{f,r,t,\lambda} \cdot c^{fue}_{f,r,t} \qquad \forall f \in FF(F) \tag{6.73}$$
$$C^{fue}_{f,r,t,\lambda,\phi} = U_{f,r,t,\lambda,\phi} \cdot \tilde{c}^{fue}_{f,\phi} \qquad \forall f \notin FF(F) \tag{6.74}$$

Non-fuel operating costs $C^{op}_{i,r,t}$ are modeled as if entirely fix, i.e. driven by the operational capacity base, and described by Eq. (6.75). Splitting operating costs into a fixed and a variable portion leads to substantially longer solution runs with virtually no changes to results.[57]

$$C^{op,con}_{i,r,t} = \sum_{\mu} K^{op}_{i,r,t,\mu} \cdot c^{op,con}_{i} \tag{6.75}$$

As current operating costs for CSP plants are high in comparison with other renewable technologies (see Fig. 2.3) and the technology is still at an early implementation stage, operating cost reductions can be expected.[58] Not accounting for them would put the technology at a serious disadvantage to more mature technologies. Therefore, a share of CSP operating costs is given the opportunity for model-endogenous cost reductions, similar to the one for

[54] Exceptions are decommissioning costs and operating costs for solarthermal power plants.
[55] Set ϕ contains the different fuel price levels.
[56] Hard coal, lignite, oil, natural gas.
[57] Tested for the base case.
[58] Given sufficient investment into the technology.

capital costs.[59] This property is implemented by representing the *learning* CSP operating cost portion $c_i^{op,red}$ as share of capital costs, which are subject to endogenous cost reductions (see sec. 6.2.6). Applicable CSP operating costs in a particular period are then the sum of the non-learning $C_{i,r,t}^{op,con}$ and the learning $C_{i,t}^{op,red}$ operating cost portions, as given by Eqs. (6.75) and (6.76), respectively.[60]

$$C_{i,t}^{op,red} = C_{i,t}^{ainv} \frac{(1+r)^{\pi_i} - 1}{r\,(1+r)^{\pi_i}} \cdot c_i^{op,red} \tag{6.76}$$

Decommissioning costs apply at the time of capacity retirement. However, to align calculation of this cost type with the approach to model only specific years until 2100, decommissioning costs are discounted back to the time of plant commissioning and transformed into annuities as shown in Eq. (6.77). Due to their occurrence at the end of a plant's useful life, their value at the time of commissioning is almost negligible in comparison with capital costs. This even holds for nuclear power plants with their comparatively high specific decommissioning costs.

$$C_{i,r,t}^{dec} = \sum_{\mu} K_{i,r,t,\mu}^{iins} \frac{c_i^{dec} r\,(1+r)^{\pi_i}}{(1+r)^{\pi_i} - 1} \tag{6.77}$$

Trans-regional power transmission costs are calculated as given by Eq. (6.78). They also only have a relatively minor impact on model results.

$$C_{r,r',t,\lambda}^{tra} = E_{r,r',t,\lambda} \cdot c^{tra} \tag{6.78}$$

6.3 Critical Assumptions

CORE is a theoretical, fundamental model. Assumptions for the numerous parameter values are based on literature reviews instead of econometric estimations. This sections covers critical assumptions used in the CORE base

[59] Operating costs associated with the power block are not subject to cost reductions.

[60] This formulation effectively reduces the incentive to prematurely retire capacity, as the share of operating costs under consideration will be added across the plant's useful life despite premature retirement. In model tests without this specific formulation for CSP, early retirement of capacity added within the model horizon never proved optimal. It is therefore not expected that the formulation will alter optimal early retirement strategy. A formulation allowing for decreasing operating costs and at the same time not affecting early retirement incentives would be computationally more expensive.

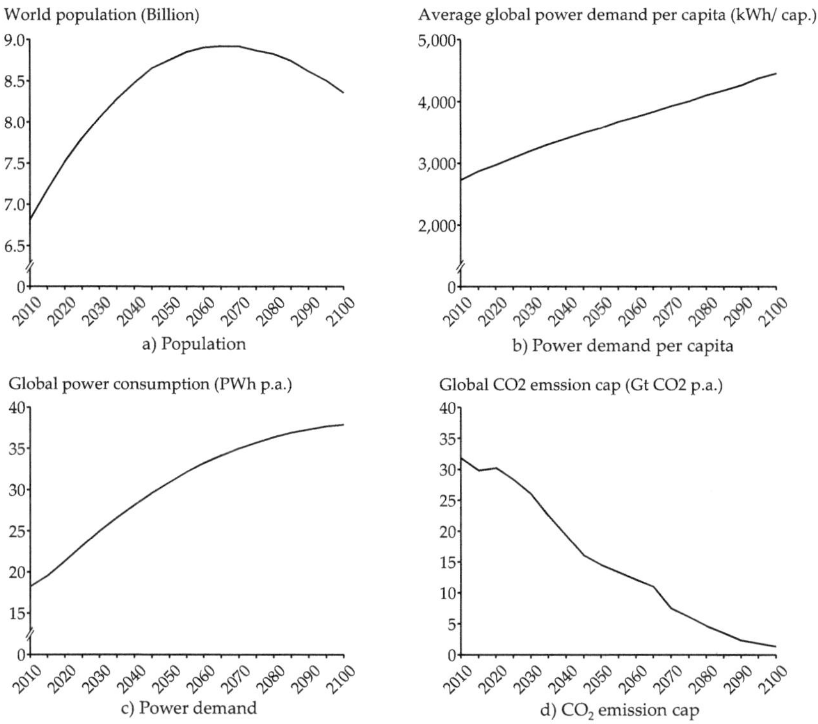

Fig. 6.7 Global power demand and carbon dioxide cap
Sources: Lutz et al. (2007), den Elzen et al. (2007), IEA (2009), own calculations

case, grouped into five main categories: Power demand, carbon cap, fuel prices, renewable potentials and technology properties.

A fundamental assumption, affecting both power demand and the carbon cap, is attitude towards climate change. As CORE is a normative model and the international community has agreed that global warming should be limited to 2°C (UNFCCC, 2010), all CORE assumptions must conform with this target. Attaining a 2°C warming target with reasonably high probability is generally associated with an atmospheric concentration of 450 ppm of CO_2-equivalents by 2100 (cf., for example, IPCC 2007; Clarke et al. 2009; Edenhofer et al. 2010).[61]

Future power demand for each of the eight model regions is calculated based on the expected development of population size and power consumption per capita. Population data is taken from IIASA's most recent forecast to 2100 (Lutz et al., 2007), power demand per capita data from the 450ppm

[61] *Overshooting* is assumed permissible, i.e. higher concentrations are allowed before 2100 as long as 450 ppm are reached by 2100.

scenario in IEA's World Energy Outlook (IEA, 2009).[62] As the per capita demand data is only available until 2030, its development has been linearly extrapolated to 2100 on a regional level.[63] A benefit from the approach using per capita demand from the IEA 450ppm instead of the reference scenario is that demand reductions due to the carbon constraint are already reflected. Therefore, the inelastic representation of demand in CORE could be retained and no assumptions on demand elasticity had to be made. Aggregated global population, demand per capita, and aggregated global demand figures are shown in Fig. 6.7.

The maximum CO_2 emission constraint must represent a reasonable trajectory to the 450ppm concentration target, which is to be reached by 2100. Several modeling exercises to identify such trajectories have been carried out in the recent past, for example, within the framework of the EMF-22 (cf. Clarke et al., 2009). Unsurprisingly, the set of existing trajectories is characterized by differences — yet most studies agree that peaking of emissions must occur in the very near future and that emissions by 2100 can only be a fraction of today's.[64] To avoid extremes, a path around the average of the reviewed studies is chosen. Until 2050, the emission trajectory underlying IEA's World Energy Outlook (IEA, 2009) is used. This also has the advantage that demand and carbon cap data stem from the same source. Beyond 2050, a medium emission scenario[65] from den Elzen et al. (2007) is chosen. Figure 6.7d shows this global CO_2 emission bound used as basis for the analysis. As CORE does not comprise all economic sectors emitting greenhouse gases, an assumption has to be made on how much of the global, energy-related emission budget is available to the power sector. The historical share of power-related CO_2 emissions was 40% in 2007 (IEA, 2009). On the one hand, power consumption is likely to increase at a faster pace than energy consumption as a whole, implying a larger power share in the future. On the other hand, the electricity sector may provide abatement options at lower costs than other sectors, which would imply a decrease in emission budget share. These two effects illustrate that even determining whether the emission share of the power sector will increase or decrease is not straightforward. For this reason the future emission allowance of the power sector is left at its historical 40% share.

Fossil fuel prices are exogenous to CORE. Prices for hard coal, lignite, oil and natural gas until 2050 have been obtained using the fundamental, stochastic model PUFF (Walther, 2009). After 2050, the regional prices have been linearly extrapolated based on the absolute average annual price changes over the period from 2020–2050. The resulting price trajectories for a sample

[62] Assumptions were made to match the data to CORE's regional structure.

[63] The regional growth from 2025–2030, in absolute terms, is assumed to continue until 2100.

[64] Several studies even identify paths including negative emissions, i.e. net extraction of CO_2 from the atmosphere, in the last decades of the century.

[65] SRES B2.

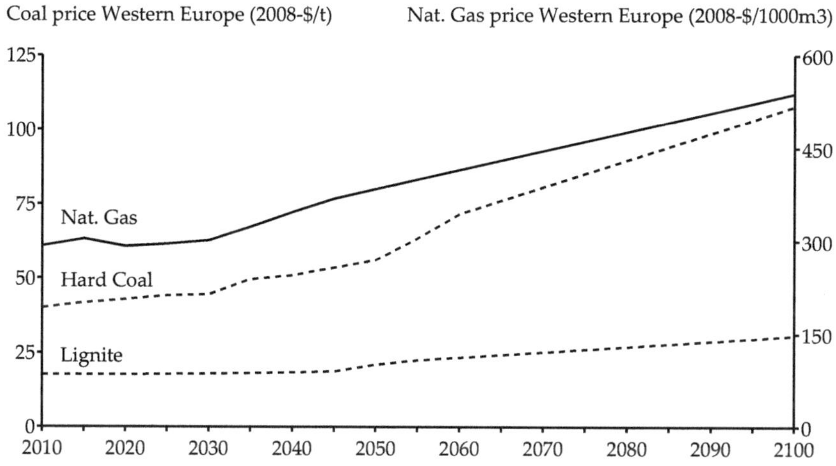

Fig. 6.8 Fossil fuel prices for Western Europe
Source: Walther (2009), own calculations

region, here Western Europe, are illustrated in Fig. 6.8.[66] Fuel prices differ substantially across regions.

Nuclear and biofuel prices depend on the amount of the particular fuel used. The supply curve for uranium at different prices levels is based on the reserve and resource information from BGR (2009). Biofuel assumptions are based on Hoogwijk (2004).[67] As the electricity sector is in competition with other sectors for low-cost biofuel, the share of total potential available to the power sector has been exogenously restricted to 20% across all model segments and regions, based on the historically observed value.

Potential data for renewable technologies has been compiled in a detailed literature review. Details on potentials for each covered technology can be found in Appendix A. Onshore wind is modeled with six potential segments[68] and is based on Hoogwijk et al. (2004), PV modeled with four potential segments and based on Hoogwijk (2004). Offshore wind potential is divided into

[66] Oil prices are not shown, since they play a minor role in CORE. Hard coal prices have floated around USD_{2008} 100 for Western Europe (Germany) in 2013, which lies above the assumptions in CORE. Based on those higher prices, the gas-coal substitution might be slightly accelerated compared to the results shown in the next section.

[67] Biofuel prices have been adjusted upwards — see Appendix A.5.3.

[68] For some regions with comparatively low potential only three segments are available.

six potential segments based on Krewitt et al. (2009).[69] CSP potential[70] is also divided into six segments based on Trieb et al. (2009) and hydro potential into four segments based on World Energy Council (2007) and World Energy Council (2009).[71]

Other assumptions, essentially related to plant specifications and costs, have also been generated in a detailed literature review (see Appendix A). CORE base case assumptions for fuel-based technologies are summarized in Table 6.4,[72] assumptions for renewable technologies except biomass in Table 6.5. Learning rates for the covered technologies are discussed in chapter 4.3. The assumed values largely correspond to the ones proposed for analysis of future cost developments in Neij (2008). A critical assumption related to experience curves is the one for floor cost.[73] Most technologies in CORE do not reach their floor cost in the base case. The notable exception is PV technology with its comparatively high learning rate of 20%, whereas all other technologies are characterized by 10% or less. The floor cost for PV are based on the Nemet's 2006 finding that even for addition of 30 TW of PV capacity, costs for PV modules would not drop below 0.63 \$/W.[74] As modules only account for a portion of PV generator capital costs, corresponding floor costs in CORE are set to 0.8 \$/W.

[69] Since the potential is not split into different availability classes, the relative split into wind classes for Western Europe presented by European Wind Energy Association (2009) has been applied to all regions as approximation.

[70] The large difference to PV potential cannot fully be explained by the comparatively higher efficiency of CSP but must also be due to methodological differences in the underlying studies. This at first glance large difference, however, does not substantially affect model results. In case PV and/or CSP are deployed, this deployment takes place in regions with high radiance, where typically not even the first potential segment is fully exploited for either technology.

[71] As the technical potential published for hydro is only divided into an economic and a technical potential group, the economic potential group has been further divided into three equal segments, where average availability, i.e. inflow, in the first segment is five percent above, availability in the third segment five percent below the one of the second segment.

[72] Assumptions for nuclear overnight investment costs were based on Olkiluoto unit 3 estimates after initial cost overruns had become public. To date, even further cost overruns are expected.

[73] Determining floor cost based on the state of technology today is a difficult, if not impossible task, as bottom-up assessments of floor cost are generally based on today's design of a technology. However, as future designs may be different, the reasons for minimal cost in today's design may no longer hold for the new design.

[74] Nemet employs an econometric model with seven explanatory variables for the historical cost reductions of PV.

Table 6.4 Specifications and costs for combustion and nuclear technologies

	H Coal	*Lign*	*GasOC*	*GasCC*	*Oil*	*Nuc*	*Bio*
Useful life (a)	35	35	30	30	30	50	30
Constr. duration (a)	3	4	1	2	2	7	2
Availability	93%	94%	92%	96%	92%	90%	90%
Thermal efficiency	46%	44%	39%	58%	43%	36%	33%*
Spec. CO_2 *(kg/*MWh_{el}*)*	717	909	513	345	651	-	-
Overn. inv. cost ($/kW)	1,700	1,900	450	700	1,600	4,000	2,700
Other op. cost ($/kWa)	70	80	20	30	40	110	90
Learning rate (%)	-	-	10%	10%	-	-	5%
Cumul. capacity (GW)	-	-	1,400	360	-	-	50

*Corresponds to used reference energy content of 4.7 MWh_{th}/t.

Table 6.5 Specifications and costs for renewable technologies

	Wind Ons	*Wind Off*	*PV*	*CSP Base*	*CSP Stor*	*Hyd New*	*Hyd Rep*
Useful life (a)	20	25	25	30	30	60	60
Constr. duration (a)	1	2	1	2	2	5	2
Storage	-	-	-	-	Yes	Yes	Yes
Overn. inv. cost ($/kW)	1,400	2,300	4,200	4,200	3,900	2,400	1,600
Operating cost ($/kWa)	45	80	40	110*	90*	55	55
Learning rate (%)	10%	10%	20%	10%	10%	-	-
Floor cost ($/kW)	600	850	800	1,000	800	-	-
Cumul. capacity (GW)	121	2.1	13.6	0.6	0.2	-	-
Max. tech. potential (PWh)	96	16	372	2,941		17	

*A portion of those costs (60 $/*kWa*) are subject to cost reductions.

6.4 Results

When interpreting model results and concluding on implications for decision makers it is important to bear the modeling approach in mind. Due to CORE's normative setup, the presented results describe how capacity and operating decisions *should* be made on a regional level to achieve a *global* optimum. Therefore, regional decisions may not be cost-optimal for the particular region but enable global cost-optimality instead.

In reality, investment and operating decisions are made by agents far below the global or regional level. As those agents aim to maximize profit and cannot appropriate all benefits from investment into still uncompetitive technologies with increasing returns, underinvestment into these technologies may occur (cf. Arrow, 1962). Governmental support schemes can then be established to resolve this type of market failure, promoting investments at the agent level. Under the normative approach followed here, governmental support schemes are not an input but an implicit output of the model. They should be designed in such a way that they enable implementation of the globally optimal strategy at the agent level. Today, a multitude of different, mostly uncoordinated support schemes exist to promote the short-term development of renewables across countries. These support schemes, or the absence thereof, will likely be the key driver of renewable power deployment in the near future, as most renewable technologies cannot yet compete with conventional ones. Therefore, CORE model results should under no circumstances be interpreted as forecast of the short-term renewable power deployment.

6.4.1 Base Case

The model results from the set of parameter assumptions considered as most likely are further referred to as *base case*. An important consideration when interpreting base case results is *penny-switching*. The expression refers to the model property that the set of activity levels, i.e. decisions, leading to the lowest objective function value are always preferred even if another set of activity levels with a marginally higher objective function value would be vastly more robust to changes in parameter values. Potentially different degrees of risk associated with the decisions are not considered. Especially in presence of endogenous technical change, a small change in parameter values may be decisive whether the large up-front investment to drive a technology to competitive cost levels should be undertaken or not. Therefore, the base case should not be regarded as *the* optimal future investment strategy into renewables but rather as a potentially optimal future strategy and as basis from which the impact of different parameters on optimal investment strategy can be studied (see sections 6.4.2 through 6.4.7). Base case results are described in two steps: Global, aggregated results followed by regional ones. A more detailed representation of results can be found in Appendix C.1.

Global Results

The optimal development of the global power generation mix under base case assumptions is shown in Fig. 6.9. Although some technologies appear to be on the decline from the start, such as hard coal, this may not be the case in absolute terms. Hard coal production, in fact, is increasing until 2020 despite a decreasing production share in the context of increasing power demand.[75] It should be noted that globally aggregated results are shown and that production mixes differ significantly across model regions — none of the regions is characterized by a technology mix mirroring the global one. In the following, several characteristics of the power sector transformation towards a higher renewable share are highlighted, as observed for base case parameterization.

(1) A large number of technologies are part of the optimal solution. Eleven of the 14 technologies available in CORE have reached cost-competitiveness

[75] Global power demand approximately doubles across the model horizon.

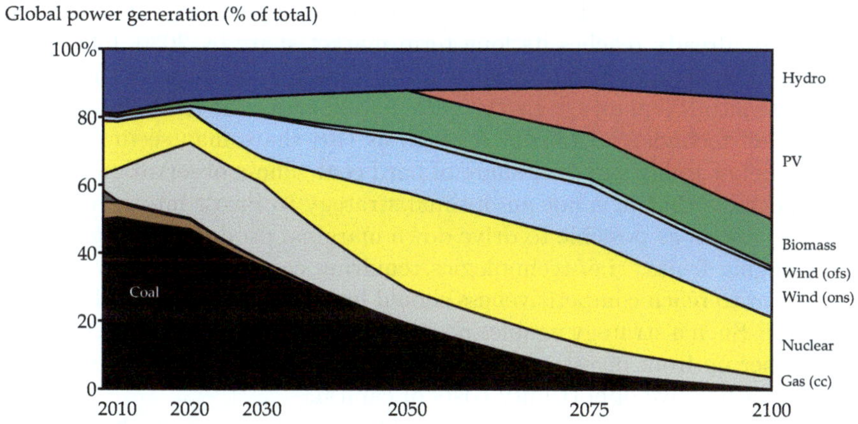

	2010	2020	2030	2050	2075	2100
Fossil	66	72	61	29	13	4
Nuclear	14	10	7	28	31	19
Renewable	21	18	32	42	56	79

Fig. 6.9 Global power generation 2008–2100 by technology
Source: Own calculations

by 2100.[76] This is particularly interesting as the presence of increasing returns from experience curves would favor dominance of one or at least few technologies (cf. Arthur, 1989), especially given several uncompetitive technologies at the beginning of the model period. The obtained results suggest that increasing returns must be sufficiently *bounded* to avoid technological dominance. An important constraint for renewable deployment is certainly the economically exploitable potential but other constraints, such as non-storability of power in conjunction with fluctuating demand and renewable supply, likely also play a role. It therefore appears that no backstop technology exists, at least not among the ones represented in the model, that can by itself resolve the challenge posed by the need to substantially reduce carbon emissions in an environment of rising demand.

(2) Although the generation mix is constantly moving towards higher renewable shares after 2020, a switching of "regime" can be observed around the middle of the century. Whereas the first half of the century is still characterized by low enough fossil fuel prices and high enough emission allowances to retain a fossil share of roughly two thirds, the second half is characterized by large share increases of renewable and nuclear technologies. Although there is already considerable renewable production in 2030, almost half thereof is still from hydro. 2075 then marks the model year, where more power is generated from renewable than from non-renewable technologies.

[76] The global power generation share of open cycle gas turbines and lignite plants in 2100 are too small to be visible on the chart; category hydro contains both new and replacement stations.

(3) The deployment of renewables takes place sequentially. Onshore wind technology already reaches its long-term market share by 2030, followed by biomass in 2050 and offshore wind in 2075. This year also marks the first investments into PV plants.[77] In 2100, PV finally arrives at the highest market share of all technologies (35%). As high as this share may seem at first, it still lies below today's market share of hard coal. These observations support the hypothesis that it is not an optimal strategy to invest into several technologies as soon as possible to drive down marginal production costs. Rather "low-hanging fruits," i.e. technologies requiring comparatively low up-front investment to reach competitiveness should be pursued first, which is onshore wind here. Such a strategy enables postponement of spending for technologies with higher up-front investment costs, ultimately reducing costs to society in presence of positive opportunity costs of capital.

(4) Global average generation costs and thus power prices are characterized by an initial decline, followed by an increase in the second half of the modeled period (see Fig. 6.10). This initial decline is mainly caused by the optimization of the regional legacy portfolios through premature retirement of redundant capacity. Examples include the decommissioning of oil plants but also of PV generators in regions with unfavorable solar regime to avoid non-fuel operating costs.[78] In the second half of the century, average system costs are increasing again. This development accompanies the buildup of renewable and nuclear technologies, which can generate power more cost effectively at this point than fossil-fuel based technologies.[79] Yet, an increase in system costs is logical, since otherwise the large-scale implementation of renewables, especially of PV, would already be optimal at an earlier point in time.

(5) The relative importance of different cost types changes over time (Fig. 6.10). Unsurprisingly, the share of fuel costs is lowest by the end of the century. This is mainly due to the comparatively low penetration of fossil technologies but also to the changed role of Biomass. The latter moves from base-load production in 2030 to mid-load in 2100. The relative share of capital costs initially decreases. This is largely driven by the addition of combined cycle gas turbine technology and end-of-life retirement of nuclear capacity, which is not directly replaced. From mid-century onwards, the share of capital costs increases again due to the increase in capital-intensive renewable and nuclear technologies. Cost reductions from learning, however, mitigate this increase to a certain extent. The share of non-fuel operating costs also increases from 2020 onwards. This is also driven by the share-gains of

[77] As will be shown in the next section, the term *solar* may be more accurate, since small parameter changes may lead to optimal deployment of CSP instead of PV technology.

[78] Non-fuel operating costs are assumed fix (see p. 159).

[79] For particular regions and load duration curve segments, after cost reductions have been experienced.

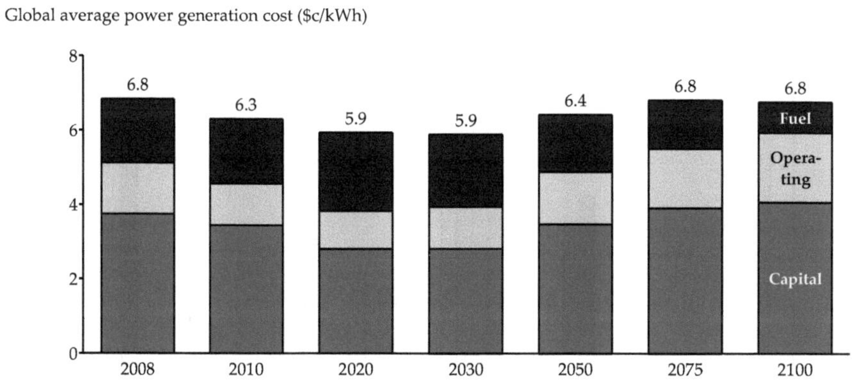

	2008	2010	2020	2030	2050	2075	2100
% Fuel	25	28	36	33	24	19	12
% Operating	20	18	17	19	22	23	28
% Capital	55	54	47	48	54	58	60

Fig. 6.10 Global average system generation costs by technology
Source: Own calculations

renewable and nuclear technologies. Yet, in contrast to capital costs, non-fuel operating costs are not subject to learning.[80]

(6) Although their positions in the merit-order remain the same, the roles of fossil and biofuel based technologies change over time (Fig. 6.11). In the initial model periods, combined cycle gas turbines operate in the mid-load segment but subsequently move towards peak. Biomass starts out in high mid-load, moves to baseload production in 2030 and subsequently reverts back to mid-load production in 2100. Coal moves from baseload to mid-load before its use is discontinued due to high specific emissions towards the end of the century. Over time renewables and nuclear with comparatively low marginal costs increase their share of power production. Graphically, this development can be imagined as a relative elongation of the flat, left-hand section of the merit-order curve and a corresponding shortening of its positively sloped, right-hand section. This positive slope becomes steeper over time.[81] The reduction in average full load hours is less pronounced for non-fuel technologies than for fuel-based ones.[82] The reduction in offshore wind full load hours can, for example, be attributed to the interaction of the differentiated renewable potentials with endogenous cost reductions. As costs are reduced further, previously uneconomic potential segments with lower productivity can be competitively exploited.

[80] With the exception of CSP technologies. Those, however, are not part of the optimal strategy in the base case.

[81] Assuming a continuous instead of a piecewise linear function.

[82] Initial output increases for wind from 2010 to 2030 are a result of the sub-optimal location of plants in the initial endowment.

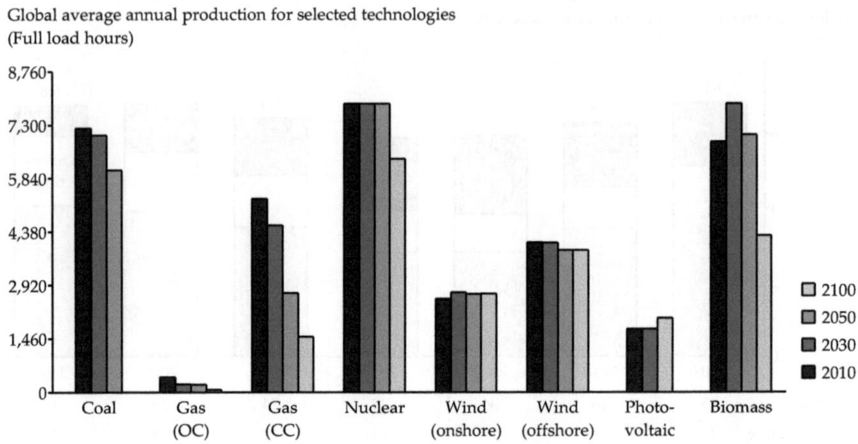

Fig. 6.11 Annual power production by technology
Source: Own calculations

(7) Levelized generation costs for fuel-based technologies increase over time (Fig. 6.12). Rising fuel prices[83] contribute to this increase in both a direct and an indirect manner. Indirectly they lead, in conjunction with other factors,[84] to lower full load hours. For coal, combined cycle gas and biomass technologies, the impact of this decrease in production on levelized generation costs is considerably larger than the direct impact from fossil fuel price increases. For non-fuel technologies, full load hours also, at most, remain constant due to expansion to sites with lower energy carrier availability. Yet, for wind and PV this effect is overcompensated by reductions in specific capital costs.[85] Generation costs for new hydro stations increase due to expansion to sites with lower output in absence of technological cost reduction opportunities.

Although the aggregated results presented in this section are well suited to provide an overview on the global result of regional investment strategies, they fail to expose the substantial differences among regional strategies, which jointly lead to global optimality.

Regional Results

In CORE, the world is divided into eight regions. Four mechanisms connect these otherwise isolated regions: Global experience curves with perfect

[83] For biomass: For higher output, biofuel potential segments characterized by higher costs must be exploited.

[84] For example, increasing CO_2 shadow prices or buildup of technologies with lower variable costs.

[85] It should be noted that offshore wind generation cost for 2010 are driven by comparatively optimistic capital cost assumptions based on early offshore implementations in shallower waters than typically found today (see also section 5.6 on this topic).

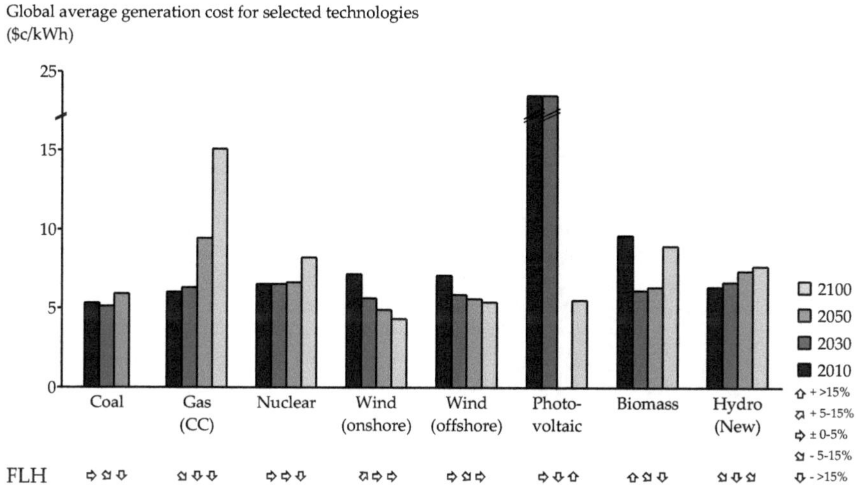

Fig. 6.12 Global average generation costs by technology
Source: Own calculations

regional learning spillover, a global carbon cap, a global pool of uranium resources and power transmission capacity among some of the regions.[86] Among others, regions differ with respect to power demand level and development, renewable potential and fuel prices. Despite the existence of regional interdependencies, the globally optimal strategy consists of very different regional technology portfolios — illustrated using sample years 2030 and 2100.

In 2030, when fuel prices are still comparatively low and the carbon emission budget comparatively large, fossil technologies dominate global power generation. Although either coal or gas technology constitutes the largest power producer in all but three regions, the global fossil dominance is mainly driven by the region Asia (Fig. 6.13). Asia is characterized by a fossil share of almost 80% and accounts for almost half of global power demand. An even higher fossil share exists in the Middle East region, mainly due to low regional gas prices. In contrast, four regions are already characterized by renewable shares above 50%. The dominating renewable technologies are onshore wind and hydro. Whereas substantial hydro capacity exists in almost all regions, large-scale wind implementations[87] are only found in three regions with favorable wind potential.

Fossil fuels have largely disappeared by 2100 (see Fig. 6.14) and the technology portfolio contains large shares of renewables. Yet the types of renewable technology implemented differ across regions. Large shares of PV can be

[86] Power transmission capacities are exogenously assigned and have a minor effect on results.
[87] Share above 20%.

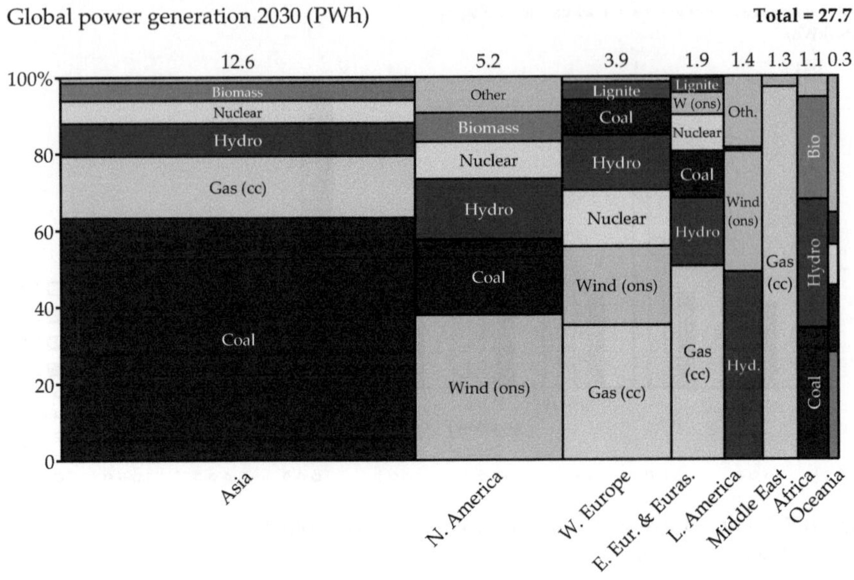

Fig. 6.13 Global power generation 2030 by region
Source: Own calculations

found in four regions but PV is not present at all in the remaining four.[88] Key driver behind these differences is the relative availability of high-productivity renewable potential.[89]

The optimal portfolio for 2100 includes large shares of intermittent renewable technologies. The example of North America in 2100 is used to illustrate the modeling of such a power system in CORE (Fig. 6.15).[90] It is clear that a power system with such high degrees of output fluctuation would also require significant investments into the transmission grid, which are not explicitly

[88] This result has implications for current German renewable policy. Today, feed-in tariffs are in place to promote the installation of photovoltaic generators in Germany. The regional CORE results, however, suggest that PV capacity addition in the model region Western Europe is not optimal for any of the model periods despite comparatively low capital costs, which are a result of large-scale deployment in other regions. The base case results therefore do not support the current policy of fostering PV diffusion in Germany.

[89] Different fossil fuel prices play a minor role here, since the stringent carbon cap severely limits opportunities for fossil power generation in 2100.

[90] Note that there is even renewable surplus capacity (wind, PV, hydro run-of-river) at virtually no marginal costs in load duration curve segment six, where highest renewable availability across the eight load segments is paired with a medium demand level. As the model does not feature ramp-up and ramp-down costs, nuclear production is decreased in load segments six and eight — this would presumably not occur in presence of ramp-up/ -down costs.

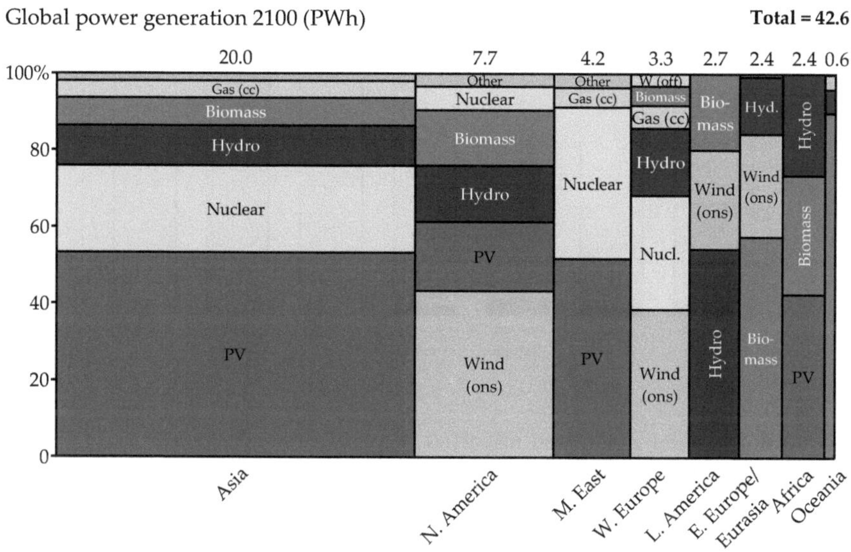

Fig. 6.14 Global power generation 2100 by region
Source: Own calculations

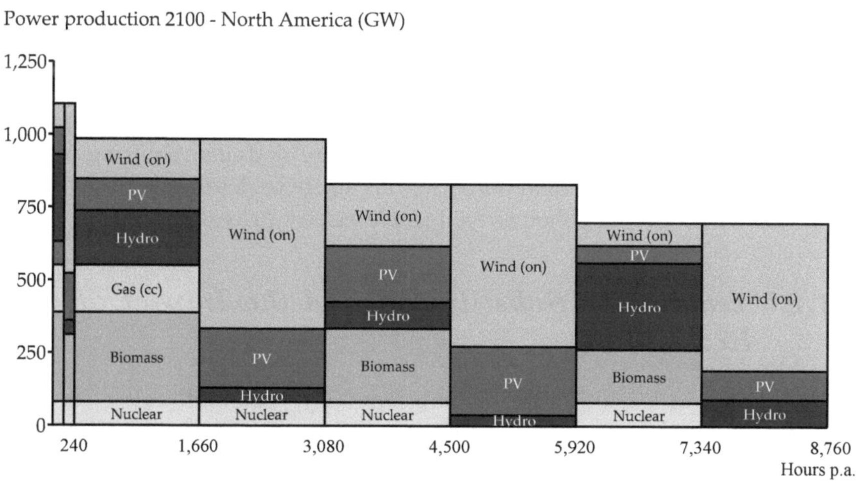

Fig. 6.15 Power generation 2100 — North America
Source: Own calculations

represented here. This issue, alongside a more detailed regional resolution to capture the usually significant distances between high-productivity renewable sites and centers of power demand, is left for further research.

Regions differ not only with respect to the optimal technology portfolio but also regarding average generation costs (Fig. 6.16). In 2100, costs in

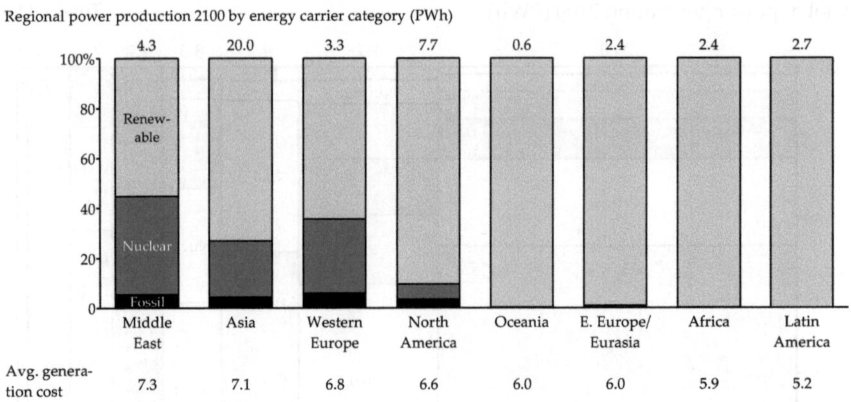

Fig. 6.16 Regional generation mix 2100 by primary energy carrier class
Source: Own calculations

the Middle East region lie 40% above the ones in Latin America.[91] The four regions with the highest costs are also characterized by higher demand and higher non-renewable shares than the four regions with lower costs. The observation further illustrates the importance of high-productivity renewable potential, enabling high shares of low-cost renewable power in the long term. The relative[92] abundance of high-productivity potential will therefore likely have a major impact on long-term power prices in a region.

In summary the regional base case results have highlighted the important role of the renewable potential. In future energy systems requiring high renewable penetrations, regional potentials are likely to determine both optimal regional technology portfolios as well as electricity prices.

6.4.2 Technology Substitution and the Impact of Cost Reductions

Small changes in learning rate assumptions may profoundly alter results of models with endogenous technical change. The easiest way of analyzing robustness of results is by means of sensitivity analysis.[93] The objective of the sensitivity analyses carried out in this section is twofold. On the one hand, minimal learning rates and/or floor costs required for large-scale implementation of initially uncompetitive technologies are identified. On the other hand, inferences are made on which technologies interact in a substituting

[91] A portion of those high costs are still artifacts of the up-front investment into PV before cost competitiveness has been reached.

[92] With respect to power demand.

[93] Uncertainty of learning rates is explicitly considered in an extension to CORE (see chap. 7).

manner.[94] For the latter objective an alternative to using learning-rate sensitivity would have been sensitivity to capital costs at the beginning of the model or average annual output. Nonetheless, learning rates are considered more relevant, since their "true" value is much less certain than the one of today's capital cost or plant output.

The endogenous technological change mechanism in CORE is not only affected by learning rates but also by floor costs and maximum cumulative capacities. To avoid the latter two interfering with learning rate sensitivity, they are also varied in line with learning rates — yet only at half of the rate applied to learning rates. For example, if the learning rate were doubled, floor costs (maximum cumulative capacity) would be decreased (increased) by 50%. Average[95] power production instead of figures for each period are used as dependent variable in the sensitivity runs to avoid time as third dimension. The results for changes in the learning rates of PV, wind, CSP and biomass are shown in Fig. 6.17.[96] The charts only show technologies, where considerable changes to power production can be observed. Put differently, technologies not displayed can be interpreted as being only marginally affected by the changes in the particular learning rate.

A general observation from all four sensitivity analyses is that optimal generation from fossil technologies is only marginally affected by changes in the learning rates of renewable technologies. This indicates a high robustness of the findings that those technologies still dominate the first half of the century whereas they are replaced by renewable technologies in the course of the second half. The key driver of this development is the existence of the global carbon cap. For large-scale PV deployment to be optimal, a minimum learning rate of approximately 16% and/or floor costs no higher than around 900 \$/kW[97] are required (Fig. 6.17a).[98] For PV learning rates below this minimum threshold, PV is substituted by CSP technology. Higher PV learning rates than in the base case[99] go in line with lower on- and offshore wind investments and, to a lesser extent, lower nuclear power production.

[94] The term substitution is used here with respect to the base case.

[95] Unweighted average across the seven model years.

[96] The category wind includes both on- and offshore wind, CSP both base plant and storage module.

[97] For simplicity, only learning rates are stated — this however, does not mean that the corresponding floor cost value does not affect the outcome of the sensitivity analysis under consideration.

[98] It may seem counterintuitive that the power production mix in Fig. 6.17 is not constant for any learning rate below the minimum rate required for implementation. This can be attributed to the model being set up as MIP with a 0.001% tolerance. For different PV learning rate values, even if there is no implementation, the branch & cut solution algorithm will follow different paths to finding a solution satisfying the tolerance. As it "picks" the first solution that fulfills the specified tolerance, this solution may not have the same optimal power mix as another one based on a different learning rate.

[99] Schaeffer et al. (2004), for example, report a rate of 23%.

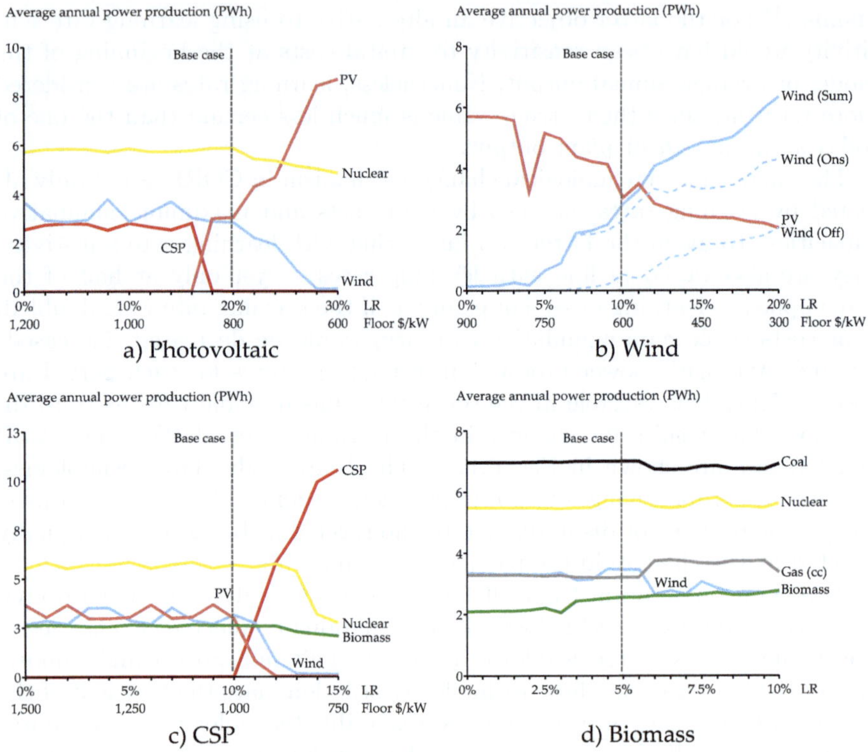

Fig. 6.17 Sensitivity to learning rate and floor costs
Source: Own calculations

Beyond a 28% learning rate, PV would completely replace wind. The author, however, is not aware of research finding such high learning rates for capital cost. These results indicate that the development of PV is not at all robust to changes in the learning rate. Although a large share of PV is reported for the base case, this share should rather be attributed to solar technology, as base case results "balance on a knife's edge" and slight changes in learning rates may tilt results towards either CSP or PV deployment.

The sensitivity analysis for CSP learning rates (Fig. 6.17c) confirms this finding. For a mere 1% increase in the learning rate of CSP and/or around 50 $/kW lower floor costs, CSP's share would be far higher than the one of PV, for a 2% increase PV would be substituted entirely. When learning rates for both technologies are varied simultaneously, this substituting behavior is confirmed (Tab. 6.6). Only given comparatively high learning rates for both technologies, PV as well as CSP technologies are deployed at large-scale. Although there are no changes to the optimal solution for CSP learning rates below the base case value of 10%, higher learning rates lead to a very different optimal strategy. Wind power is entirely replaced from a learning rate value of 14% upwards and even optimal nuclear deployment is significantly reduced,

Table 6.6 Average annual power production for CSP and PV

		CSP Base & Storage (PWh)					PV (PWh)				
PV LR		*CSP LR & floor cost ($/W)*					*CSP LR & floor cost ($/W)*				
& floor		*5%*	*7.5%*	*10%*	*12.5%*	*15%*	*5%*	*7.5%*	*10%*	*12.5%*	*15%*
cost		*1.34*	*1.21*	*1.07*	*0.94*	*0.80*	*1.34*	*1.21*	*1.07*	*0.94*	*0.80*
10%	*1.04*	-	1.1	2.8	5.9	10.5	0.8	-	-	-	-
15%	*0.93*	-	0.2	2.5	5.9	10.5	1.1	0.9	-	-	-
20%	*0.83*	-	-	-	5.9	10.5	3.0	3.1	3.7	-	-
25%	*0.72*	-	-	-	2.6	10.2	4.8	4.8	4.8	4.3	0.8
30%	*0.62*	-	-	-	-	7.3	8.7	8.7	8.7	8.7	4.3

Note: Floor cost include construction fin. charge; CSP floor cost for base plant.

mainly owed to CSP's storage capability.[100] A 14% learning rate, however, lies clearly above historically observed cost reductions and, in light of power block maturity, appears high from a real-world point of view.

The penetration of wind power is also sensitive to changes in learning rates/ floor costs (Fig. 6.17b) but to a lesser degree than penetration of solar technologies. A small amount of onshore wind power would even be deployed in the absence of learning, yet for offshore wind deployment a minimum learning rate of 9% in both onshore and offshore is required.[101] Wind appears to interact in a substituting manner with PV for both higher and lower wind learning rates than in the base case. Model results are comparatively robust to changes in the learning rate for biomass combustion technology (Fig. 6.17d). Even absence of cost reductions will not massively alter its implementation.

6.4.3 Value of Differentiated Renewable Potentials

To minimize generation costs, renewable plants should be installed at those locations with most abundant energy carrier availability first.[102] Yet, most models assume an average production level for renewable technologies across the entire technical potential, if a technical potential constraint is specified at all. In presence of endogenous cost reductions such an approach leads to an overstatement of the up-front investment required to drive down the costs of immature technologies to competitive levels. In extreme cases, this may prevent the implementation of an, in reality promising, technology altogether. For this reason marginal instead of average energy carrier availability should

[100] The higher the overall CSP share, the higher the share of storage extension becomes: In the 11% learning rate case there is virtually no storage whereas beginning from a 17% learning rate, CSP storage is higher than base plant capacity.

[101] Due to spillovers.

[102] Given similar costs across locations.

Table 6.7 Impact of renewable potential curves

	Potential segments	*2008*	*2010*	*2020*	*2030*	*2050*	*2075*	*2100*
No. of ren.	4–6	-	1	3	4	4	5	5
technologies	2	-	1	2	2	4	5	5
invested in	1	-	-	1	2	3	4	4
Share of	4–6	21	21	18	32	42	56	79
renewables (%	2	19	18	16	18	32	59	80
prod.)	1	15	15	9	8	29	61	81

Note: Both Hydro and CSP technologies counted as one technology, respectively.

ideally be used in energy system models. For reasons of data availability and to retain linearity, CORE uses approximated renewable supply curves instead of nonlinear ones.

To analyze the impact of using marginal instead of average plant availability, the model has been modified. One version uses an average plant availability across the entire technical potential, another version divides the technical potential into two segments with different availabilities. Results from these simplified models, as well as from the full model with 4–6 segments, are summarized in Table 6.7.

Results confirm the assertion that using average instead of marginal power output will substantially alter model results. Objective function values are higher the fewer segments are reflected. Moreover, both the number of technologies implemented as well as renewable power shares in the model versions with reduced segment numbers lag behind the ones from the base case.[103] In the one-segment setup, offshore wind technology is not pursued at all and onshore wind only plays a marginal role. This stands in sharp contrast to the base case, where onshore wind is the first of the new renewable technologies to be deployed at large scale based on its close-to-competitive costs at favorable locations. As differences in energy carrier availability are not considered in the one-segment setup, the up-front investment requirement for wind is clearly overstated and no implementation occurs.

Whereas the use of potential segmentation to represent different levels of supply vastly affects model results, this is not so much the case for differences in the size of these segments. When increasing each potential segment size up to double its values, model results were only moderately affected. Unsurprisingly wind, for which high-productivity potential is comparatively scarce in all regions, benefits most from larger segment sizes. PV, on the other hand, is only slightly affected since its potential is characterized by comparatively large, high-productivity segments in regions where deployment is at all

[103] Exceptions are the final years 2075 and 2100, where renewable shares of the cases with lower potential "overtake" the one from the original setup.

meaningful. In other words, if there is potential with sufficiently high irradiance in a region, there's usually a lot of it.

In summary, one finds that explicit representation of renewable power supply curves, even if only in an approximated manner is crucial when attempting to realistically assess optimal renewable deployment strategies in presence of endogenous technical change. A more detailed representation than the one with 4–6 segments in CORE could potentially further improve results.

6.4.4 Impact of Fluctuating Renewable Energy Carrier Availability

It has previously been asserted that the fluctuating availability of some renewable energy sources has a negative impact on the competitiveness of the corresponding conversion technologies (see chap. 2.2.3). This negative impact mainly stems from the need for backup capacity in times of low availability, the required strengthening of the grid and increased operational efforts to manage higher degrees of fluctuation in the power system. CORE takes the need for backup capacity into consideration when deriving optimal investment strategies, whereas the integration of the two latter effects have been left for further research.

The impact of input energy fluctuations[104] on model results has been investigated using sensitivity analysis (see Fig. 6.18). The parameter varied is the fluctuation factor $\omega_{i,\lambda}$ (see p. 140). Although the analysis has been carried out for a range of zero fluctuation to double the degree of base case fluctuation, only values up to 1.2 times the base case value are meaningful. This can be attributed to the fact that higher degrees of fluctuation would imply negative availability for some technologies and load segments under the approach taken.[105]

Results confirm the assertion from above: a higher degree of fluctuation generally leads to lower investments into the intermittent technologies PV and wind (Fig. 6.18).[106] For a theoretical absence of fluctuations, Wind and PV would be supplying baseload power. This explains, why considerably less nuclear power would be needed for low levels of fluctuation. In fact, for the case of zero fluctuation, nuclear power generation in 2100 would, in absolute terms, be lower than generation in 2008. Hydro and biomass power production also increases for higher levels of fluctuations, yet at far lower rates than the one observed for nuclear.

[104] Including both cyclical and stochastic fluctuations.

[105] As negative availability in CORE is automatically converted to zero, the implied average availability would increase for cases above 1.2 times base case fluctuation. This is the reason for the increase in production from PV beyond a fluctuation factor multiple of 1.2 in Fig. 6.18.

[106] CSP has not been tested due to its absence in the optimal base case strategy.

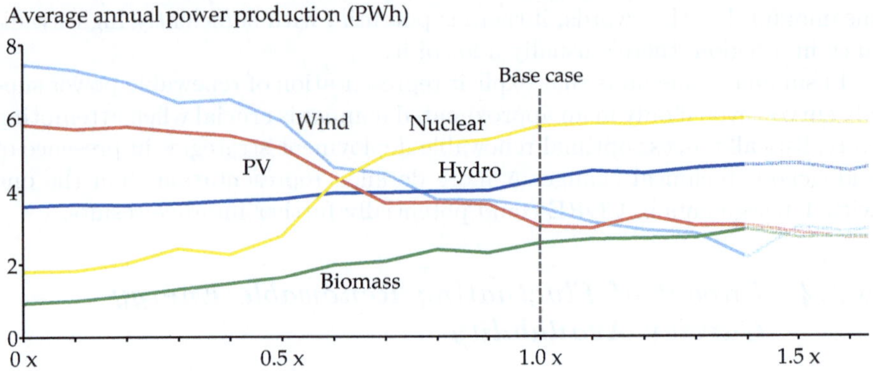

Fig. 6.18 Sensitivity to renewable energy carrier fluctuation (base case mult.)
Source: Own calculations

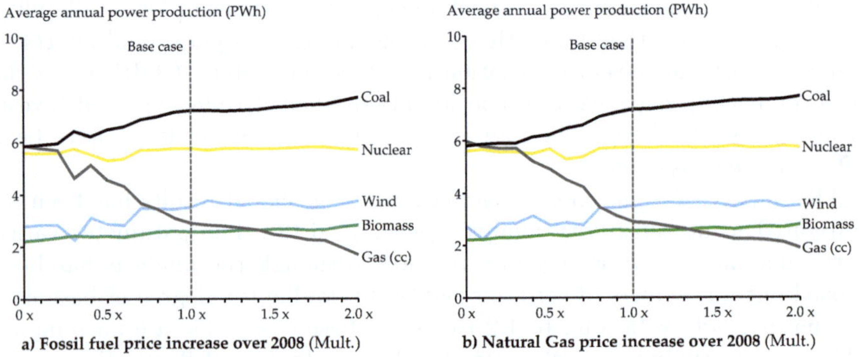

Fig. 6.19 Sensitivity to fossil prices
Source: Own calculations

6.4.5 Impact of Fossil Fuel Prices and the Carbon Cap

Fossil fuel prices only moderately impact long-term model results. For fossil fuel price sensitivity analysis, the exogenously specified price increase for all fossil energy carriers[107] over their respective 2008 starting values on the one hand, and for natural gas only on the other hand, are varied by means of a multiple (Fig. 6.19). Case 0x thus signifies no price increase at all, 1x price increases according to base case assumptions and case 2x double of the absolute price increase relative to the base case. The most important finding is that differences in fossil fuel price development hardly affect the long-term deployment of renewables. This indicates that the role of cost reductions and the carbon cap play a much more important role in the transformation of the power sector. The most eye-catching impact from changed fossil fuel price

[107] Hard coal, lignite, natural gas, oil.

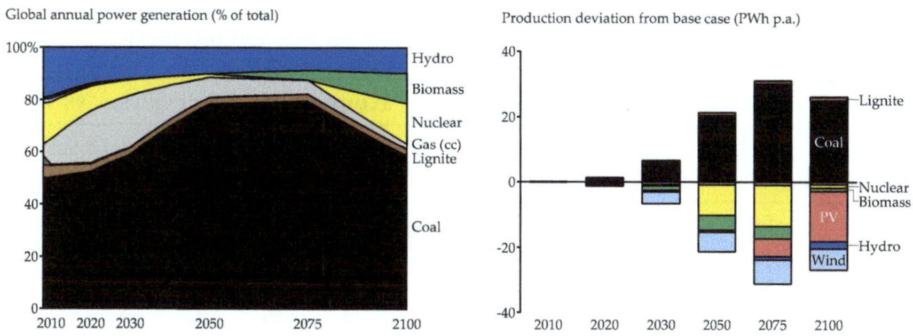

Fig. 6.20 Alternative scenario — No carbon cap
Source: Own calculations

trajectories, whether for all fossil fuels simultaneously or natural gas only, are coal-gas substitutions. The higher the relative price increase, the more gas is substituted by coal and, to a lesser degree, biomass and wind. Yet this finding is heavily dependent on the exogenously given price paths for these two energy carriers, where higher absolute price increases take place for natural gas than for hard coal.

In contrast to fossil fuel prices, the carbon cap has a massive impact on model results. An alternative scenario to the base case, where carbon dioxide emissions are not restricted, illustrates this finding (Fig. 6.20). In this scenario, hard coal penetration increases to almost 80% in 2075. Nuclear technology only becomes competitive again after coal prices have attained a high enough level towards the end of the model horizon. Hydro and biomass are the only renewables implemented and only by 2100 today's renewable share of roughly 20% is reached again. Wind and PV are most negatively impacted from the absence of a carbon cap — no implementation of those two technologies occurs at all.

The key finding from the two analyses relates to the relative importance of fossil fuel prices and the carbon cap for the long-term deployment of renewables. Whereas the carbon constraint vastly impacts optimal renewable deployment strategy, the importance of fossil fuel prices is comparatively low. Only in case of substantial fossil fuel price increases beyond the range shown in Figure 6.19 would the shift towards renewables be substantially accelerated. A fossil fuel price induced delay of renewable deployment given very low prices would only be possible in the early years of the model horizon, given the impact of the stringent carbon constraint.

6.4.6 Nuclear Power as Carbon-Free Substitute for Renewables

In public discussion nuclear technology is often referred to as potential, carbon-neutral substitute to renewable technologies. Sensitivity analyses in

Table 6.8 Impact of nuclear-specific parameters on average production

	Nuclear (% deviation vs BC)					**Renewable** (% deviation vs BC)				
Uranium	*Nuclear overnight cost ($/kW)*					*Nuclear overnight cost ($/kW)*				
potential	*2,800*	*3,400*	*4,000*	*4,600*	*5,200*	*2,800*	*3,400*	*4,000*	*4,600*	*5,200*
0.5×	-29	-36	-40	-45	-59	+17	+17	+16	+19	+27
1.0	+14	+8	±0	-21	-54	-2	-3	±0	+7	+23
1.5	+57	+51	+35	-15	-51	-22	-23	-18	+5	+22
2.0	+98	+90	+38	-12	-50	-39	-39	-17	+3	+22
2.5	+134	+120	+39	-12	-49	-55	-52	-17	+3	+21

	Fossil (% deviation vs BC)					**PV** (% deviation vs BC)				
Uranium	*Nuclear overnight cost ($/kW)*					*Nuclear overnight cost ($/kW)*				
potential	*2,800*	*3,400*	*4,000*	*4,600*	*5,200*	*2,800*	*3,400*	*4,000*	*4,600*	*5,200*
0.5	-55	+1	+4	+2	+1	+101	+82	+77	+65	+58*
1.0	-4	±0	±0	+2	+1	+11*	+14	±0	+25	+53*
1.5	-4	±0	+3	+1	+2	-3	-6	-36	+21	+68
2.0	-5	-2	-1	+2	+2	-100⋆	-74	-39	+16	+67
2.5	-5	-1	-1	+2	+1	-100	-100	-39	+13	+44*

* CSP is implemented at large-scale but with lower average production than PV
⋆ CSP substitutes PV entirely

this section are directed towards testing this assertion as well as base case result robustness to changes in parameters directly impacting nuclear competitiveness. Overnight capital costs, which are not subject to learning,[108] as well as uranium reserves and resources are varied across sensitivity scenarios (Tab. 6.8).

For scenarios characterized by nuclear capital costs below or equal to the ones in the base case (4,000 $/kW), uranium resources are the key factor determining nuclear deployment. The more resources are available the higher power production will be. However, already for base case capital costs of 4,000 $/kW an increase in resources by more than a factor of 1.5 does not translate into a significant increase in nuclear power production. When capital costs cross a certain threshold, somewhere between the base case value of 4,000 $/kW and 4,600 $/kW, they become the main determinant of nuclear deployment, which is lower than in the base case.

Higher levels of nuclear capacity crowd out non-nuclear capacity. If the assertion that nuclear is a substitute to renewables were true, one would expect this crowded out capacity to be mainly renewable. In presence of a

[108] Empirical studies have not found evidence supporting specific capital cost reductions for nuclear power plants (Neij, 2008).

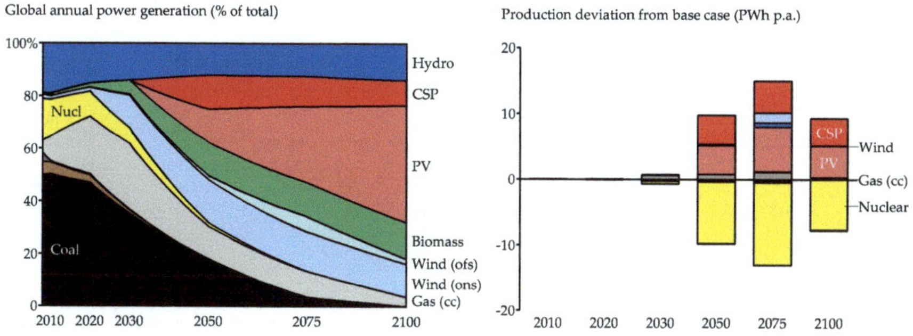

Fig. 6.21 Alternative scenario — No nuclear investments
Source: Own calculations

carbon cap this is definitely the case in CORE.[109] Fossil production is only marginally affected by different levels of nuclear deployment, whereas the opposite is true for renewables.

The renewable technology most affected by changes in nuclear deployment is PV. In most of the sensitivity scenarios, PV is deployed despite improved nuclear competitiveness — though large-scale implementation may be delayed. An exception are the three scenarios with lowest capital costs and highest uranium resources, where no PV implementation occurs. Interestingly, nuclear deployment in this framework also has an impact on CSP technology. For all scenarios marked with an asterisk in Table 6.8, a considerable amount of power[110] is produced from CSP in addition to PV. In one particular scenario, CSP even replaces PV entirely, given low capital costs for nuclear and doubling of uranium resources.

The observations from this sensitivity analysis therefore generally support the notion that, in presence of a stringent carbon cap, nuclear acts as a substitute to renewable technologies. For an additional perspective on this issue, a model scenario, where investments into new nuclear capacity have been prohibited, has been calculated (see Fig. 6.21). The main beneficiaries of such a setup are, from 2050 onwards, solar technologies and, to a lesser extent, combined cycle gas to compensate for the increasing fluctuations within the system.[111] The findings from this scenario therefore also support the idea that nuclear indeed interacts with renewables in a substituting manner, given a stringent carbon bound.

[109] For an exogenously given carbon price in lieu of the carbon cap, both fossil and renewable capacity were crowded out in approximately equal shares.

[110] More than 5 PWh per year, on average.

[111] CSP implementation alongside PV largely occurs due to presence of the growth constraint. However, even in absence of a growth constraint some CSP capacity is implemented.

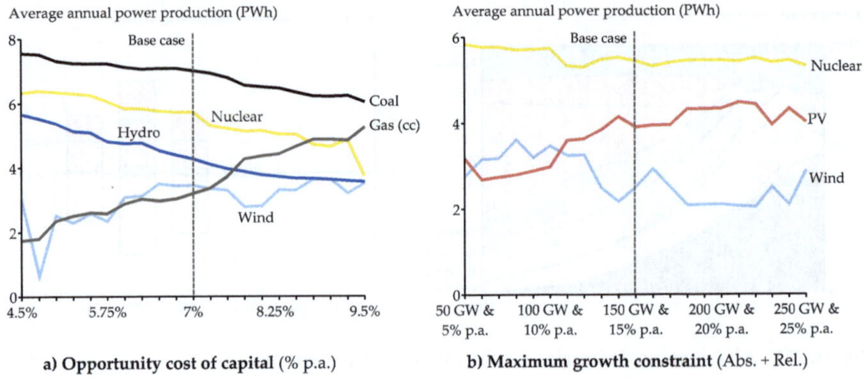

Fig. 6.22 Sensitivity to discount rate and maximum growth constraint
Source: Own calculations

6.4.7 Influence of Interest Rate and Growth Constraint

Capital intensity of the power mix increases over the modeled time horizon, mainly driven by the introduction of renewable power generation. An important factor impacting the competitiveness of capital-intense over less capital-intense technologies are opportunity costs of capital. Results from sensitivity analysis suggest that higher opportunity costs of capital lead to lower penetrations of coal, nuclear and hydro (Fig. 6.22a), which is altogether not surprising. Conversely, the comparatively less-capital intense combined cycle gas technology benefits most from an increases in opportunity costs of capital. The same holds for wind power, but to a lesser extent. This effect can partly be attributed to wind's higher share of non-fuel operating over total costs than found for other renewables, such as PV or hydro.

The maximal growth constraint only exerts a moderate influence on model results (Fig. 6.22b). The technologies it affects most are PV, wind and nuclear. As one would expect, the stricter the growth constraint, the more wind is deployed in lieu of PV. Yet, this effect can largely be observed for growth constraint assumptions below the base case value — above base case values the impact of the constraint on model results is minor. The latter observation also demonstrates that the growth restriction in the base case is not unreasonably strict, as more relaxed values for the growth constraint would not change results by much.

6.5 Key Findings and Implications

In this chapter the partial equilibrium power system model CORE has been used with two main objectives. On the one hand, to examine the impact of including marginal renewable potential supply curves into a bottom-up system model with endogenous learning. On the other hand, to investigate properties of the globally-optimal deployment strategy for renewable power technologies over the 21st century.

Concerning the methodological objective, results indicate that explicit representation of renewable potential, which is differentiated by productivity,[112] will considerably alter model results. Use of marginal instead of average productivity promotes renewables in a similar manner as niche markets do, effectively reducing total up-front investment costs into a technology. Higher degrees of differentiation in the technical potential will thus lead to faster deployment of renewable technologies. It is consequently suggested that system models aiming to realistically reflect renewable deployment explicitly consider renewable resources differentiated by technology, region and productivity.

In the optimal investment strategy derived in CORE, the first and the second half of the century are characterized by very different optimal technology portfolios. Whereas fossil technologies still dominate in the first half, the second half is characterized by a fast buildup of renewables and, to a lesser extent, nuclear. Interestingly, despite the presence of increasing returns and numerous non-competitive technologies at model start,[113] the technology portfolio in 2100 is quite diverse. The detailed representation of renewable potentials is presumably the main driver behind this property — increasing returns are clearly *bounded*. Despite a diverse technology portfolio in 2100, implementation of different technologies does not occur simultaneously but rather sequentially. Technologies with comparatively low up-front investment costs, such as onshore wind, are developed first, whereas technologies with larger gap to competitiveness, such as PV, are postponed. Model results therefore suggest that it may be reasonable to postpone promotion of some technologies until low-cost potentials from other technologies have been exhausted.

Although in the initial hypothesis climate change, cost reductions and increasing fossil fuel prices were proposed as main drivers of future renewable competitiveness (see chap. 2), model results demonstrate that the impact of fossil fuel prices in presence of a carbon cap and endogenous cost reductions play a relatively minor role. Climate change and cost reductions therefore remain as the two key drivers for the transition towards higher shares of renewables with fossil fuel prices only playing a less important role.

Average system costs over time increase, suggesting that the movement towards renewables comes at a cost to society, although a continued use of fossil technologies would also entail cost increases from the depletion of the corresponding resources as well as implicit costs associated with greenhouse

[112] Represented as differences in renewable energy carrier availability or biofuel price.

[113] This will often lead to dominance of one or few technologies in case of insufficient bounding; this property is also often referred to by the terms lock-in or lock-out, depending on the perspective taken.

gas emissions. Regional technology portfolios as well as the corresponding average generation costs differ widely. Regional renewable potential endowments constitute the main driver of those regional differences in long-term electricity costs.

In the long-term, model results suggest that at least one solar technology is required to fulfill increasing power demand while simultaneously lowering global greenhouse gas emissions. In CORE, the two solar technologies PV and CSP act in a highly substituting manner and only for very few parameter combinations it is optimal to incur the up-front investment costs for both technologies. Model results are not robust to changes in cost-related parameters for the two technologies, in particular learning rates, and small changes may tip implementation decisions towards one or the other technology. Nuclear technology also has a non-negligible influence on the choice between them. For example, for comparatively high penetrations of nuclear, there may be cases where CSP technology with its ability to reallocate some power production to peak times, is a more suitable technology to complement high nuclear shares in the power mix.

Model results have implications for optimal renewable support policy. Two implications relate to the timing and location of governmental support spending: (1) Costs to society will be lower, if government support is first focused on those technologies, where the highest benefits can be expected at comparatively low costs. The fostering of other technologies should be postponed until their corresponding economic potentials have been exhausted and the conditions for cost-optimal investment timing into the next technology are fulfilled.[114] Earlier than cost-optimal investment into a particular technology may be chosen to minimize risk, for example, from a technology not following an expected cost reduction trajectory — yet this approach will increase expected costs to society. Those incremental costs can be viewed as "insurance premium" in case society is assumed to be risk-averse. (2) Implementing a still uncompetitive technology in regions with comparatively high-productivity potential will result in lowest cost to society. For example, if plant output in one region is double the one of another region, the nominal up-front deployment support costs in the first region will, ceteris paribus, be half the one of the second region.[115] For technologies, where multi-billion dollar up-front investments are required, implementation at high-productivity sites may be an approach to consider.[116]

[114] In CORE, the technologies with the lowest up-front investment requirements are the ones picked for earlier installation, followed by the ones with higher gaps to competitiveness. However, picking technologies solely based on up-front investment costs may not always yield the optimal investment sequence, as benefits associated with the investment must also be taken into consideration.

[115] Given sufficiently high power demand and equal prices for power.

[116] Note that not all up-front investment costs are related to deployment of the technology in the field. Costs for R&D efforts, for example, are likely far less dependent on potential and thus the region, where they are carried out.

These implications should, however, only be extended to current, local policies with extreme caution, bearing the limitations of the model in mind. First, CORE implicitly assumes full transparency, perfect international cooperation and perfect markets. Second, the setup as a partial equilibrium model ignores interactions with other sectors of the economy outside the power sector, particularly interactions with other energy subsectors. Third, by choice of the technologies available for investment, the model implicitly requires a certain movement towards renewables in the future. Other carbon-free technologies, such as carbon capture & storage, fissile nuclear alternatives, nuclear fusion or fuel cells may also play an important role in the long-term development of the power sector.[117] Finally, the model approach assumes perfect foresight and thus absence of risk. This last limitation is explicitly addressed in the next chapter.

References

Arrow, K.J.: The Economic Implications of Learning by Doing. The Review of Economic Studies 29(3), 155–173 (1962)

Arthur, W.B.: Competing Technologies, Increasing Returns, and Lock-In by Historical Events. The Economic Journal 99(394), 116–131 (1989)

Barreto, L.: Technological Learnin. In: Energy Optimisation Models And Deployment Of Emerging Technologies: Dissertation. Dissertation, Swiss Federal Institute of Technology, Zurich (2001)

Barreto, L., Kypreos, S.: Multiregional technological learning in the energy systems MARKAL model. International Journal of Global Energy Issues 17(3) (2002)

BGR. Energierohstoffe 2009. BGR, Hannover (2009)

Boiteux, M.: Peak-load pricing. Journal of Business 33, 157–179 (1960)

Clarke, L., Edmonds, J., Krey, V., Richels, R., Rose, S., Tavoni, M.: International climate policy architectures: Overview of the EMF 22 International Scenarios. Energy Economics 31(suppl. 2), S64–S81 (2009)

de Vries, B., van Vuuren, D.P., Hoogwijk, M.M.: Renewable energy sources: their global potential for the first-half of the 21st century at a global level: an integrated approach. Energy Policy 35(4), 2590–2610 (2007)

den Elzen, M.G.J., Meinshausen, M., van Vuuren, D.P.: Multi-gas emission envelopes to meet greenhouse gas concentration targets: Costs versus certainty of limiting temperature increase. Global Environmental Change 17(2), 260–280 (2007)

Edenhofer, O., Knopf, B., Barker, T., Baumstark, L., Bellevrat, E., Chateau, B., Criqui, P., Isaac, M., Kitous, A., Kypreos, S., Leimbach, M., Lessmann, K., Magne, B., Scrieciu, S., Turton, H., van Vuuren, D.P.: The Economics of Low Stabilization: Model Comparison of Mitigation Strategies and Costs. The Energy Journal 31(special issue 1) (2010)

European Wind Energy Association. The economics of wind energy, Brussels (2009)

GE Energy. The 1.5 MW Wind Turbine (2009)

[117] These technologies have not been explicitly represented since they either are in a very early developmental stage and/or are expected to face high degrees of public resistance for other than cost reasons.

Goldemberg, J. (ed.): World energy assessment: Energy and the challenge of sustainability. UNDP, New York (2000)
Hoogwijk, M.: On the global and regional potential of renewable energy sources: Doctoral thesis. Doctoral thesis, Utrecht University, Utrecht (2004)
Hoogwijk, M., de Vries, B., Turkenburg, W.: Assessment of the global and regional geographical, technical and economic potential of onshore wind energy. Energy Economics 26(5), 889–919 (2004)
IEA. Energy Technology Perspectives: Scenarios & Strategies to 2050. IEA Publications, Paris (2008)
IEA. World Energy Outlook 2009. IEA Publications, Paris (2009)
IPCC. Contribution of Working Group III to the Fourth Assessment Report of the Intergovernmental Panel on Climate Change. Cambridge Univ. Press, Cambridge & New York (2007)
Krewitt, W., Nienhaus, K., Kleßmann, C., Capone, C., Stricker, E., Graus, W., Hoogwijk, M., Supersberger, N., Winterfeld, U., Samadi, S.: Role and Potential of Renewable Energy and Energy Efficiency for Global Energy Supply. Dessau-Roßlau (2009)
Loulou, R., Labriet, M.: ETSAP-TIAM: the TIMES integrated assessment model Part I: Model Structure. Computational Management Science 5(1-2), 7–40 (2008)
Lutz, W., Sanderson, W., Scherbov, S.: IIASA's 2007 Probabilistic World Population Projections: IIASA World Population Program Online Data Base of Results 2008 (2007)
Mattsson, N.: Internalizing technological development in energy system models: Licentiate. Licentiate. Chalmers University, Gothenburg (1997)
Mattsson, N., Wene, C.-O.: Assessing New Energy Technologies Using an Energy System Model With Endognized Experience Curves. International Journal of Energy Research 21(4), 385–393 (1997)
Messner, S.: Endogenized technological learning in an energy systems model. Journal of Evolutionary Economics 7(3), 291–313 (1997)
Neij, L.: Cost dynamics of wind power. Energy 24(5), 375–389 (1999)
Neij, L.: Cost development of future technologies for power generation — A study based on experience curves and complementary bottom-up assessments. Energy Policy 36(6), 2200–2211 (2008)
Nemet, G.F.: Beyond the learning curve: factors influencing cost reductions in photovoltaics. Energy Policy 34(17), 3218–3232 (2006)
NREL. National Solar Radiation Data Base (1991–2005 Update): Typical Meteorological Year 3, Albuquerque Intl Arpt [Class I] (2010)
Poponi, D.: Analysis of diffusion paths for photovoltaic technology based on experience curves. Solar Energy 74(4), 331–340 (2003)
Rout, U.K., Blesl, M., Fahl, U., Remme, U., Voß, A.: Uncertainty in the learning rates of energy technologies: An experiment in a global multi-regional energy system model. Energy Policy 37(11), 4927–4942 (2009)
RTE. Historique des consommations journalières en puissance (2010)
Schaeffer, G.J., Alsema, E., Seebregts, A.J., Beurskens, L., Moor, H.H., van Sark, W., Durstewitz, M., Perrin, M., Boulanger, P., Laukamp, H., Zuccaro, C.: Learning from the Sun: Analysis of the use of experience curves for energy policy purposes: The case of photovoltaic power. Final report of the Photex project. Petten (2004)
South Dakota State University. Climate and Weather (2010)

Trieb, F., O'Sullivan, M., Pregger, T., Schillings, C., Krewitt, W.: Characterisation of Solar Electricity Import Corridors from MENA to Europe: Potential, Infrastructure and Cost, Stuttgart (2009)
UNFCCC. Outcome of the work of the Ad Hoc Working group on long-term Cooperative Action under the Convention: Draft decision -/CP.16 (2010)
van der Zwaan, B., Rabl, A.: The learning potential of photovoltaics: implications for energy policy. Energy Policy 32(13), 1545–1554 (2004)
Walther, W.: Long-term price uncertainities of fossil primary fuels and implications for the electricity industry. Cuvillier, Göttingen (2009)
Weber, C.: Uncertainty in the Electric Power Industry: Methods and Models for Decision Support. International Series in Operations Research & Management Science, vol. 77. Springer, New York (2005)
World Energy Council. 2007 Survey of Energy Resources, London (2007)
World Energy Council. Survey of Energy Resources: Interim Update 2009, London (2009)

Chapter 7
Implications of Uncertainty for Renewable Power Deployment

Future rates of cost reduction, fossil fuel prices, the global carbon bound and other key factors impacting renewable power development are highly uncertain. For this reason, uncertainty is introduced to the previously deterministic CORE model by means of a stochastic program. After the discussions of objective, approach and current state of research, the required amendments to the model's formulation are described in section 7.2. Results are then presented in section 7.3 for uncertain learning rates, carbon bound, fossil fuel prices, renewable potentials as well as uncertainty in several parameters simultaneously. The chapter concludes with a summary of findings and implications.

7.1 Objective and Approach

Despite general consensus that different developments in the future will lead to increased renewable power generation, the magnitude of those developments is ex-ante unknown. In the deterministic framework from the previous chapter, parameters describing those developments have been assumed perfectly known. Yet, the introduction of a global carbon cap, for example, would require agreement on a global emission budget, allocation thereof to individual countries as well as rigorous national emission reduction according to the agreed budgets. In light of these preconditions, it is almost certain that actual future carbon emissions will deviate from the corresponding assumption made in the deterministic model — unfortunately it is virtually impossible to determine by how much. For other developments a deviation from the assumptions made in CORE can be expected as well. Experience curves, for example, have been used to forecast future cost developments. But not only is it impossible to predict whether costs will continue to develop at the historically observed learning rates but also may these observed learning rates be "inaccurate" due to methodological and data limitations. It is therefore comprehensible that several authors call for explicit representation of uncertainty in

F. Wagner, *Renewables in Future Power Systems*,
Green Energy and Technology,
DOI: 10.1007/978-3-319-05780-4_7,

energy system models with endogenous cost reduction mechanisms (e.g. Alberth, 2008; Berglund and Söderholm, 2006; Grübler et al., 1999).

Investment decisions in the power sector impact technology portfolios and cost structures for several decades into the future. For this reason it is necessary to take alternative, widely varying future developments of the most important cost drivers into account at the timing of the investment decision. Accordingly, it is the objective of the analyses in this chapter to examine the impact of parameter uncertainty in several important cost drivers on the future deployment of renewable power technologies. A stochastic energy system model is applied for this purpose. Apart from insights regarding the impact of uncertainty on particular technologies and the group of renewables as a whole, a methodological contribution of the model is the explicit consideration of effects from shifts in experience curves at stochastic stage transitions.

Two approaches commonly applied to introduce uncertainty in models are sampling techniques, e.g. Monte-Carlo, and stochastic programming.[1] A major difference between the stochastic programming and the sampling approach is that normative results are obtained for the first, descriptive results for the latter. Under the stochastic programming approach, investment decisions can be optimized in order to "hedge" for alternative future states of the world. Using a stochastic programming approach thus better fits the overall objective of this work to analyze *optimal* renewable deployment strategy. This benefit, however, comes at the expense of vastly higher computational requirements than for a sampling approach.[2]

Research based on bottom-up energy system models featuring uncertainty and technological change appeared on the scene about a decade ago, following the initial implementations of linearly approximated experience curves in a deterministic setting (Messner, 1997; Mattsson and Wene, 1997). Over the last few years, models featuring uncertainty in technological change have largely followed top-down and hybrid approaches, as policies to counteract climate change rather than technology assessment moved into focus. Therefore, most publications relevant for the bottom-up model presented in this chapter have appeared in the early 2000's. This bottom-up model based research can roughly be divided into sampling- and non-sampling based approaches.

The sampling based research has largely been developed at the IIASA.[3] Prominent examples are Grübler and Gritsevskyi (1997), Gritsevskyi and Nakićenović (2000), Grübler and Gritsevskyi (2002) or, more recently, a working paper by Shittu (2008). Whereas Grübler and Gritsevskyi (1997) still employ a stylized model with three technologies,[4] Gritsevskyi and Nakićenović (2000) consider 100 technologies characterized by some degree of spillovers

[1] Further methods exist, for example, chance constraint programming as used by Held et al. (2009) or straightforward scenario analysis.

[2] For additional information on this issue see sec. 3.5.2.

[3] Mostly based on the MESSAGE model.

[4] Mature, Incremental and Radical; the latter two are characterized by uncertain, model-endogenous cost reductions.

within defined "technology clusters." Future technology costs are represented by log-normal distribution functions, for which both expected value and variance decline with increasing installations of the corresponding technology.[5] All of these models assume risk averse decision makers, may it be a global planner or several different agents, such as in Grübler and Gritsevskyi (2002).

The key findings on the impact of uncertainty for this group of models relate to investment timing and the benefit of technological diversification. Whereas the diffusion of the radical technology in Grübler and Gritsevskyi (1997) is delayed when introducing learning uncertainty, the advent of carbon-free technologies is accelerated by a decade in Shittu's 2008 analysis, when uncertainty is brought into the picture. Gritsevskyi and Nakićenović (2000) find that "different structures of the energy system emerge with similar overall cost" and recommend that support policies in light of uncertain learning be directed towards several technologies characterized by spillover effects among each other instead of fully concentrating investment on one technology. Similarly, Shittu (2008) finds that for the optimal development of the energy system under uncertain learning "a more diversified technology portfolio is revealed."

Enhancing the ERIS and GENIE models, respectively, Barreto (2001) and Mattsson (2002) used stochastic programs to assess the impact of uncertainty in bottom-up models with endogenous technological change. Whereas Barreto uses a two-stage model with uncertainty resolution at a pre-specified point in time, stage transition is triggered endogenously in Mattsson's three stage model. The latter approach fits well with the ambition to model learning entirely endogenous, i.e. investment into a technology is required to learn about future cost reduction possibilities. Yet it has the serious drawback that for each learning-enabled technology an additional stochastic stage as well as integer variable must be added to the model.[6] It is clear that the *curse of dimensionality* will quickly render such models computationally infeasible.[7] Moreover, simultaneous introduction of other types of uncertainty with resolution at specified points in time, would lead to asynchronous stage transitions and thus further complication of the model.

Both authors highlight that investment into several different technologies is an effective hedging strategy against uncertain learning rates. Barreto finds that "uncertain learning rates may drive the model to follow a more prudent

[5] Due to the high computational requirements of the model in Gritsevskyi and Nakićenović (2000), the authors reduce the number of infinite possible futures into 500 so-called technology dynamics. These are "more generic characterization of future developments, specifying "a set of uncertain, generic relations." For each of these "dynamics" they sample 250 scenarios in parallel.

[6] In case of information updating occurring once over the model horizon. For multiple information updates, the number of required stochastic stages further increases proportionally to the number of updates.

[7] For this reason Mattsson (2002) only includes uncertain learning for two technologies and highlights that the "methodology is feasible on a demonstration level" only.

and gradual path of investments into learning technologies," yet also states that the model reacts highly sensitive to the transition probability values. For example, in his stochastic base case, PV investment levels lie below the ones from his deterministic base case, after slight modification of transition probabilities, above. It is also possible that the presence of floor costs may have played a role in the deviation of the "PV hedging path" in the stochastic setup from the deterministic one. Mattsson also qualifies his findings, highlighting that the investment into both learning technologies instead of concentration on one of them is likely caused by the fact that there are not enough technologies within the model "to provoke competition" among them for learning opportunities. Moreover, since indirect spillovers between those two learning technologies are possible, there is an inherent incentive for co-evolution.

More recently Rout et al. (2009) and Kumbaroglu et al. (2008) have examined the impact of uncertainty in bottom-up models with endogenous cost reductions. They approached the topic using deterministic scenario analysis and backwards solution of a model featuring non-stationary processes for uncertainty representation, respectively. As the corresponding approaches and results are of lower relevance to the research in this book, the interested reader is referred to the original publications for details.[8]

In line with the objective to examine the globally optimal renewable deployment strategy, i.e. taking a normative point of view, a stochastic programming over a sampling approach is followed for the model in this chapter. In light of the otherwise very long stage duration and to enable multiple information updating, a three-stage-setup is chosen. Information updating occurs twice at pre-specified points in time, notably from model year 2015 to 2030 and again from 2050 to 2075. This introduces conditional probabilities, reflecting improved information decision makers have on future developments based on actual observations. For example, if a high learning rate materializes in the second stochastic stage, it is considered quite unlikely that it will drop again to a very low level in the third stage.

An alternative to exogenously fixing stage transitions at pre-specified points in time would have been to trigger stage transition endogenously, comparable to the approach used by Mattsson (2002). Yet the associated computational burden would have been prohibitive, in particular due to information updating occurring more than once. Moreover, although endogenous stage transition triggers are conceptually convincing for learning rate uncertainty,[9] such an approach would be as arbitrary for carbon bounds, renewable potential and fuel prices as an approach based on pre-specified points in time.

[8] Rout et al. (2009) employs a deterministic scenario approach. Kumbaroglu et al. (2008) set up a recursive model with uncertainty being represented as non-stationary stochastic processes.

[9] Presuming that the largest share of cost reductions indeed originates from endogenous sources, such as economies of scale or learning by doing, and not from spillovers.

The global investor is assumed to be risk-neutral. On the one hand, since assumptions on the degree of risk aversion in a setup of global welfare optimization are difficult to make and, on the other hand, since consideration of risk aversion could make methodological interpretations more difficult by "masking" properties visible in a risk neutral setting. As the number of possible states of the world is finite, optimal decision making under this model effectively follows Bayes' rule.

With respect to the methodological setup, the model comes closest to the one presented by Barreto (2001). However, a model feature the author has not encountered in existing literature is the explicit correction for unrealistic discontinuities at stage transitions in the formulation of the model (see chaps. 5.4.1 and 7.2). This correction is only required in case of exogenously specified stage transitions, where the cumulative manufactured capacity of a technology at the point in time of stage transition is not known ex-ante.[10]

7.2 Model Description

The stochastic program is implemented by extending the deterministic CORE model described in the previous chapter and will further be referred to as SCORE (Stochastic CORE). At each stochastic node, branching into two new states of the world occurs (see Fig. 7.1). The second stage is thus made up of two, the third stage of four states of the world. For parameter values the program follows a recombining tree. For variables, however, the tree does not recombine, as the parameter value in node ab/ba can be reached via different "paths". Although a setup with more stages and stronger branching at each node would be desirable, implementation thereof is unfortunately prevented by the associated computational burden. The impact of parameter uncertainty on the deployment of renewable technologies is analyzed for:

- Learning rates (PV, CSP, all learning-enabled renewables)
- Fossil fuel prices (natural gas, natural gas and coal)
- Carbon dioxide bound
- Renewable potential
- Multiple (learning rates, carbon bound, fossil fuel prices, potential)

In cases, where uncertainty is represented for several parameters simultaneously, e.g. learning rates for several technologies, the correlation in the development of those parameters is exogenously specified to avoid the disproportionately high increase in possible states of the world and retain the basic setup described above. Such an increase would render solution of the model infeasible within a reasonable time frame. Additional simplifications versus the deterministic model version include the reduction of learning curve

[10] In a setup as the one used by Mattsson (2002), where stage transition is triggered at an ex-ante known capacity, no adjustments are necessary.

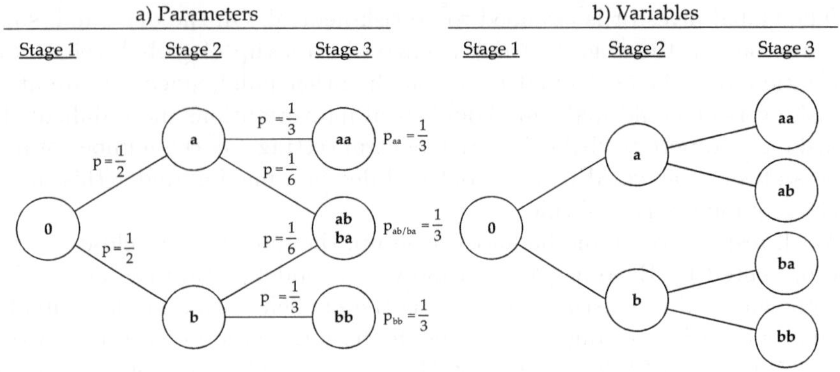

Fig. 7.1 Stochastic tree for SCORE
Source: Own representation

segments to four and reduction of the number of model years by one.[11] Even under this setup, the MIP solution process[12] on an Intel Core i5 computer with 16 gigabytes of memory requires between three and fourteen wall clock days, depending on parameterization. The simplifications mandate the computation of a new deterministic reference case to be compared to the stochastic results.

The global planner is considered risk neutral. Transition probabilities are chosen in such a manner that, from a first stage perspective, expected parameter values from the stochastic equal the ones from the deterministic model (see Fig. 7.1). From first to second stage the uncertain parameter moves to either a *high* (node *a*) or a *low* (node *b*) value.[13] From a first stage perspective, all three parameter states of the world in stage three are equally likely. From a second stage perspective, however, either the high (low) path continues, corresponding to node *aa* (*bb*), or the parameter value moves back to the expected value in node *ab*/*ba*. This setup reflects the belief that after a certain parameter value has already been demonstrated for some time, it is rather unlikely that a drastic change happens later on. For example, after node *a* (high path) is realized in stage two the value will not drop again to the low path represented by nodes *b* and *bb*.

The following subsections describe the changes in model formulation to convert the deterministic CORE into the stochastic SCORE model. Implementation of uncertainty for fossil fuel prices, carbon bound and renewable potential are comparatively straightforward. The implementation of learning

[11] Model years in SCORE: 2008, 2015, 2030, 2050, 2075, 2100.
[12] For a tolerance of 0.001%.
[13] Generally the *high* (*low*) path corresponds to 1.5 (0.5) times the deterministic parameter value. For learning rate uncertainty it is important to apply this value to the learning rate instead of the learning index to ensure preservation of the expected value (given floor costs are not reached).

rate uncertainty, however, requires additional modifications. The approximations of the non-linear cumulative cost curves have to be adjusted for unrealistic cost discrepancies at the cumulative capacities associated with stage transitions. This adjustment process has previously been referred to as *vertical shift* of the cost curve (see section 5.4.1).

7.2.1 Non Learning-Related Modifications

Let $N = \{0,\ a,\ b,\ aa,\ ab,\ ba,\ bb\}$ denote the set of stochastic nodes, which can be further divided into three subsets associated with the respective stages one to three.

$$\begin{aligned} N^{S1} &= \{0\} \text{ for stage 1,} \\ N^{S2} &= \{a, b\} \text{ for stage 2 and} \\ N^{S3} &= \{aa, ab, ba, bb\} \text{ for stage 3} \end{aligned} \tag{7.1}$$

Transition from one node to the next is possible along the paths illustrated in Figure 7.1. For the defined setup, model periods 2008 and 2015 thus make up the first stage, 2030 and 2050 the second and 2075 and 2100 the third one.

As initial step of the stochastic enhancement, dimension n is added to all model variables and equations with the exception of the objective function. Additional modifications are required for all equations incorporating a time lag. The need for this arises from the existence of more than one model period per stochastic stage. A variable in period t referring to a previous period may either carry the same or a different value for n as the one in period $t-1$, depending on whether the reference extends across stage borders. Equation (7.2) is used as example for the stochastically enhanced formulation, corresponding to Equation (6.20) from the deterministic model.

$$\begin{aligned} K^{op}_{i,r,t,\mu,n} = \ & X_{i,r,t,\mu,n} - V^{act}_{i,r,t,\mu,n} \\ & + \begin{cases} K^{op}_{i,r,t-1,\mu,n^{pre}(n)} & \forall t = t^{1st}(n) \\ K^{op}_{i,r,t-1,\mu,n} & \forall t \neq t^{1st}(n) \end{cases} \end{aligned} \tag{7.2}$$

$t^{1st}(n)$ denotes the first period of a particular node, $t^{las}(n)$ the last period of this node. Here, the respective periods are 2008, 2030, 2075 and 2015, 2050, 2100 for $t^{1st}(n)$ and $t^{las}(n)$, respectively. $n^{pre}(n)$ represents the node directly preceding n, $n^{pre}(n^{pre}(n))$ the node preceding n two stages back. Modifications analogous to the ones shown in Eq. (7.2) are also required for other equations referencing a previous period.[14]

[14] Equations (6.14), (6.18), (6.19), (6.24), (6.26), (6.27), (6.32) and their variations, such as versions without dimension μ for technologies not modeled using potential bands.

Variables tracking investment and their corresponding capital cost annuities, such as $\dot{X}_{i,r,t,t',\mu,n}$, also require modification. For those, accurate allocation of time periods and stochastic nodes is ensured as shown in Equation (7.3).

$$\left.\begin{aligned}\dot{X}_{i,r,t,t',\mu} = \quad & X_{i,r,t',\mu,n} \\ & + X_{i,r,t',\mu,n^{pre}(n)} \qquad \forall\, s(t) - s(t') = 1 \\ & + X_{i,r,t',\mu,n^{pre}(n^{pre}(n))} \; \forall\, s(t) - s(t') = 2\end{aligned}\right\} \forall\, \hat{\tau}_t - \hat{\tau}_{t'} < \pi_i + \tau_t \tag{7.3}$$

with parameter $s(t)$ indicating the number associated with period t.[15]

The first term to be stored in Equation (7.3), $X_{i,r,t',\mu,n}$, is the one for investment in the current stage and thus node. The following two terms refer to investments in previous stages. Equations modified in the same manner are (6.45) and (6.51).

The objective function requires modification as well. The summands in the deterministic objective function, given by Eq. (6.1) and represented by [...] in Eq. (7.4), are modified to introduce probability-based weighting of the different states of the world.

$$\sum_n p_n \cdot [\ldots] \tag{7.4}$$

7.2.2 Learning-Related Modifications

Experience curves are power functions. Thus, curves characterized by different learning rates but passing through the same point, which is described by a particular cumulative capacity and specific investment costs, will be characterized by different shapes.[16] This property leads to undesirable as well as unrealistic discontinuities in cumulative investment costs if learning rates change across exogenously specified stochastic stage transitions. Figure 7.2 illustrates such a discontinuity for an approximated cumulative cost curve. Here, a stage transition occurs at a cumulative capacity of roughly 13 TW, where the learning rate decreases.[17] Consequently, cumulative investment costs are lower in point A than in B. Since investment costs are calculated as difference between cumulative investment costs of the current and the previous period, the sample discontinuity would lead to an overstatement of specific costs in the first period of the new stochastic stage. The periods thereafter are not affected, as long as there are no further stage transitions with changes in learning rate. Taken to the extreme, the presence of such a discontinuity could lead to negative investment costs despite capacity

[15] Possible values for $S = \{1, 2, 3\}$.

[16] First derivatives will differ for the same cumulative capacity value.

[17] The applicable cumulative capacity at stage transition is determined model-endogenously.

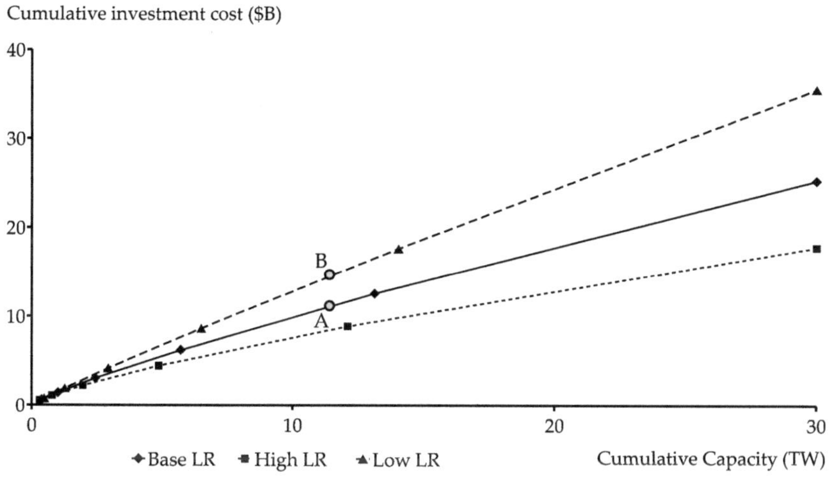

Fig. 7.2 Cumulative cost curves for different learning rates
Source: Own representation

additions in a particular period.[18] This clearly illustrates the need to correct for the effect of this discontinuity.

In the analytically solved stochastic model from chapter 5 the issue is resolved by vertically shifting the learning curve at stage transition. However, this operation was applied to the learning curve directly and not to the implied cumulative cost curve. More important, this shift had been formulated as a nonlinear relationship. Since the piecewise linear nature of the objective function in SCORE needs to be preserved in order to reliably identify a globally optimal solution, this approach is not possible here. Instead, a *vertical shift* of the approximated cumulative cost curve is implemented in a linear manner. Figure 7.3 illustrates such a shift for a sample increase in learning rate. The resulting combined cumulative cost curve has a kink in point B. As the model is set up as three-stage stochastic program, potential vertical shifts have to be considered for both stage transitions. The required formulation modifications are implemented in two major steps:

(1) First, the parameters describing the approximated cumulative cost curve must be adjusted. Figure 7.3 illustrates that the capacity breakpoints for cumulative cost functions based on different learning rates must coincide. Yet, this is not the case if the segmentation procedure used in the deterministic model version, described by Eqs. (6.63) through (6.70) and Figure 7.2, were applied. Although this alignment of segment borders may not be optimal with respect to approximation error minimization, it is required to enable a linear formulation of the vertical shift.

[18] For an increase in learning rate and sufficiently low capacity additions.

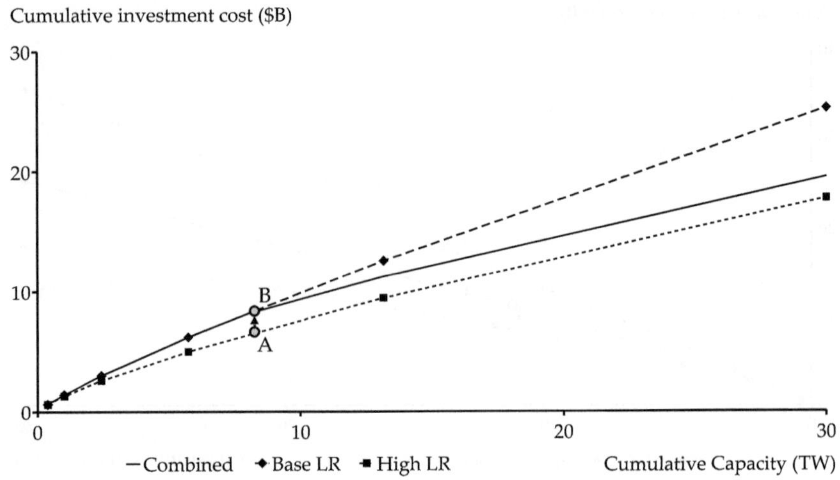

Fig. 7.3 Combined cumulative cost curves for change in learning rate
Source: Own representation

The cumulative capacity breakpoints calculated in Eqs. (6.69) and (6.70) are based on the learning rate assumptions for the first stage ($n = 0$). These breakpoints must now remain valid for all stochastic nodes. Therefore, cumulative investment cost[19] at the segment breakpoints for stochastic nodes beyond $n = 0$ need to be recalculated to match the capacity values from stage one. The presence of floor costs complicates this matching process. In CORE, floor costs were confined to the last learning curve segment, if reached at all. Since capacity breakpoints are fixed for all states of the world in SCORE, the capacity at which floor costs are reached, $\hat{\kappa}_{i,n}^{flo}$, will be different from the capacity delimiting the last and the second-to-last segment.[20] For example, $\hat{\kappa}_{i,n}^{flo}$ will be reached somewhere before the beginning of the last segment in case of a learning rate increase. For a decrease, it will fall within the last segment or not be reached at all before the specified maximal cumulative capacity $\hat{\kappa}_{i,n}^{Max}$ is attained. Figure 7.4 illustrates the case of a shift in the capacity at which floor costs are reached resulting from a decrease in learning rate. $\hat{\kappa}_{i,n}^{flo}$ is reached within the last segment instead of exactly at the transition from the second-to-last to the last one. Cumulative costs at the pre-defined segment capacity breakpoints are then calculated according to Eq. (7.5), which takes the issue arising from presence of floor costs into account.

[19] Required to derive the segmental slopes and y-intercepts of the approximated curves.

[20] Given a change in learning rate between stages takes place.

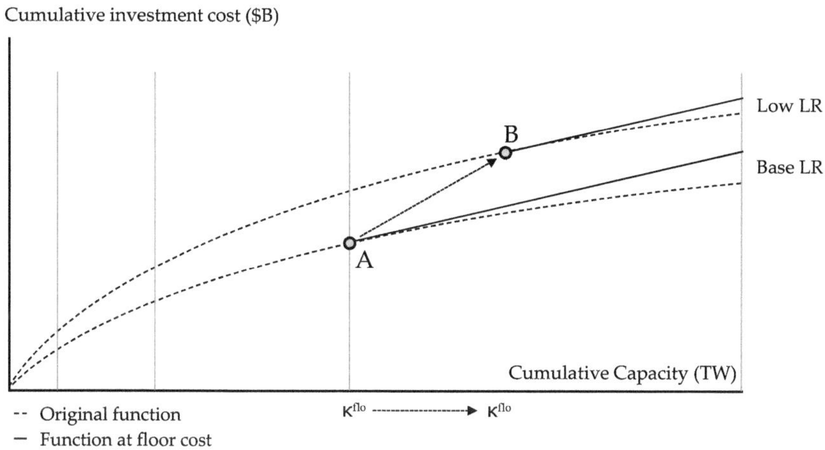

Fig. 7.4 Shift of cumulative capacity at floor costs (illustrative)
Source: Own representation

$$\hat{c}^{tinv,seg}_{i,\chi,n} = \begin{cases} \frac{c^{1st}_{i,n}}{1-\gamma_{i,n}} \left(\tilde{\chi}_{i,\chi,n}\right)^{(1-\gamma_{i,n})} & \forall i \mid \hat{\kappa}^{flo}_{i,n} \geq \tilde{\chi}_{i,\chi,n} \\ \frac{c^{1st}_{i,n}}{1-\gamma_{i,n}} \left(\hat{\kappa}^{flo}_{i,n}\right)^{(1-\gamma_{i,n})} + c^{flo}_{i,n} \left(\tilde{\chi}_{i,\chi,n} - \hat{\kappa}^{flo}_{i,n}\right) & \forall i \mid \hat{\kappa}^{flo}_{i,n} < \tilde{\chi}_{i,\chi,n} \end{cases} \tag{7.5}$$

The calculation depends upon the relationship between capacity at which floor costs are reached $\hat{\kappa}^{flo}_{i,n}$ and the upper capacity limit of the segment under consideration $\tilde{\chi}_{i,\chi,n}$. In case floor costs are not reached within the segment under consideration, cumulative costs at segment breakpoints are calculated as previously. If floor costs are reached within a particular segment, the second case applies. Cumulative costs at the end of a segment are then the sum of the cumulative costs corresponding to the capacity, where floor costs are reached, and the costs from investments at floor cost. The slopes and y-intercepts of the particular cumulative investment cost curve segments before a potential vertical shift are calculated as before. Their calculation is described by Eqs. (6.71) and (6.72).

The vertical shift of an approximated cumulative cost curve at a specific point can be formulated as a linear function. Hence, in addition to the slope and y-intercept for cost curve segments before a vertical shift, a slope $\alpha^{vs}_{i,\chi,n}$ and y-intercept $\beta^{vs}_{i,\chi,n}$ describing the vertical shift as linear function, are introduced, as given by Eqs. (7.6) and (7.7). The latter are nothing else than the difference between the slope and y-intercept, respectively, of the current and the preceding node.

$$\alpha^{vs}_{i,\chi,n} = \alpha_{i,\chi,n^{pre}(n)} - \alpha_{i,\chi,n} \tag{7.6}$$

$$\beta^{vs}_{i,\chi,n} = \beta_{i,\chi,n^{pre}(n)} - \beta_{i,\chi,n} \tag{7.7}$$

(2) The second major modification step is to adjust the model equation calculating cumulative costs, as shown in Eq. (6.49), which is in turn the basis for the calculation of investment costs in a particular model year, given by Eq. (7.8). As vertical shifts of all preceding stages, not only the immediately preceding one, need to be considered and a node and time-period referencing is required, the complexity of the cumulative cost calculation increases in comparison to the deterministic setup.

$$\begin{aligned}
\hat{C}^{tinv}_{i,t,n} = & \sum_{\chi}\left(\alpha_{i,\chi,n} \cdot \hat{\kappa}^{seg}_{i,t,\chi,n} + \beta_{i,\chi,n} \cdot A_{i,t,\chi,n}\right) \\
& + \sum_{\chi}\Big(\alpha^{vs}_{i,\chi,n} \cdot \hat{\kappa}^{seg}_{i,t^{las}(n^{pre}(n)),\chi,n^{pre}(n)} \\
& \qquad + \beta^{vs}_{i,\chi,n} \cdot A_{i,t^{las}(n^{pre}(n)),\chi,n^{pre}(n)}\Big) \qquad \forall s(t) \geq 2 \qquad (7.8) \\
& + \sum_{\chi}\Big(\alpha^{vs}_{i,\chi,n^{pre}(n)} \cdot \hat{\kappa}^{seg}_{i,t^{las}(n^{pre}(n^{pre}(n))),\chi,n^{pre}(n^{pre}(n))} \\
& \qquad + \beta^{vs}_{i,\chi,n^{pre}(n)} \cdot A_{i,t^{las}(n^{pre}(n^{pre}(n))),\chi,n^{pre}(n^{pre}(n))}\Big) \quad \forall s(t) \geq 3
\end{aligned}$$

The first of the three main summands in Eq. (7.8) corresponds to the cumulative cost calculation in the deterministic model version. Here, this expression is used to calculate cumulative costs based on the learning rate applicable in the currently active stage. Summands two and three represent the differences to be added to or subtracted from this cumulative cost value, previously referred to as vertical shifts of the curve. For example, the second summand would add or subtract the vertical shift applying for stage one to two transition in a hypothetical two-stage model setup. In a setup with more than two stages, the difference applied to cumulative costs before vertical shifts is a sum of more than one vertical shift. For the three stage model, the second summand referring to the previous period, thus yields the value of the shift between stages two and three, the third summand the value of the shift between stages one and two.

The described setup now enables representation of uncertainty for learning rates, fossil fuel prices, the carbon bound and renewable potential. The resulting model is characterized by more than 210,000 rows, 210,000 columns, 800,000 non-zero variables and 640 discrete columns.

7.3 Results

Over the following subsections the impact of different types of parameter uncertainty on the optimal renewable deployment strategy within the framework of the SCORE model is described. After a brief review of parameterization, stochastic results are presented and compared to the deterministic reference case. In addition to the probability-weighted mean of the stochastic results, the values for the different stochastic nodes[21] are shown as well.[22] Detailed results in tabular form can be found in Appendix C.2.

Some limitations of the model, in addition to the ones already noted for the deterministic CORE, should be born in mind when interpreting results. (1) Representation of uncertainty is highly simplified. The model is divided into only three stochastic stages and branching into only two new states of the world takes place at each node. The duration of the stages are several decades long, whereas information updating in the real world occurs continuously. In case of multiple uncertain parameters, the latter are grouped into scenarios to avoid an increase in the number of stochastic nodes. Approximated experience curves are only divided into four distinct segments. (2) In absence of empirical data, parameter values for the different states of the world and the corresponding transition probabilities have been exogenously assigned. (3) Risk-neutrality is deliberately assumed, which may, however, complicate comparability of results with other publications, where decision makers are modeled as risk averse. (4) Finally, although the MIP is solved to a very small tolerance, it is possible that slightly better results than the one presented as optimal here exist, especially for model year 2100.[23]

7.3.1 Uncertain Learning Rates

Three different parameter configurations are used to examine the impact of learning rate uncertainty on optimal renewable deployment strategy. PV and CSP learning rates, each in isolation, as well as a configuration where learning rates of several renewable technologies are uncertain.[24] In the latter case, learning rate developments are perfectly correlated, e.g. if learning rates are high for onshore wind, they are also high for PV.

Examination of the uncertainty in the learning rate of PV is of particular interest, since differences may lead to vastly different optimal investment strategies: either making PV the dominant technology in 2100 or preventing its implementation altogether. Figure 7.5 displays results for this case of an uncertain PV learning rate (detailed results can be found in App. C.2).

From an overall perspective, the deployment of PV is, on average, accelerated in presence of uncertainty. Whereas investment does not commence before 2075 in the deterministic case, it already starts in 2050, if node *a* is realized in stage two. In stage three, however, only in three of the four stochastic nodes (*aa, ab, ba*) PV is pursued at all. If a low learning rate trajectory *b* in stage two continuous into stage three (*bb*), investments are fully

[21] a, b, aa, ab, ba, bb.

[22] The presented variables are either power generation or cumulative investments across all model years.

[23] Results for 2100 have a comparatively small impact on the objective function value due to discounting.

[24] Wind (on- and offshore), biomass, PV, CSP (base plant and storage).

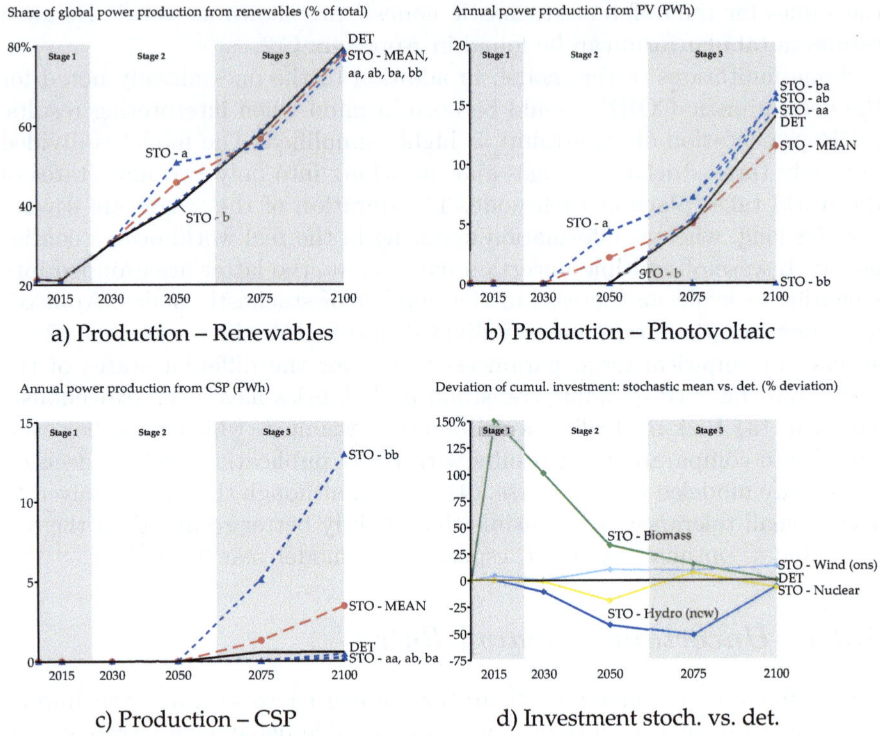

Fig. 7.5 Results for uncertain PV learning rate
Source: Own representation

directed towards CSP. This fact illustrates the important role of information updating in the model. If node a is realized in stage two, it is certain that at least a medium path PV learning rate will be realized in stage three. In this case it is optimal to already begin with PV deployment in stage two. In contrast, if node b materializes, there is a high likelihood that PV learning will continue to remain at the low path rate of 10%. In this case it is optimal to postpone investment into one of the solar technologies until uncertainty is resolved at the beginning of stage three. Investment into CSP would therefore not be optimal in the current setting.[25]

In line with the results for PV, the transition towards renewables is, on average, moderately accelerated in presence of an uncertain PV learning rate. Whereas the share of renewables in 2050 clearly lies above the deterministic case in node a, it approximately equals it for the low learning rate in node b. This acceleration occurs largely at the expense of nuclear technology, whereas

[25] Consequently the use of the stochastic model over a deterministic scenario analysis would not improve decision making with regard to optimal CSP investment under this setup.

the power share from fossil technologies in the stochastic setting remains approximately equal to the one in the deterministic reference case.

Optimal investment into other technologies than PV and nuclear is also affected by the presence of PV learning rate uncertainty (see Fig. 7.5d). Investments into new hydro plants are considerably delayed — yet this delay is made up for again in the last model year to arrive at almost the same value for cumulative investments as in the deterministic case.[26] The opposite holds true for biomass, for which deployment is, on average, accelerated over earlier models periods.

For the other two parameter configurations, i.e. uncertain CSP learning rate and uncertain learning rates for several renewables, the same overall effect on the share of renewables can be observed as for PV learning rate uncertainty (see App. C.2.1 and C.2.3): the deployment of renewables is, on average, accelerated so that they command a moderately higher share of power generation in 2050 than in the deterministic case. For uncertain CSP learning, CSP implementation already starts by 2030 in node *a*, i.e. given a high CSP learning rate is demonstrated in stage two. In node *b* no CSP implementation takes place but PV investments are brought forward to 2050 since the opportunity for CSP to follow a high learning rate path in stage three no longer exists, implying that there is no danger of regretting the up-front investments into PV after uncertainty has been fully resolved in stage three.

In summary, the most important effect from the introduction of learning rate uncertainty has been the moderate acceleration of renewable power deployment, driven largely by earlier investments in one of the solar technologies. In contrast, two technologies characterized by constant costs and long useful plant lifetimes, new hydro and nuclear, were delayed in presence of uncertain learning rates. Despite the changes in optimal investment timing, the power generation landscape in 2100 again looked very similar to the one in the deterministic reference case for all three model configurations with uncertain learning.

7.3.2 Uncertain Carbon Bound

Already in the deterministic model version, results were highly sensitive to global carbon cap assumptions. This holds particularly towards the end of the model horizon, when allowed greenhouse gas emissions only constitute a small share of the amount emitted today. In the stochastic setup, the carbon bound is specified as either 1.5 times (high, node *b*) or 0.5 times (low, node *a*) of the deterministic value. Consequently absolute deviations in the carbon cap

[26] To a certain degree the reduction of the relative deviation in Fig. 7.5d is also due to an increasing denominator in the calculation (cumulative investments from the deterministic case).

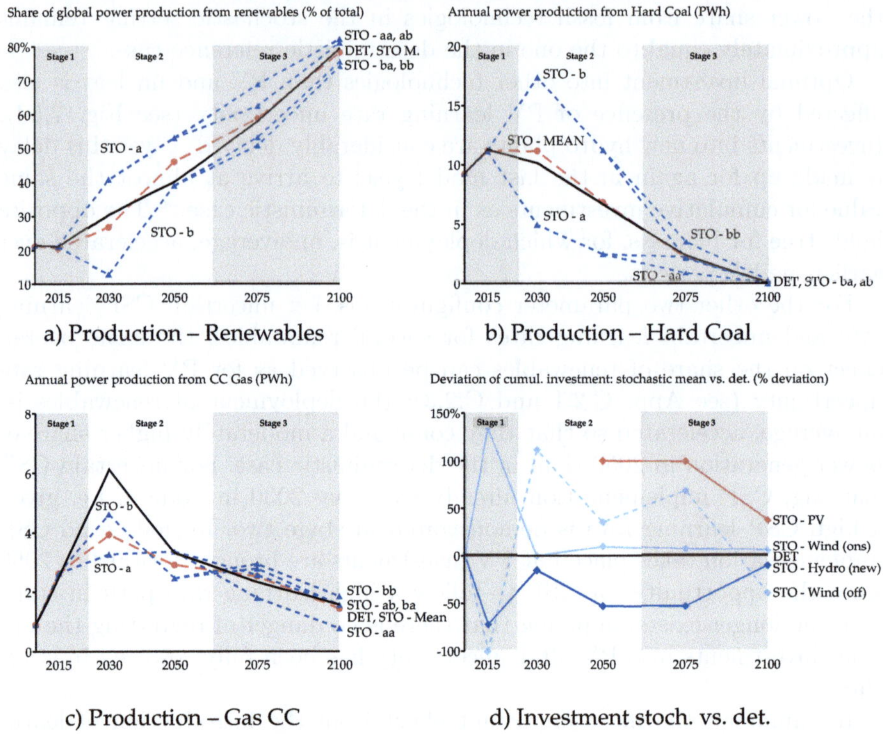

Fig. 7.6 Results for uncertain carbon bound
Source: Own representation

between the different states of the world are higher in earlier model periods, starting with 2030.

The most obvious observation under this configuration is a large discrepancy in results depending on whether a high or low cap value materializes in stage two (Fig. 7.6). As one would expect, power generation from hard coal in 2030 is almost doubled in case of a high cap and almost halved for a low one (see Fig. 7.6b). This development is also clearly reflected in the renewable share: deployment is greatly accelerated in case of a low cap and postponed for a high one (see Fig. 7.6a). On average, however, both the share of production from coal as well as the renewable power share are similar to the one in the deterministic case.

The introduction of carbon bound uncertainty has an interesting effect on production from combined cycle gas turbines (Fig. 7.6c). For any state of the world realized, production is lower than in the deterministic case. In case of a high cap, gas is replaced by lower-cost hard coal technology in 2030 and, to a lesser extent in 2050. In case of a more stringent bound, nuclear and wind, in 2050 also PV, substitute power production from gas. On average, the introduction of carbon cap uncertainty, therefore clearly has a negative effect on generation based on natural gas here.

Introduction of carbon cap uncertainty also impacts early renewable technology investments. For 2015, investments into new hydro stations and offshore wind are redirected into onshore wind. A possible explanation for this observation is that investments into technologies, which have a higher dependence on the state of world realized in stage two, are postponed in favor of investments into technologies that will be pursued for any realization (Fig. 7.6d). This way, regretting the investment, if an unfavorable state of the world is realized, can be avoided. Once uncertainty is then partly resolved in 2030, the delays are made up for again to a large degree.

In summary, the introduction of uncertainty in the carbon cap will, on average, not lead to large changes in the optimal deployment of renewables. Yet, the optimal development of the energy system will vary considerably depending on the realization of the applicable carbon cap in the mid-term. On average, PV and offshore wind, benefit from uncertainty introduction.[27] New hydro plants, as in the previous section, as well as combined cycle turbines play a less important role than in the deterministic setup.

7.3.3 Uncertain Fossil Fuel Prices

As fossil fuel price formation is not endogenous to the model, gas and coal prices constitute another important factor impacting optimal investment in the power sector, whose future values are highly uncertain. The two parameter configurations tested here are, on the one hand, uncertain natural gas prices in isolation and, on the other hand, uncertain hard coal and natural gas prices together. Similarly to the runs with multiple learning rates, the development of hard coal and gas prices in the second configuration is perfectly correlated. Nodes *a* and *aa* are characterized by high prices, *b* and *bb* by low ones.

Similar to the results from the previous sections, the technology affected directly by the introduction of natural gas price uncertainty benefits most: combined cycle gas turbines (see Fig. 7.7). On average, gas-based power production under gas price uncertainty lies significantly above production in the deterministic reference case. Yet, also similarly to carbon bound uncertainty, there are large variations in production between the stochastic nodes in stage two (see Fig. 7.7c). Whereas production in node *b* with low gas prices is more than three times the one in the deterministic reference case, the opposite holds for the one in node *a* with high gas prices. Production levels only fully return to the deterministic value in 2100, when the impact of the carbon cap renders cost advantages from low fuel prices irrelevant. Other technologies, whose production is being impacted considerably by gas price uncertainty,

[27] Except for model year 2100.

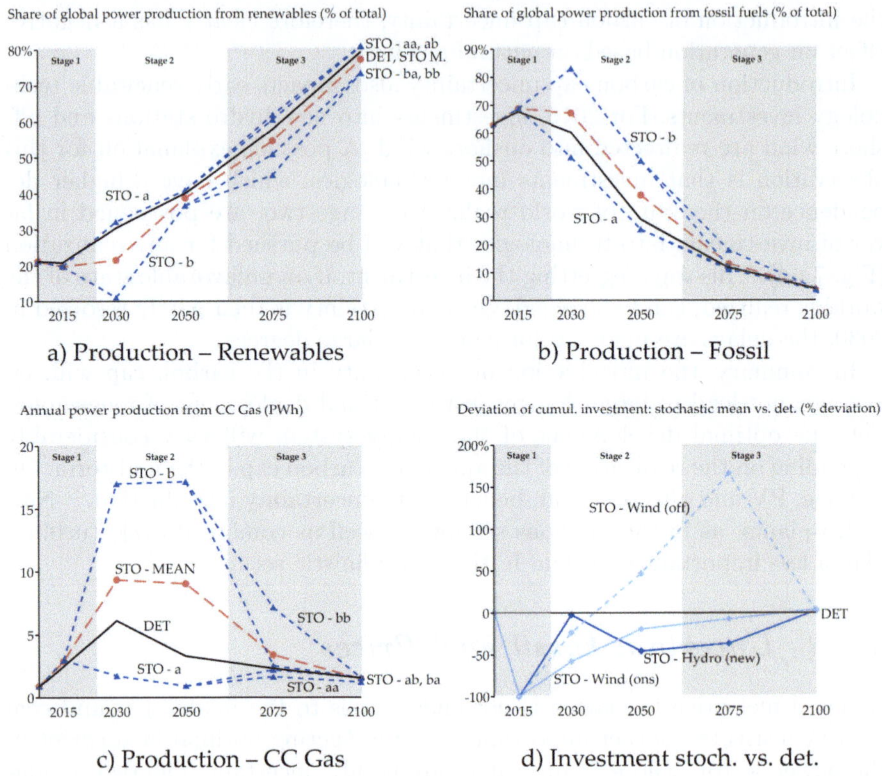

Fig. 7.7 Results for uncertain gas price
Source: Own representation

are coal, nuclear and biomass. Unsurprisingly, those dispatchable technologies interact in a substituting manner with gas based conversion technologies.[28]

The higher average share of gas-based production also translates into temporarily higher shares of fossil generation. The other side of the coin is a moderately postponed deployment of renewable power. In fact, the renewable technologies impacted most negatively are onshore wind and new hydro. Wind and hydro installations are reduced to zero in stage one to avoid regretting those investments in case low gas prices materialize in stage two. Cumulative investments into onshore wind and new hydro then trail investments of the deterministic case across the entire model horizon and only "catch up" again by 2100. In contrast to uncertainty in learning and carbon bound, PV is not positively impacted by the presence of uncertainty here and technology deployment does not start before 2075, as in the deterministic reference case.

Results for the case of combined coal and gas price uncertainty are similar to the ones for gas (see App. C.2.5). In periods before 2050, fossil technology

[28] Arguably the degree of dispatchability varies substantially among those technologies.

capacity additions even exceed the ones from the gas-only case at the expense of renewables. Renewable shares, however, then catch up at an even higher rate than in the gas-only case to approximately reach the value from the deterministic case again by 2100. Despite the positive correlation of coal and gas prices, gas on average still substitutes some coal production in comparison to the base case but to a lesser extent than in the setup with price uncertainty for gas only.

7.3.4 Uncertain Renewable Potential

Uncertainty in the renewable potential differs somewhat from the other types of uncertainty featured in this chapter. Whereas the future value of the carbon cap is uncertain, depending to a large degree on decisions of economic and political actors, the renewable potential for a particular reference technology will not change over time. Uncertainty in the renewable technical potential thus rather originates from lack of transparency than from actual developments in the future. Further research efforts could remedy or at least mitigate this issue.[29]

The stochastic run is set up so that the renewable potential in each of the segments of the technical potential function is either large (*a, aa*), medium (*ab, ba*) or small (*b, bb*), where large is 1.5 times and small 0.5 times the value of the deterministic reference case. For the computational reasons already mentioned earlier, sizes of the potential segments are perfectly correlated across technologies, e.g. if potential for onshore wind is large, potential for PV is also large.[30]

From an overall perspective, uncertain renewable potential has a moderately negative effect on the deployment of renewable power. On average, some of the investments into renewables may be postponed from 2030 to 2050 (Fig. 7.8a). The actual investment amount, however, depends on the materialized stochastic node: whereas a slight acceleration of renewable deployment versus the deterministic case can be observed for large potentials, a substantial postponement can be observed for small potentials. The technologies affected most by realization of low potentials are wind and hydro, for which potential with high energy carrier availability is scarce across all model regions.

At first glance surprising, uncertainty in the renewable potential has a positive impact on the deployment of solar technologies, which substitute other renewables. PV capacity is added earlier and power production even

[29] Changes in technical potential occur through technological performance improvement, e.g. increases in plant efficiency. Efficiency of the renewable technologies in this model are, however, held constant.

[30] The sizes of the different potential segments for the same technology are perfectly correlated as well, i.e. all segments are either larger than, smaller than or equal to their original size.

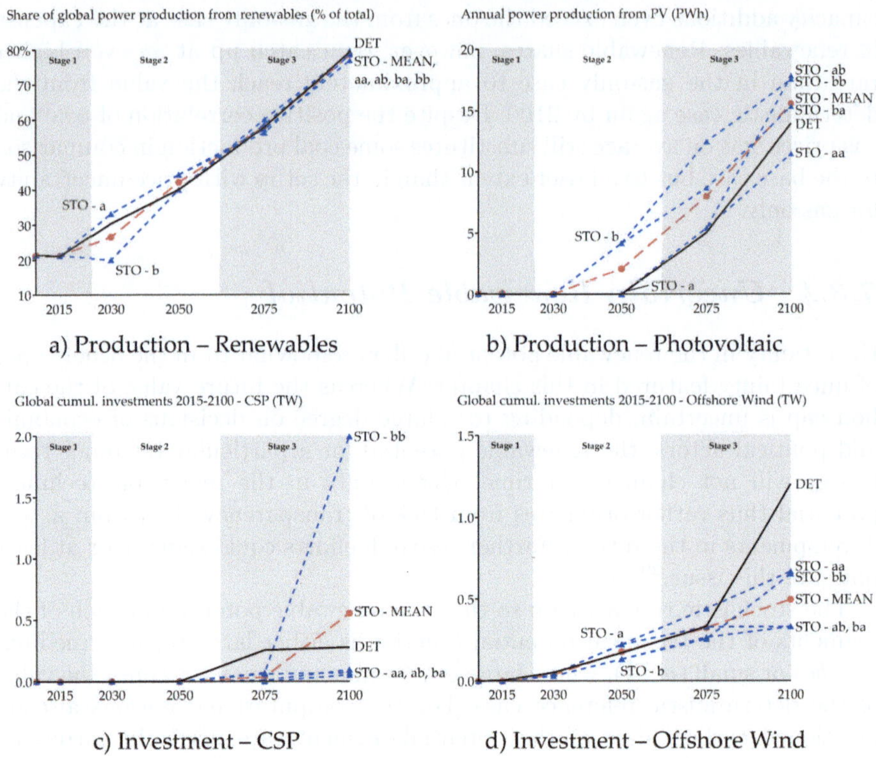

Fig. 7.8 Results for uncertain renewable potential
Source: Own representation

remains higher than in the deterministic reference towards the end of the model horizon. In case of sustained small technical potential in the third stage, substantial CSP capacity is even added alongside PV. These observations can be explained by the shape of the technical potential function. In regions, where sites are characterized by sufficiently high energy carrier availability to make solar power generation cost-competitive,[31] a large number of such high-irradiance sites exist. Consequently, the slope of the corresponding solar potential curves is comparatively flat and even if only half of the potential of the deterministic case is available, there is still enough potential with high irradiance for large scale deployment. In contrast, the slopes of the potential curves for other renewable technologies are comparatively steep and only at sites "at the beginning" of the curve power can be produced at competitive cost.

In summary, the key observation from the introduction of renewable potential uncertainty is that on average a moderate postponement of renewable

[31] For example, Africa or the Middle East.

deployment and partial substitution of wind, hydro and biomass through solar technologies takes place.

7.3.5 Multiple Uncertainties

In reality, a multitude of uncertain factors impact the cost-optimal investment strategy in the power sector simultaneously. It is obviously not possible to capture all of them within a model. Here a stochastic run was configured to include uncertain learning rates for all renewable technologies,[32] hard coal and natural gas prices, renewable potential and the carbon cap. It would certainly be most interesting to examine the impact of multiple uncertain parameters within a setup allowing for all different state-of-world combinations among the possible parameter developments, implying multiple branches per stochastic node. Unfortunately computational limitations render this impossible in SCORE. Instead, branching into two new nodes, as previously, is retained by introducing a *favorable* (*a*, *aa*) and an *unfavorable* (*b*, *bb*) parameter development path for renewables.[33] In light of this greatly simplified setup the interpretation of results is difficult and therefore confined to a mostly descriptive presentation.

On average, the deployment of renewables is moderately accelerated, however, with extreme differences depending on the state of the world being realized in stage two. In 2050, for example, the share of renewable power ranges from below 20% in node *b* to above 70% in node *a*. It is likely that the forced perfect correlation of parameter values is driving this large difference. Although some degree of correlation can be expected in the real world, results in a model setup with lower degrees of correlation might be less extreme.

The presence of multiple uncertainties here has a positive impact on gas and PV based power production. Production from combined cycle gas turbines is more than twice the one in the deterministic case, presumably driven by the combination of low gas prices and a comparatively lax carbon bound (see Fig. 7.9). PV development is, on average, greatly accelerated, with implementation in node *a* starting as early as 2030. Interestingly, for a sustained favorable path (*aa*), PV implementations fall below their deterministic value in stage three. Here the shares of biomass, onshore wind but also nuclear production are greatly increased over the deterministic reference.

The build-up of onshore wind is moderately postponed in stage two but then higher overall production is attained towards the end of the modeling horizon. As for model configurations with only one uncertain parameter, implementation of the long-lived hydro and nuclear technologies is postponed. This effect is, once again, most pronounced in 2050, presumably to wait with long-term investment decisions until uncertainty is resolved in the next model period.

[32] Except hydro, where no cost reductions take place.

[33] Favorable: High learning rates, high renewable potential, high fossil fuel prices, low carbon cap; unfavorable: opposite.

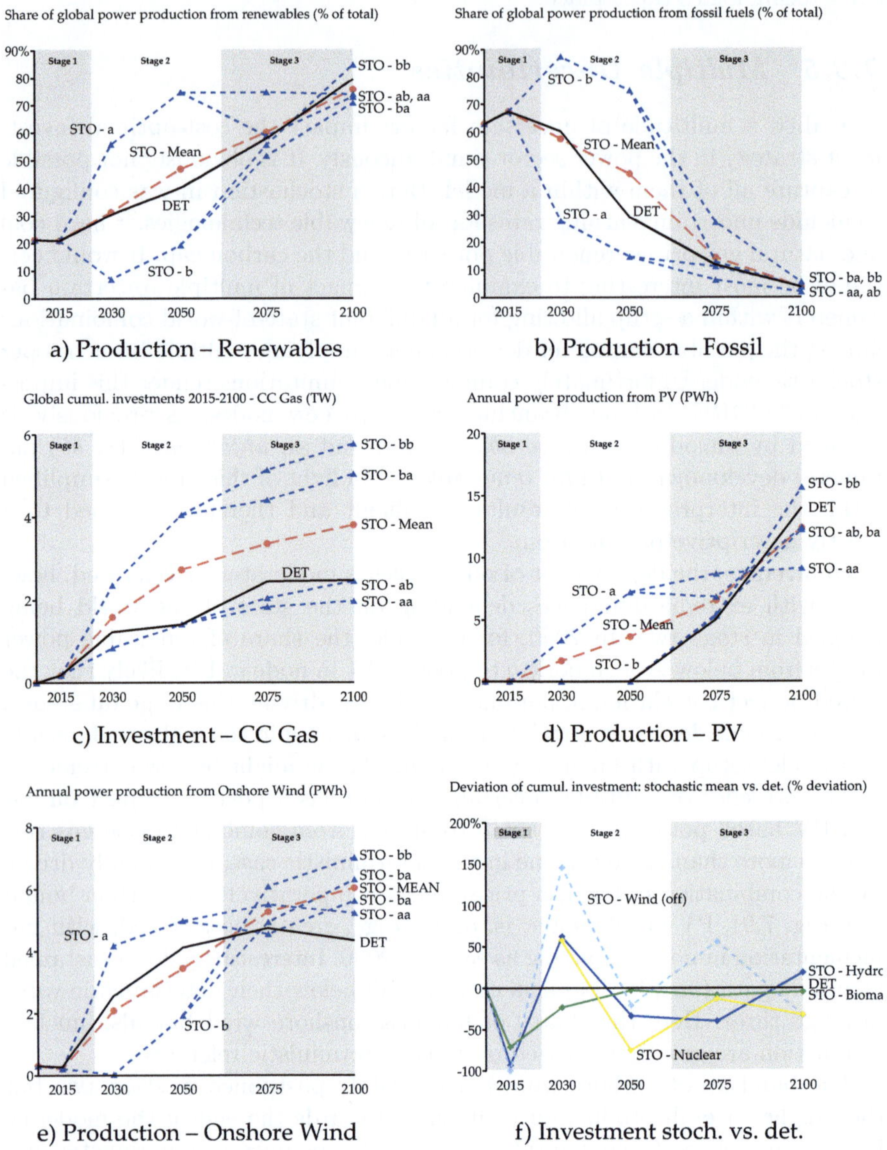

Fig. 7.9 Results for multiple uncertainties
Source: Own representation

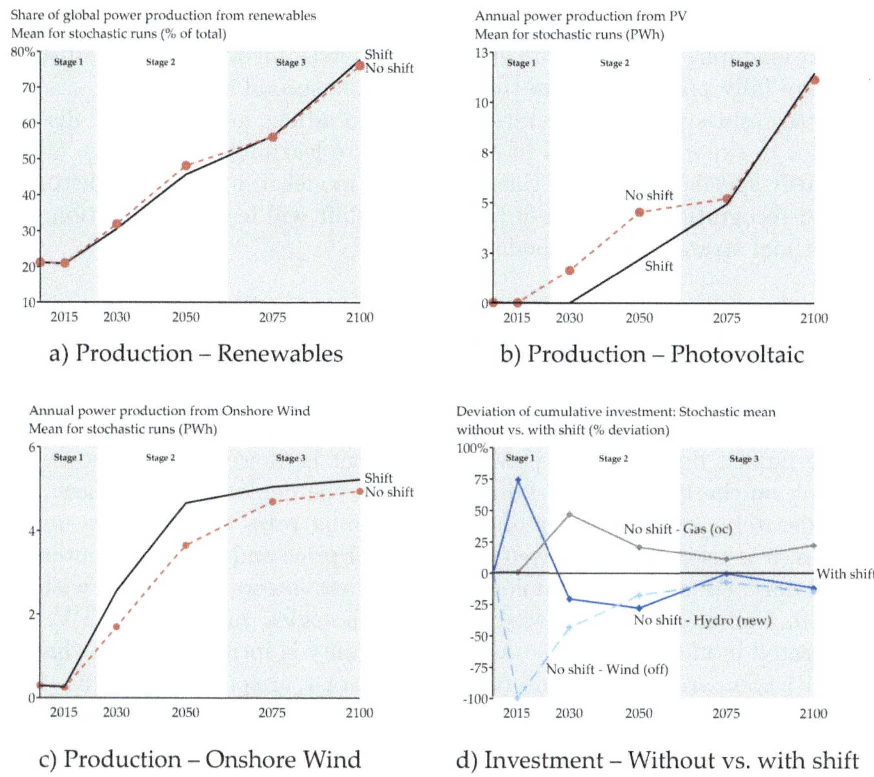

Fig. 7.10 Results without vertical shift
Source: Own representation

7.3.6 Vertical Shift Omission

In the previous section it has been asserted that omitting the vertical shift of the cumulative cost curve from the model's formulation will distort results. To demonstrate that this is indeed the case, a model version not reflecting this vertical shift has been set up and solved. Parameterization equals the one for PV learning rate uncertainty (sec. 7.3.1). The results of the model version reflecting and the one not reflecting the vertical shift are then compared to illustrate the impact of vertical shift omission (see Fig. 7.10).

When considering renewables as a group, only minor differences can be observed between the results of the two model versions. Yet, when reviewing results at the technology level, differences become apparent. Although results in both model version re-approach their deterministic counterparts towards the end of the model horizon, the build-out of PV is even more accelerated in the incorrect setup without vertical shift than already observable for the one with vertical shift. In contrast, power production from onshore wind in the version without shift trails the one in the version with shift across the entire model horizon. Other technologies are affected as well — hydro investments are, for example, accelerated in stage one, whereas investments into offshore wind are fully postponed from the first into the second stage.

These results further illustrate the need to adjust for unrealistic discontinuities in experience curves in a setup, where learning rates change at exogenously specified points in time. Although model results are not distorted beyond recognition, omission of the vertical shift will lead to suggestion of a suboptimal strategy by the model.

7.4 Key Findings and Implications

Introduction of uncertainty to the model had an impact on optimal investment timing for all uncertain parameters examined. The direction of this timing impact on renewable power deployment is, however, ambiguous, depending on the type of uncertainty. Whereas the deployment of renewables is moderately accelerated for uncertain learning rates and multiple uncertainties, it is moderately delayed for fossil fuel price and renewable potential uncertainty. For specific technologies, in contrast, general tendencies with respect to timing can be observed. Solar technologies, in particular PV, are accelerated in most of the cases when uncertainty is introduced.[34] Technologies with comparatively long useful life, in contrast, are postponed. This effect is most pronounced in model periods preceding a stage transition, likely to avoid regretting the long-term commitment in case of an unfavorable development for the competitiveness of those technologies in the next stage. This holds true for new hydro stations across all setups and, to a lesser extent, for nuclear power. Although power production from those technologies in the first stage is still largely determined by the initial capacity endowment, optimal investment levels already show considerable deviations from the deterministic base case by 2015 in expectation of the different possible future states of the world.

Despite impacts on optimal investment timing, technology shares reapproach their deterministic values again towards the end of the model horizon, i.e. technologies whose development had been postponed earlier on "catch up," growth of previously accelerated technologies "slows down." Introduction of uncertainty therefore appears to primarily have a timing impact here instead of provoking a substantial, sustainable change to the power system beyond the current century. A major driver behind this development is the stringent carbon bound, which steers the power sector towards a low emission structure in the long term. Yet, it is important to highlight that these findings only hold on average. Along the different possible paths parameter values may follow, optimal invest strategies differ substantially. Given an

[34] PV is accelerated for all types of uncertainty examined, except for uncertain fossil fuel prices.

uncertain carbon cap, for example, optimal annual production from hard coal in 2030 ranges from less than five to more than 15 PWh.

In case a model with explicit uncertainty representation and risk-neutral decision making, such as the one presented here, is considered more realistic than its deterministic counterpart, optimal deployment of a technology, whose deployment depends heavily on cost reductions,[35] should occur earlier than suggested by the deterministic model. Yet despite this finding, even under explicit consideration of uncertainty, PV or CSP deployment today would not be optimal from a cost minimization perspective.

From a methodological point of view, results support the assertion that adjusting for the vertical shift of the experience curve in models with learning uncertainty resolution at fixed points in time is necessary. Otherwise, the investment timing suggested by the model solution will not be optimal, as demonstrated here for the case of uncertain PV learning rates.

In comparison to previous research on learning rate uncertainty, the model results here most closely match the results of Mattsson (2002) and Shittu (2008), who witness early deployment of learning technologies in case of uncertainty. In the models presented by Barreto (2001) and Grübler and Gritsevskyi (1997), on the other hand, introduction of learning uncertainty postponed the advent of the corresponding learning technology. Comparison of the results obtained with the ones from literature is, however, only possible to a limited extent, as the corresponding models assume risk aversion.

The inclusion of risk aversion in the SCORE model would be a logical next step, as well as the simultaneous examination of multiple uncertainties in a setting, where parameter developments are not perfectly correlated. This way, interactions among the different types of uncertainty could be studied. For the latter, ways must be found to render the approach computationally feasible, whether this be by choosing a different methodology for the stochastic enhancement or a simplification of the underlying deterministic optimization model. Finally, the strong influence of uncertainty in the renewable potential highlights the need for further efforts in the detailed quantification of renewable potential with consistent methodologies applied to a comprehensive set of renewable technologies. These efforts could eliminate or at least greatly reduce uncertainty in renewable potential due to limited transparency.

References

Alberth, S.: Forecasting technology costs via the experience curve - Myth or magic. Technological Forecasting and Social Change 75(7), 952–983 (2008)

Barreto, L.: Technological Learnin. In: Energy Optimisation Models And Deployment Of Emerging Technologies: Dissertation. Dissertation, Swiss Federal Institute of Technology, Zurich (2001)

[35] For example, PV or CSP.

Berglund, C., Söderholm, P.: Modeling technical change in energy system analysis: analyzing the introduction of learning-by-doing in bottom-up energy models. Energy Policy 34(12), 1344–1356 (2006)

Gritsevskyi, A., Nakićenović, N.: Modeling uncertainty of induced technological change. Energy Policy 28(13), 907–921 (2000)

Grübler, A., Gritsevskyi, A.: A Model of Endogenous Technological Change through Uncertain Returns on Learning (R&D and Investments): Paper presented at the International Workshop on induced Technical Change and the Environment, Laxemburg, June 26–27 (1997)

Grübler, A., Gritsevskyi, A.: A Model of Endogenous Technological Change through Uncertain Return on Innovation. In: Grübler, A., Nakićenović, N., Nordhaus, W.D. (eds.) Technological Change and the Environment, pp. 280–319. Resources for the Future and IIASA, Washington D.C. (2002)

Grübler, A., Nakićenović, N., Victor, D.G.: Dynamics of energy technologies and global change. Energy Policy 27(5), 247–280 (1999)

Held, H., Kriegler, E., Lessmann, K., Edenhofer, O.: Efficient climate policies under technology and climate uncertainty. Energy Economics 31, S50–S61 (2009)

Kumbaroglu, G., Madlener, R., Demirel, M.: A real options evaluation model for the diffusion prospects of new renewable power generation technologies. Energy Economics 30, 1882–1908 (2008)

Mattsson, N.: Introducing uncertain learning in an energy system model: a pilot study using GENIE. International Journal of Global Energy Issues 18(2-4), 253–265 (2002)

Mattsson, N., Wene, C.-O.: Assessing New Energy Technologies Using an Energy System Model With Endognized Experience Curves. International Journal of Energy Research 21(4), 385–393 (1997)

Messner, S.: Endogenized technological learning in an energy systems model. Journal of Evolutionary Economics 7(3), 291–313 (1997)

Rout, U.K., Blesl, M., Fahl, U., Remme, U., Voß, A.: Uncertainty in the learning rates of energy technologies: An experiment in a global multi-regional energy system model. Energy Policy 37(11), 4927–4942 (2009)

Shittu, E.: Stochastic Modeling of Endogenous Technological Change (2008)

Chapter 8
Summary and Outlook

The main objective of this work has been to examine the future deployment of renewable power from a normative point of view. A particular focus has been put on identifying particular properties characterizing the cost-optimal transition towards a renewable power system, analyzing the key drivers behind this transition with a particular focus on technological cost reductions, and investigating the implications of uncertainty associated with those drivers. Bridging the gap between the two strands of literature of renewable potential, on the one side, and energy system modeling with endogenous technological change, on the other side, counts among the methodological contributions of this book. Moreover, specific considerations for the formulation of models featuring variable learning rates over time have been described and the effect of uncertainty for optimal investment strategies in an environment with bounded increasing returns has been explored.

A model-based approach has been followed to address those topics. First, the methodological questions regarding correct model formulation in case of variable learning rates over time as well as optimality of technological concentration versus diversification under uncertain learning rates have been examined employing a *stylized, analytically solved nonlinear program* with two stages (see chapter 5). It features two substitute technologies and strictly increasing returns due to endogenized experience curves. As the latter also render the model non-convex, the analytical solution process follows an enumerative approach, under which solutions for all meaningful local optima have been derived. Those solutions have then been compared among each other to identify the global optimum. The *large-scale, numerically solved power system model CORE* has then been employed to examine the optimal future deployment strategy for renewable power (see chapter 6). It has been set up as bottom-up, mixed integer program and reflects aspects of particular relevance for the competitiveness of renewable technologies in detail. Those aspects include endogenized experience curves, detailed renewable potential curves and intermittency but also technology specific aspects, such as detailed storage modeling for CSP plants or learning spillovers between

F. Wagner, *Renewables in Future Power Systems*,
Green Energy and Technology,
DOI: 10.1007/978-3-319-05780-4_8,

on- and offshore wind. Experience curves as well as renewable potential curves have been linearly approximated to avoid non-convexity of the model and to harmonize the model's setup with the available data, respectively. Implications of uncertainty have then been examined using the *three-stage stochastic program SCORE*, which is effectively an extension of CORE (see chapter 7). A stochastic programming approach has been chosen in support of the normative perspective taken for all analyses in this book. The model has been applied to investigate the impact of introducing different types of parameter uncertainty, such as uncertain learning rates or carbon bound, on the optimal renewable deployment strategy.

A key result of the CORE model is a perspective on the future *optimal deployment strategy for renewables.* For the base case, a technological "switching of regime" can be observed around the middle of the century. Whereas its first half is still characterized by fossil fuel dominance, renewables dominate in the second half. Transition towards a largely renewable power system therefore does take place under cost optimization (see chapter 6). The main drivers behind this development are the carbon dioxide emission cap and cost reductions for renewables along the experience curve. Increases in fossil fuel prices, in contrast, play a less important role. In fact, in absence of an emission restriction, the transition towards renewables would not take place and, instead of an almost 80% global renewable penetration in 2100, only 20% would be optimal. Despite presence of increasing returns, the optimal global technology portfolio is quite diverse towards the end of the century. This can largely be explained by the fact that increasing returns are bounded, in particular from the restrictions imposed by the explicitly considered, differentiated renewable potential. The investment strategy meeting the stringent carbon emission reduction target in a cost-optimal fashion includes a solar technology towards the end of the century and solar technology, although not at all pursued during the first half of the century, even constitutes the largest power producer by 2100. Its penetration, however, still lies below today's penetration of coal, also illustrating the movement towards a more diverse technology portfolio. Not only solar but also other technologies' penetrations differ vastly across world regions. Those different regional technology mixes also go in line with large deviations in average generation costs. Whereas power in some regions with comparatively generous renewable potential endowments fully comes from renewable technologies in 2100, other regions with less favorable renewable energy carrier supply still operate conventional plants leading to considerably higher average production costs. The optimal renewable deployment strategy is also characterized by a sequential rather than simultaneous implementation of initially uncompetitive technologies. Whereas onshore wind, for which up-front investments to reach competitive cost levels are comparatively small, is pursued first, implementation of solar technologies, for which high up-front investments are required, is postponed to the second half of the century. The rationale behind this strategy is that "low hanging fruits" are picked first and only when plant locations with sufficiently favorable conditions are occupied, investment into the next most attractive technology should take place. If risk neutrality is assumed, as done throughout the analyses in this book, the sequential technology implementation approach is superior to a simultaneous one in presence of positive opportunity costs of capital.

Current national renewable support policies, which are aimed at mitigating market failure from presence of increasing returns, do not fully conform with the findings described above. Germany, for example, currently fosters wind as well as PV deployment. Even when ignoring the aspect of optimal timing, the implementation of PV in Germany appears not to be part of the cost-optimal strategy. Although PV contributes large amounts of power for the cost-optimal strategy from the CORE model, no implementation takes place in model regions characterized by similar radiation regimes as found in Germany — and this despite comparatively low capital costs below 1,000 \$/kWp towards the end of the century. Under the optimal strategy, capital cost reductions from deployment of PV are realized in regions with comparatively favorable solar regime. As output from plants in those regions is often more than twice the one found in regions with a solar regime comparable to the one of Southern Germany, up-front investment costs to drive the technology to competitive cost levels, are, ceteris paribus, also only half as high.[1] Deployment in regions with favorable energy carrier supply may consequently result in savings of billions of dollars. These observations illustrate that agreement on a stringent international carbon cap and implementation of a corresponding, competitive international trading mechanism may be prerequisites for a globally cost-optimal power technology deployment. Inferences from model results on real world renewable support policies, however, need to be qualified in light of the model's limitations. Those relate, among others, to the partial instead of general equilibrium approach, the pre-selection of technologies available for investment, model simplifications, such as the representation of the technological change process in form of experience-curves or the implicit assumption of perfect foresight. Probably the greatest simplification is the neo-classical approach of cost-minimization as sole objective. In chapter 4 it has been illustrated that costs are only one aspect in the process of technological diffusion and substitution. Other elements also vastly affect this process of change, such as institutions or decision maker preferences.

Implications of uncertainty on optimal renewable strategy are ambiguous (see chapter 7). Whereas introduction of uncertainty to most of the examined model parameters has a substantial impact on optimal investment *timing*, the *long term penetration* of renewables in the stochastic setting always re-approaches its deterministic counterpart. This development is largely driven by the stringent carbon cap. With regards to optimal timing, parameter uncertainty in renewable learning rates, on average, moderately accelerates the deployment of renewables, whereas uncertainty in fossil fuel prices postpones renewable deployment. For particular technologies, more general observations

[1] In reality, price levels for power deviate across regions.

can be made. Deployment of technologies with long useful economic life is generally postponed. This is most extreme for new hydro power plants, which are postponed for all considered parameter uncertainties. The effect is particularly pronounced preceding a stage transition, where the decision maker postpones investment to avoid regretting the long term commitment in case a state of the world materializes that would render pursuing a different technology less costly. Introduction of uncertainty also leads to investment reallocations among renewable technologies. For uncertain carbon bounds, for example, new hydro and offshore wind investments in the first stage are redirected towards "less risky" onshore wind. Across different uncertain parameters, it also appears that technologies directly affected by uncertainty benefit most. In the risk-neutral setting used, this is likely driven by the fact that the opportunities from favorable development of the parameter under consideration outweigh the disadvantages associated with a negative development — in the latter case other substitute technologies can replace the technology directly affected by uncertainty. Overall, introduction of uncertainty appears to mainly affect optimal investment timing under the chosen setup. Even in this risk-neutral setting, a certain degree of "hedging" can be observed before stage transitions, where a portion of technologies still implemented in the deterministic case are substituted by technologies with shorter useful life or lower gaps to competitiveness.

CORE combines *differentiated renewable potential* restrictions with a *bottom-up modeling framework featuring experience curves*, therefore effectively merging those two strands of literature. The results demonstrate that considering potential, differentiated by region and productivity in system models with endogenized experience curves, vastly affects model outcomes (see chapter 6.4.3). A more diversified technology portfolio emerges when differentiated potentials are considered, owed to more stringent bounding of the increasing returns brought about by endogenous cost reductions. The more detailed the regional potential data is differentiated by productivity, i.e. the more segments are used to approximate the nonlinear supply curve, the more technologies are implemented across the model horizon.[2] Moreover, the deployment of renewables is accelerated for higher degrees of differentiation, as the now smaller segments with higher average energy carrier supply at the beginning of the supply curve act in the same manner as niche markets, effectively reducing the up-front investments required to bring costs to competitive levels. Based on these results, it is the clear recommendation that future bottom-up modeling efforts, aiming to realistically reflect renewable competitiveness, consider potentials differentiated by energy carrier availability.

Another methodological finding, the author did not identify in existing literature, pertains to *model formulation for changing learning rates* over

[2] Up to the maximum amount of six used for the base case model runs.

time.[3] At stage transitions, where cumulative capacity[4] values are determined model-endogenously and therefore not known ex-ante, a change in the learning rate will lead to an unrealistic discontinuity in the experience curve. The model formulation is therefore adjusted to explicitly correct for this discontinuity. Formulation of the adjustment term, called *vertical shift* in the previous chapters, differs depending on whether a nonlinear or linear model is used (see chapters 5.4.1 and 6.2.6, respectively). Two SCORE model runs with and without this adjustment clearly illustrate that the adjustment is not only required to ensure academic accuracy but also that ignoring it will impact model results. It is therefore recommended that models featuring experience curves with changes in learning rates at specified points in time reflect this *vertical shift* to avoid distortion of results.

An interesting question relating to optimal technological investment strategies in models with endogenous learning under uncertainty is the one of *concentration versus diversification* before uncertainty resolution, i.e. optimality of concentrating investments on one technology versus spreading it over several ones. Several pieces of research have identified optimality of diversification strategies, yet also note that these findings may not be generally valid. The results from the analytical investigations in chapter 5 suggest that presence of endogenous cost reductions via experience curves under uncertainty, given technologies are perfect substitutes and increasing returns are not bounded, is not a sufficient condition for optimality of a diversification strategy. This is demonstrated by comparing the six analytical solutions, where investments are concentrated on one or the other technology, with the relevant solution, where any mix between the two substitutes is permissible. As this demonstration is done using numerical experiments, a generally valid proof is left for further research.

The question of optimality of investment concentration versus diversification would in a broader sense constitute an interesting topic for further research. Specific research topics could include a generally valid proof of the finding from above or exploring optimal strategies under risk aversion and in models with recourse, e.g. by introducing construction lead times or uncertainty in constraints for a model similar to the one used in this book. Furthermore, the case of bounded increasing returns could be investigated, e.g. deriving conditions under which a diversification would be optimal, if so at all.

Regarding the topic of optimal renewable deployment strategies, which are derived using large-scale system models, several further topics of research could build upon the work carried out. A major shortcoming of CORE is its limitation to the power sector. Therefore, analyses based on models retaining a similar technological explicitness of the power sector, but also considering

[3] Time-dependent uncertainty resolution is used over a model-endogenous one to avoid asynchrony of stage transitions for learning rates and other parameters characterized by time-dependent uncertainty resolution.

[4] Independent variable of the experience curve.

interactions with the remainder of the energy sector, sectors apart from energy or even a climate module could offer additional insights. Another topic would be the broadening of the technology portfolio available for investment, in particular with respect to alternative carbon free technologies, such as CCS. Moreover, the impact of intermittency on optimal strategy could be investigated following a more elaborate approach than the one used in CORE as well as one explicitly reflecting grid-related aspects. There is also a need for further research regarding renewable potential data. Value could be created by deriving differentiated regional potential curves for a large set of technologies using the same methodology. This would also reduce the degree of uncertainty from renewable potential transparency, when deriving optimal renewable investment strategies. With respect to the impact of parameter uncertainty on optimal renewable deployment, a next step could be to extend the SCORE model to feature risk-aversion as well as investigating the effect of considering different types of parameter uncertainty in a simultaneous manner without parameter developments being perfectly correlated.

No matter how sophisticated economic models may become over time, they will still not be capable of fully predicting the actual evolution of the power sector. Apart from the sector's highly complex nature, a high degree of political interference impacts its development over time, mainly aimed at issues not efficiently regulated by market mechanisms alone. Many of those issues, for example, preferences for supply security or the willingness of a society to invest in pollution mitigation, can hardly be meaningfully quantified and represented in system models. Yet despite those shortfalls, the models applied in this book demonstrate that fundamental changes must occur, even from a cost optimization perspective, to transform today's power sector into a more sustainable one over the course of this century.

Appendix A
Renewable Power Technologies: Description, Generation Cost and Potential

This appendix provides further details on the renewable technologies considered in the CORE model: hydro, wind, PV, CSP and biomass. In addition to a technology description, current levelized generation costs for sample locations are calculated and renewable potential literature is reviewed for each of these technologies. The term *current* refers to the year 2010. Levelized power generation costs are calculated according to Equation (A.1).

$$C = \frac{i \cdot (K + E)}{\left(1 - \frac{1}{(1+i)^{\pi}}\right) \cdot h} + \frac{O}{h} + F + P, \tag{A.1}$$

where π represents the useful life of the plant in years, i the annual interest rate in percent and h the expected annual production, expressed in terms of full load hours.[1] Cost types considered are:

i. Overnight capital costs (K) in \$/kW
ii. Construction financing costs (E) in \$/kW
iii. Non-fuel operating costs (O) in \$/kWa
iv. Fuel cost (F) in \$/kWh
v. CO_2 certificate costs (P) \$/kWh[2]

Overnight capital and construction financing cost annuities are based on a real interest rate i of 7%. Construction financing costs[3] are calculated assuming a linear buildup of cumulative investments across the construction duration and annual compounding of interest. Generally, decommissioning costs are incurred at the end of a plant's useful life. Yet, these costs are ignored since they only constitute a marginal share of total costs when discounted back to the time of plant commissioning. Operating costs are stated in annualized form. The assumed values for overnight capital and operating

[1] Full load hours $= \frac{\text{Annual power generation (Wh/a)}}{\text{Rated capacity (W)}}$.
[2] Assumed as 25 $\$/t_{CO_2}$.
[3] Value at plant commissioning date.

costs are based on a literature review. This data is generally stated in Euro or US Dollar. US Dollars are used here, since most cost data in the energy sector, especially fuel prices, are quoted in this currency.[4] Cost data from reviewed literature, which has been modified, is marked with an asterisk in the corresponding charts. Modifications are usually currency conversions but may also include the conversion of operating cost data into an annualized value.

An important aspect when determining the current cost position of renewable technologies are cost reductions, for example, due to learning or economies of scale. To ensure meaningful comparison of capital costs, only recent literature data has been considered for technologies, where considerable cost reductions have been observed, notably PV, CSP or wind.

Besides costs, the other key factor impacting levelized generation costs is annual power production. In contrast to most conventional plants, renewable plants generally need to be constructed at sites with local availability of the primary energy carrier. The supply of the primary energy carrier determines possible power production and differs across locations. Generation costs in this appendix are therefore calculated for sample locations.

The concept of the *renewable potential* is widely used to describe renewable energy carrier availability for a defined area (see chapter 2.2.2 for an introduction). The potential data presented in this appendix focus on the technical and economic potential categories and have been taken from literature. For each covered renewable technology, a reference technical potential, differentiated by world region and productivity, is presented.[5]

A.1 Hydro Power

A.1.1 Overview

Hydro power is one of the most mature power generation technologies and already played a key role in the electrification process more than a century ago. The electricity for the first long distance, high-voltage transmission demonstrated at the electro-technical exhibition at Frankfurt a.M. in 1891, for example, had been generated by a hydro power station. It has since played an important role in global power generation and today still accounts for 16%

[4] Currency conversion poses a considerable challenge due to the appreciation of the Euro over the US Dollar across the last decade. To reduce the impact of year-on-year differences in publication date on the applicable exchange rate but still preserve the relationship of publication date to current exchange rate level, a four-year rolling annual average exchange rate up to the year of publication is used (see Tab. A.1). The consideration of exchange rates in years prior to the publication date is considered reasonable, since most data has presumably been gathered before the year of publication.

[5] In case of data availability.

Table A.1 Average annual EUR/USD conversion rates (Oanda.com, 2010)

Year	Exchange rate (EUR/USD)
2000	0.92492
2001	0.89658
2002	0.94590
2003	1.13208
2004	1.24386
2005	1.24539
2006	1.25622
2007	1.37074
2008	1.47134
2009	1.39463

of global production (see chapter 2.1) with a worldwide nameplate capacity of approximately 850 GW (World Energy Council, 2009).

Hydro power stations generate electricity by first converting the potential and kinetic energy of water into mechanical energy via a turbine, which is subsequently converted into electricity by means of a generator. The rated power of a particular station in Watt, P, is given by Equation (A.2) (Giesecke et al., 2009).

$$P = \eta \cdot \rho \cdot g \cdot Q \cdot h, \tag{A.2}$$

where η represents the efficiency of the entire plant in percent, ρ the density of water in $\frac{kg}{m^3}$, g the gravitational acceleration in $\frac{m}{s^2}$, Q the water throughput in $\frac{m^3}{s}$ and h the head[6] in meters. The key drivers determining the rated power of a station are consequently the head, water throughput and plant efficiency, since gravitational acceleration and the density of water can be considered constant.[7]

Hydro stations can be grouped into two main types: *run-of-river* and *storage* stations.[8] Run-of-river stations are characterized by a low head and thus mainly draw on the kinetic energy of the flowing water for power production. As the availability of the primary energy carrier is fairly stable in the short term, they produce base load power. The pondage power station is a subtype of the run-of-river station. It is capable of damming up water for a few hours, which is then released in times of high demand. Storage stations are characterized by a high head and a water storage reservoir with a natural water inflow. As water throughput can be controlled, they are generally used for mid- and peak-load power production. A particular subtype is the pumped storage power station, where two reservoirs are located at different altitudes.

[6] Height of fall of the water.

[7] Strictly speaking, gravitational acceleration and the density of water may vary slightly, depending on location and water temperature.

[8] Classification based on water pressure or height of the head are also common.

During hours of electricity overproduction from other technologies, water is pumped from the lower to the higher reservoir to be released again in times of high demand.

Power generation from hydro stations is subject to fluctuating energy carrier availability. Water throughput fluctuates only slightly in the short-term but shows considerable differences over the annual cycle and often also across years. In Central Europe, spring and summer are characterized by higher water throughputs than fall and winter, driven by the melting of snow in spring and high precipitation in summer. This directly impacts the power output from run-of-river plants, whereas stations with large reservoirs are capable of delivering a fairly constant output level across the year. Those reservoirs attain their lowest filling level towards the end of winter and their highest one by the end of summer. Year-on-year fluctuations in precipitation influence annual output of hydro power plants. The total annual precipitation, for example, has deviated by $\pm 25\%$ for the city of Munich, Germany, around the respective long-term average between 1961 and 1995 (Kaltschmitt et al., 2006).

Hydropower projects are in most cases not solely pursued for the sake of establishing power generation capacity. Among others, they are used to reduce the risk of floods, to ensure firm water supply to domestic, industrial or agricultural users, to improve navigability of a river and to serve as recreation facilities. An important challenge for hydro station construction is sustainability, the goal of which can be defined as ensuring "that development preserves the resource for future use and does not exceed the environment's absorptive capacity of negative impacts" (Frey and Linke, 2002). Examples for these issues include the conversion of an aquatic ecosystem into a terrestrial one, alteration of downstream ecosystems, potential disruption of fish migration and relocation of the population living within an area to be flooded as reservoir. These issues usually lead to controversies between supporters and opponents of large-scale projects.

A.1.2 Generation Costs

Capital costs for hydro stations strongly depend on conditions at the site of implementation. This may explain the large bandwidth of values found in literature (see Fig. A.1). Hydro power technology can be considered mature and significant cost reductions, as observed for second and third generation renewable technologies, are not expected. Capital costs for run-of-river stations and storage plants are comparable, yet the share of civil in comparison to mechanical and electrical engineering costs is higher in storage than in run-of-river stations (IEA, 2003, p. 37). The opportunity to reuse some hydro station components, especially civil ones, after the end of the station's useful life for the construction of a replacement plant, leads to different capital costs for new and replacement stations.

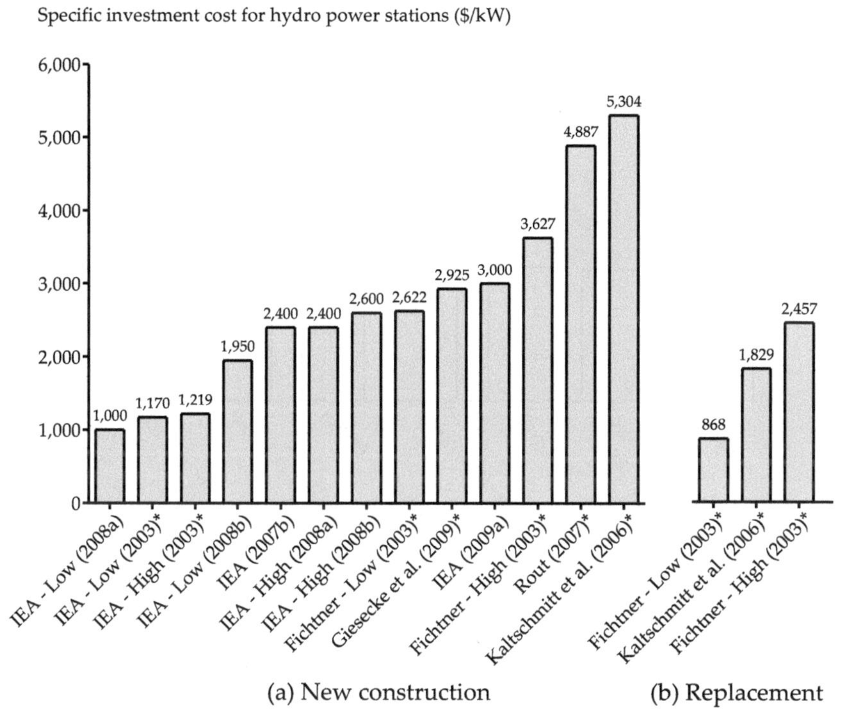

Fig. A.1 Overnight capital costs for hydro stations

Specific investment cost values identified in literature range from 1,000 up to 5,400 $/kW, with a large number of reported values in the range of 1,900 to 3,000 $/kW. For the calculation of generation costs 2,400 $/kW are used for new constructions. Generation costs for replacement stations are based on a value of 1,600 $/kW.[9]

The differences among reported operating costs, shown in Figure A.2, are not as pronounced as the ones for capital costs. For generation cost calculation, operating costs of 55 $/kWa are used here. Construction duration for new plants heavily depends upon the size of the station. Stations of around 100 MW tend towards a four-year construction duration but smaller stations of size 10 MW capacity require only approximately half of that time. Modernization of non-civil parts of a station may often be carried out within a year's time (Fichtner, 2003, pp. 8–9). Construction of very large projects with capacities above 100 MW may take even longer. The construction of the

[9] The large spread of values for replacement stations can be attributed to the underlying assumptions for the high and low Fichtner estimates. The lower bound is based on a low cost scenario for an existing plant modernization, the high bound on an unfavorable cost scenario for an entirely new plant, where only the reservoir exists already.

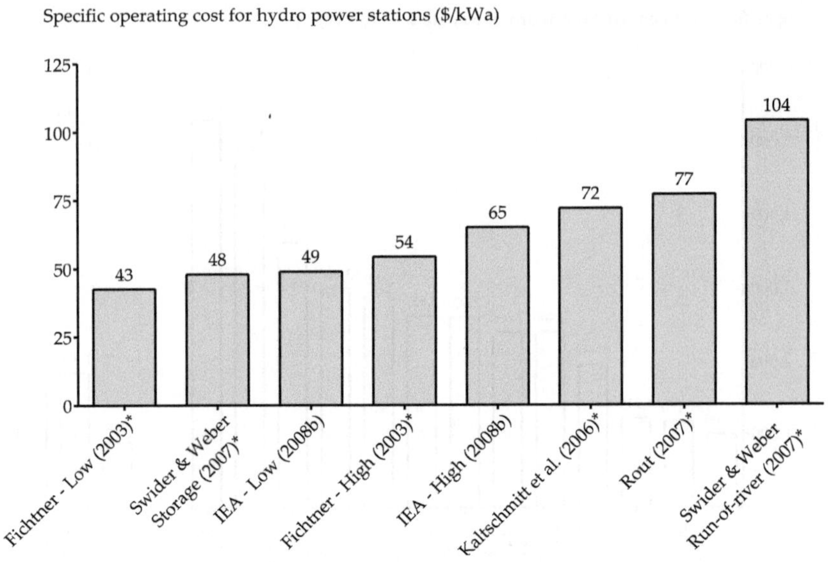

Fig. A.2 Annual operating costs for hydro stations

Three Gorges project in China with 18.2 GW capacity, for example, has taken more than a decade to complete. For generation cost calculation, a five year construction period is used for new and a two year period for replacement stations. Construction financing costs then amount to 457 and 114 $/kW for new and replacement plants, respectively, at the time of plant commissioning. Hydro power plants are characterized by a comparatively long useful life. The civil components can be used for up to 80 years, whereas the remainder of the plant has to be replaced after approximately 40, some electrical components even earlier (Fichtner, 2003; Kaltschmitt et al., 2006). An average life of 60 years is used here for generation cost calculation.

Overall, total capital costs including construction financing amount to 2,857 and 1,723 $/kW for new and replacement hydro stations, respectively. Together with operating costs, this yields total annual costs of 260 and 178 $/kWa (see Tab. A.2). Production of 4,700 full load hours p.a. (FLH/a) for a station in Latin America and 2,900 FLH/a for a European station have been used in the calculation of generation costs. For new hydro power stations those then lie in the range of 5–9 ct/kWh, placing them at the upper end of current wholesale electricity prices. Generation costs for replacement stations, with 4–6 ct/kWh, fall well within the range of current wholesale prices.

It is important to note that the calculated generation cost values are for stylized hydro plants. Run-of-river-stations are generally characterized by higher annual outputs than storage stations. This would results in comparatively lower generation costs for run-of-river than for storage stations.

Table A.2 Generation costs for stylized new and replacement hydro stations

New Hydro Station			Replacement Hydro Station		
Cost assumptions					
	$/kW	$/kWa		$/kW	$/kWa
Investment	2,400	172	Investment	1,600	115
Construction fin.	457	33	Construction fin.	123	8
Operation		55	Operation		55
Other specifications					
Useful life (a)		60	Useful life (a)		60
Constr. period (a)		5	Constr. period (a)		2
FLH Latin America (h/a)		4,730	FLH Latin America (h/a)		4,730
FLH Europe (h/a)		2,890	FLH Europe (h/a)		2,890
Generation cost					
Latin America:		5.5 ct/kWh	Latin America:		3.8 ct/kWh
Europe:		9.0 ct/kWh	Europe:		6.2 ct/kWh

However, a competitiveness comparison between these two types of plants solely on the basis of generation costs would not deliver meaningful results, as storage plants produce peak- and mid-load power, which are of higher value than the base-load power from run-of-river-stations. Additional costs from output fluctuations for run-of-river plants should only weakly increase the calculated generation costs, since those fluctuations are rather long- than short-term. Storage power stations are capable of compensating cyclical supply fluctuations.

A.1.3 Potential

There is consensus that even if all of the remaining global hydropower potential were developed the resulting capacity would not be sufficient to satisfy global electricity demand (cf., for example, Bartle 2002). This fact is illustrated in Fig. A.3 showing the technical potential for hydropower by region. One third of worldwide economic potential[10] has been developed so far, which would consequently at most allow for a tripling of current capacity. Hydropower currently makes up 16% of global power generation, consequently a tripling of capacity would only lead to less than a 50% share.

[10] Defined here has "capability that can be exploited within the limits of current technology under present and expected local economic conditions" (World Energy Council, 2007, p. 277) and not further specified in terms of generation costs or earnings.

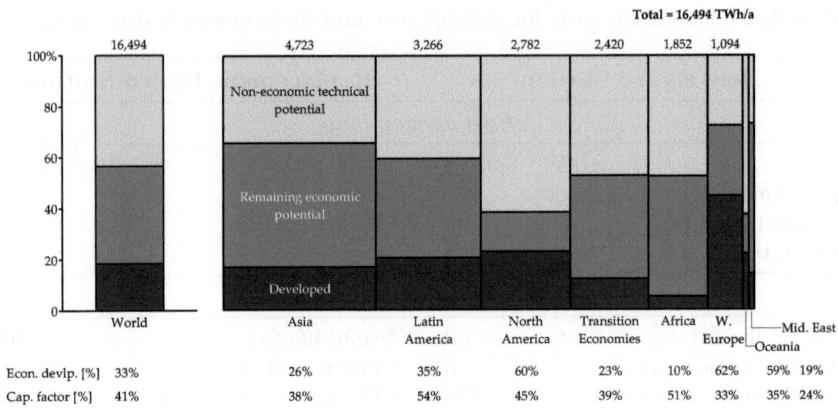

Fig. A.3 Global technical potential for hydro power
Source: World Energy Council (2007, 2009), own calculations

Potential and current degree of implementation vary across regions. In North America and Western Europe approximately two thirds of economic potential have already been exploited, whereas other regions still have large shares of their economic potential available for development. Average capacity factors differ considerably across regions.[11] They range from only 24% for the Middle East to 54% for Latin America. These deviations may be caused by actual differences in annual water throughput. However, different mixes of run-of-river versus storage plants across regions could also be a factor responsible for these differences in capacity factors, since run-of-river stations are generally characterized by higher capacity factors than storage stations. Unfortunately, the data does not allow for a final conclusion on this issue.

A.2 Wind Power

A.2.1 Overview

Power generation from wind dates back to the 19th century. In 1887, James Blyth constructed a wind turbine to electrify his cottage in Scotland. Shortly thereafter, Charles Brush constructed a 12 kW DC turbine with a rotor made up of 144 wooden blades. Despite these early experiments it took until the oil crises of the 1970s before wind was seriously considered as option for grid-connected power generation. It then further took until the 1990s before substantial grid-connected capacity was put in place, except a brief, subsidy-induced hype in the 1980s in California. Since the mid-1990s the wind power

[11] Implied capacity factors are based on current reported capacity and actual production (World Energy Council, 2009, pp. 47–49).

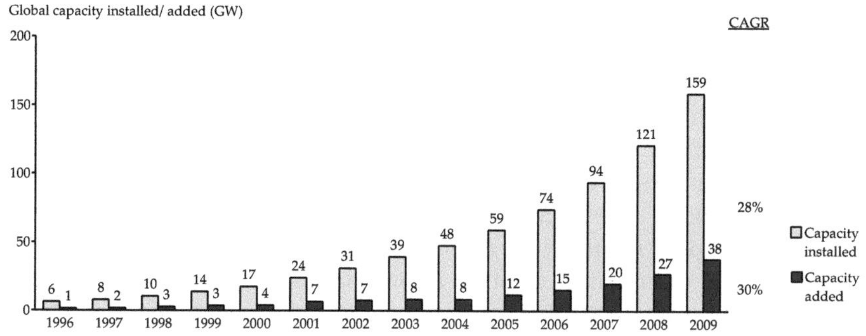

Fig. A.4 Global wind power capacity and additions 1996–2009
Source: GWEC (2010)

market has been characterized by a rapid expansion, experiencing roughly 30% year-on-year growth rates (see Fig. A.4). This growth has been fueled by governmental support under different national schemes. Today's turbines are quite different from the ones in Blyth's and Brush's time. The rotor of a typical wind turbine today is made up of three fiber-reinforced composite blades, rated turbine power mostly lies in the 1.5–2.0 MW range and AC electricity is fed into the grid. Wind turbines are often erected together as wind parks to realize economies of scale, for example, from joint grid connection and easier access for maintenance. The vast majority of installations to date occur onshore but about 1.5% of new capacity in 2009 has already been added offshore.

Wind turbines convert the wind's kinetic energy passing through their rotor into mechanical energy, which is then converted into electrical energy via a generator. The rated power P of a wind turbine in Watt is given by Eq. (A.3) (Jarass and Obermair, 2009, pp. 28ff.).

$$P = \eta \cdot c_P(v) \cdot \frac{\rho}{2} \cdot v^3 \cdot \pi r^2, \tag{A.3}$$

where η represents the efficiency of the generator and, if applicable, the gearbox in percent, c_P the power coefficient in percent, ρ the density of air in $\frac{kg}{m^3}$, v the velocity of wind in $\frac{m}{s}$ and πr^2 the swept area of the rotor with effective rotor blade length r in meters. Power coefficients c_P of modern turbines already reach values of 50% (Jarass and Obermair, 2009, p. 31) with a theoretical maximum value of $c_{P,\max} = 59.3\%$.[12]

Equation (A.3) highlights the importance of wind speed v for actual power generation, impacting turbine power to its cube. It further illustrates that increasing rotor diameter is the most promising lever to increase power output for a given wind speed pattern, system efficiency and power coefficient. In fact, rotor diameters have grown from about 20 to more than 120m over the

[12] $c_{P,\max}$ is known as the Betz coefficient (Betz, 1926).

20 years 1985–2005 (European Wind Energy Association, 2009, p. 39). This corresponds to an increase in rated turbine power by a factor of 36. In line with rotor diameters, hub heights have increased to take advantage of higher wind speeds at higher altitude.

Calculation of the average annual output for wind turbines is not straightforward. Two inputs are required: The location-specific annual frequency distribution of wind speeds and the turbine-specific power curve, which describes turbine power output as a function of wind speed. The annual wind speed frequency distribution is often approximated using a fitted Weibull or Raleigh distribution. Multiplying the wind speed frequency distribution with the power curve yields the frequency distribution of annual power output. The integral of this function then yields the theoretical annual power production. Actual production lies 5–20% (European Wind Energy Association, 2009, p. 55) below this theoretical value due to array inefficiencies, caused by the interferences among the turbines of a wind farm and turbine downtime.

Wind power production is characterized by a high degree of cyclical and stochastic fluctuations. Depending on system penetration, this may require addition of backup generation or storage capacity and reinforcements of the transmission grid. For additional information on this topic the interested reader is referred to chapter 2.2.3.

A.2.2 Generation Costs

As for most renewable plants, capital costs make up the largest share of annual costs for wind power. For onshore wind parks the wind turbines account for three fourths of overnight capital costs with the remaining fourth mainly made up of grid connection, foundation and erection costs (Jarass and Obermair, 2009, p. 179). In offshore parks, grid connection accounts for a larger share of capital costs.

Three issues complicate the identification of representative capital costs for on- and offshore plants from literature apart from the challenges posed by the appreciation of the Euro. First, there has been a surge in turbine input material prices from 2004 onwards, with steel and copper price increases of more than 50% from 2004 to 2008. Second, the turbine market has been severely supply-constrained in the years following 2005, as wind turbine manufacturers were not able to add manufacturing capacity in line with the exponentially increasing demand, ultimately leading to higher turbine prices. Finally, substantial cost reductions for wind turbines can be observed (see chapter 4) and thus only recently published data incorporates the reductions experienced to date. Using current capital cost data (see Fig. A.5) as starting point may consequently lead to an overstatement of future costs when turbine input prices are back to normal levels and higher turbine manufacturing capacity is in place.

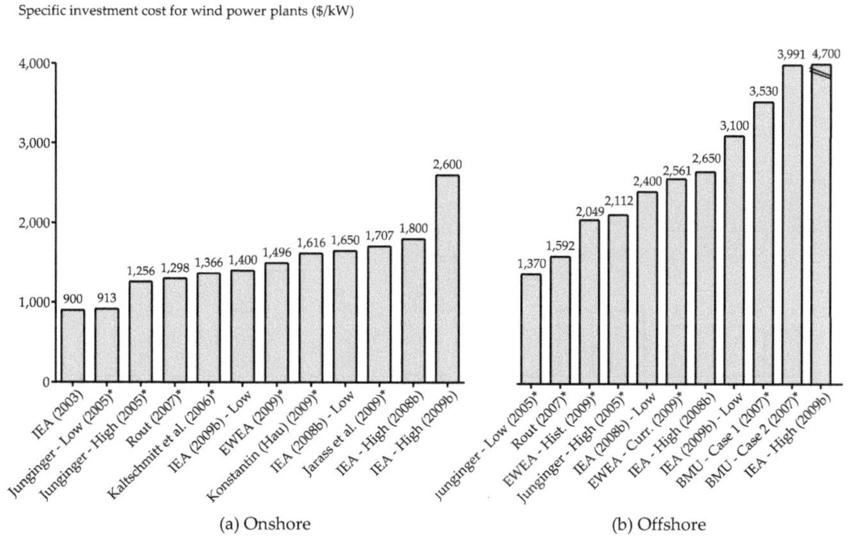

Fig. A.5 Overnight capital costs for wind power plants

Although the capital cost data in Fig. A.5a span a broad range, the recent increase in specific costs is apparent. The cost data points for onshore wind upwards 1,400 $/kW are exclusively from publications in 2008 and 2009, whereas all data points with lower specific capital costs are from earlier years. One might therefore argue that a value below 1,400 $/kW would best represent wind's true mid-term capital costs, considering input price appreciation and turbine supply shortages as temporary phenomena and presuming that turbine manufacturing costs must have decreased over the last five years. Yet, as the rapid growth of wind power is not expected to cease in the short term despite the financial crisis of 2008 and supply shortages may persist, a more conservative stance is taken and a value of 1,400 $/kW is used for the calculation of onshore wind generation costs.

The situation is similar for offshore capital costs, presented in Fig. A.5b. Current EWEA (2009) cost estimates of 2,100 €/kW are roughly 25% higher than their reported historical costs based on the Horns Rev 1 and Nysted offshore wind parks of 1,680 €/kW. These parks were commissioned in 2002 and 2003, respectively. Capital cost values from literature span a wider range than for onshore wind, which may be attributable to the comparatively small number of actually commissioned projects and thus available data points. For generation cost calculation purposes a value in between EWEA's historical and current estimates, 2,300 $/kW, will be used.

Operating costs found in literature (Fig. A.6) are also characterized by large differences. The highest value identified is more than triple the lowest one, although operating costs should be only slightly affected by the surge in

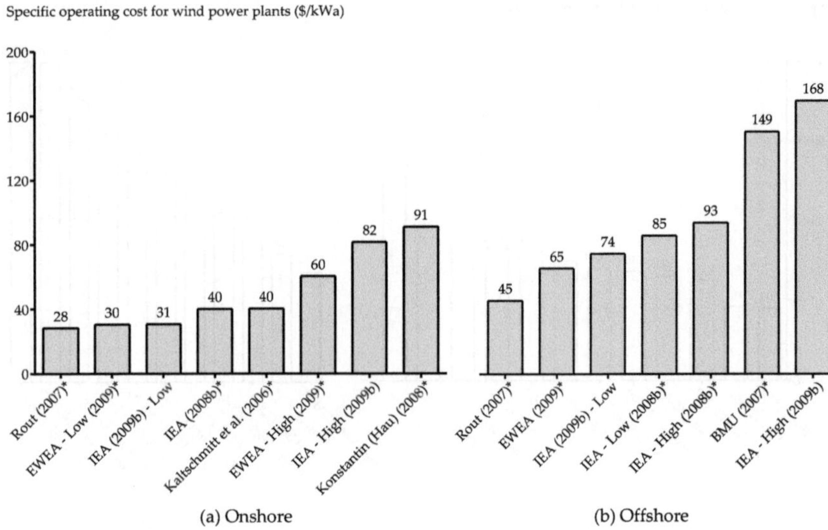

Fig. A.6 Annual operating costs for wind power plants

turbine input prices and supply shortages. To avoid extreme values, 45 $/kWa for onshore and 80 $/kW for offshore wind parks are used for reference.

In comparison with most conventional power plants, wind parks can be constructed quickly. One year is generally sufficient for the construction of an onshore windpark. For offshore parks, longer construction durations must be assumed, especially since rough weather may impede construction activities during winter months. The recent construction of Germany's first offshore windpark, Alpha Ventus, required roughly two years, from August 2007 to the first feed-in in August 2009 (Deutsche Offshore-Testfeld und Infrastruktur GmbH & Co. KG, 2010). In line with the construction duration of Alpha Ventus, two years are used as reference for generation cost calculation. These construction durations imply construction financing charges of 49 and 164 $/kWa for on- and offshore parks, respectively.

In line with most literature sources (cf., for example, Kaltschmitt et al. 2006, p. 329), an economic life of 20 years for onshore plants is used, although a useful life of up to 25 years can be considered realistic (European Wind Energy Association, 2009, p. 35). For offshore parks, lifetimes of up to 30 years may be realistic, however, in line with European Wind Energy Association (2009, p. 35), 25 years will be used. Since few turbines have so far reached the end of their useful life when compared to the total number of turbines in operation, empirical data on plant lifetimes is still rare for onshore and even inexistent for offshore technology.

For calculation of onshore wind power generation costs two sample sites are used. First, a site with exceptionally high 2,900 FLH/a, which one could find in some coastal locations in Europe or in the Great Plains in the US.

Table A.3 Generation costs for stylized on- and offshore wind farms

Onshore wind farm			**Offshore wind farm**		
Cost assumptions					
	$/kW	$/kWa		$/kW	$/kWa
Investment	1,400	137	Investment	2,300	211
Construction fin.	49	5	Construction fin.	164	14
Operation		45	Operation		80
Other specifications					
Useful life (a)		20	Useful life (a)		25
Constr. period (a)		1	Constr. period (a)		2
FLH Excellent site (h/a)		2,900	FLH Typical site (h/a)		3,300
FLH Good site(h/a)		2,300			
Generation cost					
Excellent site:		6.4 ct/kWh	Typical site:		9.2 ct/kWh
Good site:		8.1 ct/kWh			

Second, a site characterized by 2,300 FLH/a, which can still be considered a good site for inland continental Europe.

The resulting current generation costs for onshore at sites with favorable wind regime lie in the range of 6–8 ct/kWh, offshore generation costs at approximately 9 ct/kWh. After adding another 0.5–2.0 ct/kWh attributable to variability[13] (see sec. 2.2.3), it becomes apparent that wind power still lies above current baseload wholesale prices. This also explains why the vast majority of installed capacity is located in countries with attractive governmental support schemes.

A.2.3 Potential

The global technical potential for onshore wind has been the subject of several publications (e.g. Grubb and Meyer 1993; Goldemberg 2000; Hoogwijk et al. 2004; Archer and Jacobson 2005; de Vries et al. 2007; REN21 2008; Krewitt et al. 2009). They are characterized by a wide range of technical potential estimates, ranging from 53 to 127 PWh/a with differences in assessment methodologies and assumptions. The onshore technical potential data provided by Hoogwijk et al. (2004), in total 96 PWh/a globally, will be used for further reference. Since it has been assessed in a bottom-up process with global coverage, it has the advantage of not only stating an overall technical potential but also differentiating the technical potential with respect to average wind speed. This information is important for wind power

[13] These costs do not yet include potentially required grid reinforcements.

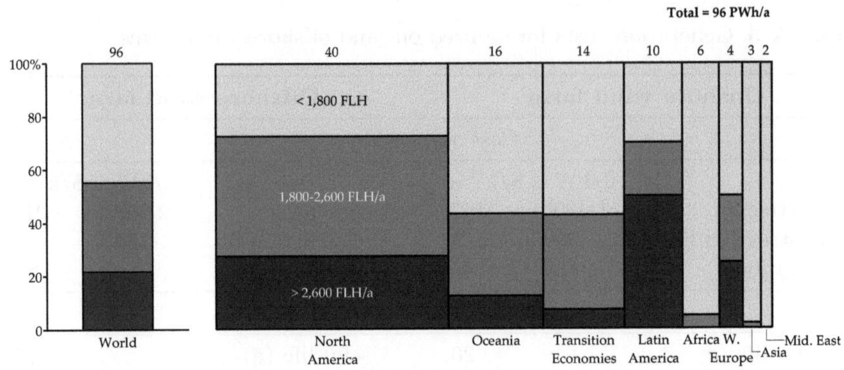

Fig. A.7 Global technical potential for onshore wind power
Source: Hoogwijk et al. (2004)[15]

competitiveness assessment, since plants at sites with higher average output will, ceteris paribus, be characterized by lower generation costs than plants at sites with lower output. Basing generation costs on average instead of marginal wind speeds would understate the competitiveness of wind power, since locations with higher wind speeds will be occupied first, presuming a profit maximizing investor.

Even for the lowest of the assessed technical potentials, a twofold of today's power demand could theoretically be satisfied from wind power.[14] The technical potential, however, is characterized by large differences in wind speed. Figure A.7 shows the technical reference potential differentiated into three power production categories. Only about 20% of technical potential can be attributed to sites with annual production above 2,600 FLH/a. Since output from plants at those sites is more than double the one from plants at sites in the lowest of the three categories, it is safe to say that at most 10% of technically suitable sites are characterized by an output above 2,600 FLH/a. Moreover, these sites with favorable wind regime are distributed unevenly across regions. Whereas a large number of high-output sites can be found in North America, the technical potential for Asia is comparatively low and further characterized by low power output. This is particularly interesting in light of the rapid growth of wind power capacity additions in China since 2007.

The global technical potential for offshore wind at sites within 50 kilometers off the shore and in water no deeper than 40m (Krewitt et al., 2009)

[14] For such a high wind penetration, intermittency would likely become a constraining factor.

[15] Full load hours reflect array efficiency losses and downtimes. Full load hour category thresholds calculated based on published generation costs breakpoints and corresponding cost assumptions. Category thresholds rounded to the nearest multiple of 100.

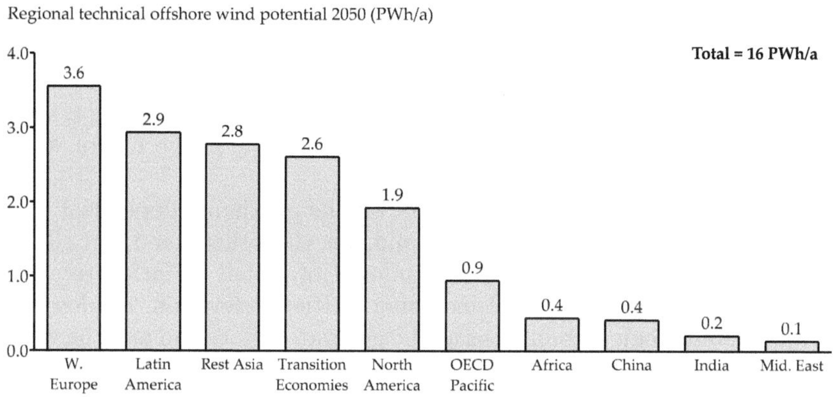

Fig. A.8 Regional technical potentials for offshore wind power
Source: Krewitt et al. (2009)

is significantly lower than the potential for onshore wind. Values for the global technical potential from different older studies range from only 3.9 to 5.2 PWh/a (REN21, 2008). Here the more recent assessment by Krewitt et al. (2009) stating a global technical potential of 15.9 PWh/a is used for reference, as it is globally comprehensive and uses more appropriate energy density assumptions than the older studies. For offshore, the comparatively small region of Western Europe is characterized by the in absolute terms greatest technical potential (see Fig. A.8).

A.3 Photovoltaic Power

A.3.1 Overview

The discovery of the photovoltaic effect dates back to 1839, when Becquerel observed that exposure to light increased power generation of an electrolytic cell. In 1876 Adams and Day discovered that a piece of Selenium, exposed to light, would produce an electric current. They named this phenomenon the photoelectric effect, which was renamed into photovoltaic effect in the 1920s in honor of Alessandro Volta. The first solar cell made from silicon and capable of powering everyday appliances was then developed in the early 1950s by Pearson, Chapin and Fuller (cf. Wagner, 2010, pp. 1–3).

Contemporary photovoltaic generators are made up of two main parts: The solar *modules* and the *balance of system* (BOS), a collection of different auxiliary components. The solar module consists of several interconnected solar cells covered by protecting and anti-reflecting devices. Within the solar cells energy from solar radiation (Photons) is converted into a direct electric current drawing upon silicon's semiconductor properties. The key BOS component, the inverter, then converts the direct into an alternating current. Solar cells made of crystalline silicon currently dominate the PV market, accounting for 85–90% of total installations (IEA, 2010c) but other technologies either already play an important role in the commercial market, such as thin film, or have the potential for playing a key role in the future, such as concentrating PV.

Energy yield from a PV generator is, on the one hand, dependent on the amount of global radiation received and, on the other hand, on the generator's technical specifications. The amount of global radiation received is influenced by numerous factors including latitude, elevation, shadowing effect and meteorological conditions at the generator's location and the vertical angle and azimuth of the solar module's surface. In almost all locations it is beneficial to install the solar module at a vertical inclination instead of horizontally to increase the amount of radiation received. Some generators are equipped with one- or two-axis tracking systems to further increase yield. Less than 20% of the radiation reaching the solar generator can ultimately be converted into electricity. Efficiency for commercial crystalline and thin film systems fall into the range of 13–20% and 6–12%, respectively (IEA, 2010c). Module efficiency is dependent on temperature, with higher temperatures corresponding to lower efficiency. Since higher solar radiation generally correlates with higher temperatures, an increase in radiation will not yield a correspondingly large increase in power output. Put differently, the marginal yield will decrease for increasing radiation. In practice, actual AC power output is lower than the theoretical one calculated based on the PV system's rated power, efficiency and radiation received. The ratio of actual to theoretical output is often referred to as *performance ratio* or *quality factor*. Values around 0.75 are commonly found in literature but some crystalline cells may attain values of 0.85 and above.

Similar to wind, the market for PV generators has been characterized by rapid growth over the last few years (Fig. A.9) with annual growth rates exceeding 50%. The global, grid-connected capacity at the end of 2009 amounted to approximately 21 GWp. Put into perspective, this still corresponds to less than 10% of power generation capacity from wind.[16] Installations are concentrated in three countries with sufficiently attractive support schemes: Germany, Japan and Spain with respective global capacity shares of 47%, 16% and 12% (REN21, 2010). The case of Spain illustrates the high dependence of PV deployment on subsidies: capacity additions in 2009 dropped to almost zero from a record high of 2.4 GW in the previous year after a revision of the support scheme.

In the past, PV technologies have often been criticized for requiring more energy during their manufacturing process than they would return in operation. These claims are no longer valid for contemporary PV generators, which are characterized by energy payback times of approximately two years given

[16] Actual power yield for wind is approximately twice as high as the one for PV per respective nameplate capacity unit at a Central European location.

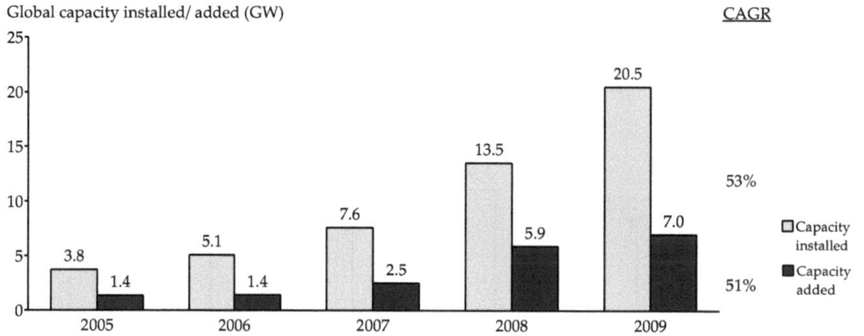

Fig. A.9 Global photovoltaic capacity and additions 2005–2009
Source: REN21 (2010)

an output of 1,500 FLH/a (IEA, 2010c). In Germany, where most of current installations take place, energy payback times would consequently be around three years.

As for wind, the intermittent output of PV generators poses a challenge for large-scale system integration. Output of a PV generator reaches a high around noon and drops to zero at night. Since the higher output during daytime correlates with higher power demand, a certain share of PV can be added to a power system without adding storage or backup capacity. Stodola and Modi (2009) assess this maximal share of PV without storage/backup for the US at 8%. Similarly, Denholm and Margolis (2007) find that excess power would be produced in a system without power storage, where 10–20% of power come from PV. Stochastic fluctuations of global radiation during daytime further complicate PV integration. Chapter 2.2.3 offers a more detailed discussion of this topic.

A.3.2 Generation Costs

PV modules usually account for slightly more than half of turnkey investment costs, with BOS and grid connection accounting for the remainder. PV modules but also BOS components continue to experience considerable cost reductions. Most literature sources agree on a historically demonstrated PV capital cost reduction of about 20% for a doubling of cumulative manufactured capacity (see chapter 4.3). For this reason, only recently published capital cost data can be meaningfully used to calculate current generation costs. The data in Fig. A.10 reflects the fact that cost reductions have occurred and are expected to continue into the future. Whereas a forecast for 2015 costs (IEA, 2008) is located at the lower end of the range, data from 2006 (Kaltschmitt et al., 2006) is located towards its higher end. Portals such as pvXchange.com or solarbuzz.com provide current module cost development indexes, which also reflect these capital cost reductions, in particular

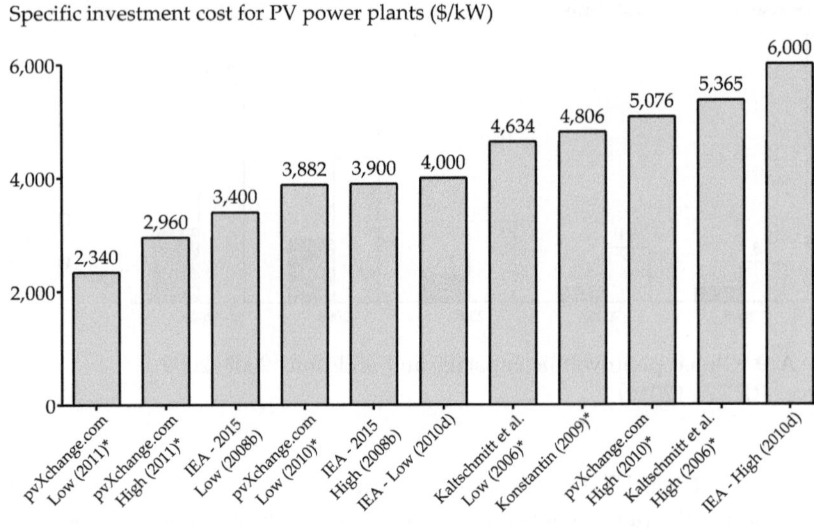

Fig. A.10 Overnight capital costs for photovoltaic generators

after the manufacturing capacity and silicon supply bottlenecks of the mid-2000s have been resolved. pvXchange (2012) states wholesale module costs of 1.12 €/Wp[17] for December 2011. Applying the respective turnkey cost surcharge between 0.5 and 0.9, turnkey capital cost would lie in the range of 2,300 to 3,000 $/kWp. These figures, however, may have to be taken with a grain of salt. Due to substantial global manufacturing capacity additions until 2011 paired with a decrease in module demand in 2011, it is unclear whether these prices are a fair representation of PV's capital cost or only a temporary low. For example, module prices have decreased by 34% over 2011 which would imply a substantially higher learning rate than the 20% which has been observed over the last decades. For generation cost calculation turnkey overnight costs of 3,000 $/kWp are used.[18]

Operating costs for PV modules include insurance and maintenance costs. The latter, however, are relatively small in comparison with technologies characterized by moving parts. Generally replacement parts are also accounted for under operating costs. Here the replacement of the inverter, usually required before the end of the generator's useful life, constitutes a major expense. Although frequently found in literature, the approach of expressing operating costs as constant share of capital costs is problematic. Such an approach implies that operating cost reductions occur at the same pace as reductions for capital cost. Although this assumption may hold for spare parts, it appears

[17] For large-scale applications.

[18] Capital cost of 4,200 $/kWp (2010 figures) are used in the models covered in chapters 6 through 7 to ensure consistency with the respective cumulative manufactured capacity figures.

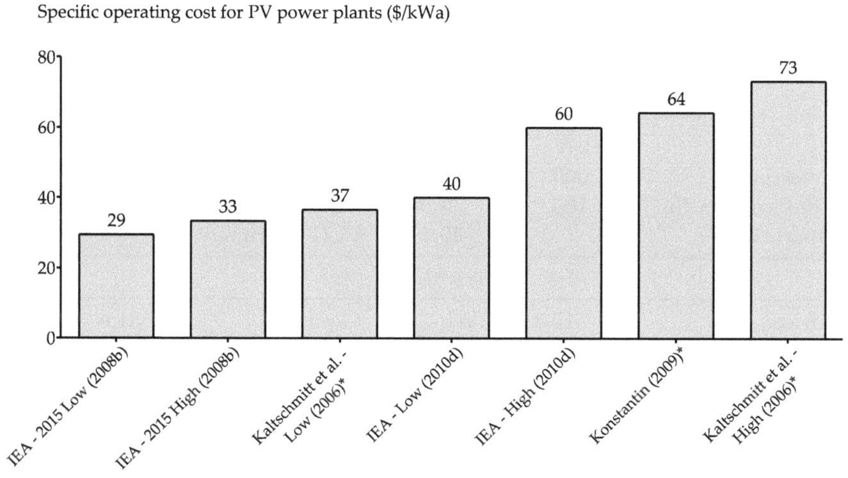

Fig. A.11 Annual operating costs for PV generators

unrealistic for insurance and labor costs. Therefore, the operating costs in Fig. A.11 are expressed as absolute values. 40 $/kWa are used as reference value.

PV plants are characterized by a useful life of over 20 years with decreasing output over time. Performance at 90% and 80% of nameplate capacity during the first and second decade of operation, respectively, is usually guaranteed by module manufacturers. Inverters often require replacement within the second decade of operation. The annualized costs for this replacement are usually included in operating costs. Useful economic lifetimes in literature range from 25 to 30 years and it is expected that plant lifetime will increase in the future. IEA (2010c) targets a useful life of 40 years for modules manufactured in 2050. For current generation cost calculations, a lifetime of 25 years is used. Construction durations for new generators are comparatively short, partly due to the modular nature of the technology and a duration of one year is used for generation cost calculation. The resulting financing charges for this period amount to 147 $/kWp.

For generation cost calculation, two sample locations are used. A desert site in Africa with an output of 2,000 FLH/a and a site in Southern Germany with 1,000 FLH/a, where a significant amount of current capacity addition occurs. At the desert site with high solar radiation generation costs below 20 ct/kWh (see Tab. A.4) can be achieved. Most current implementations, however, are concentrated in countries with comparatively unfavorable solar regime. Power output in Southern Germany of around 1,000 FLH/a can be expected, implying generation costs in excess of 30 ct/kWh. Obviously, even the generation costs at the favorable site are far above contemporary wholesale costs, even before the consideration of additional costs related to intermittency.

Table A.4 Generation costs for stylized PV generators

Cost assumptions			*Other specifications*	
	\$/kW	\$/kWa	Useful life (a)	25
Investment	3,000	267	Constr. period (a)	1
Construction fin.	105	9	FLH: Africa (h/a)	2,000
Operation		40	FLH: Germany (h/a)	1,000
		Generation costs		
Africa:	15.8 ct/kWh		Germany:	31.6 ct/kWh

A.3.3 Potential

The global technical potential for PV power is a many-fold of current global electricity production and roughly a 3–4 fold of technical wind power potential. Hoogwijk (2004) arrives at a global technical potential of 327 PWh/a. It is used for further reference, in particular since Hoogwijk (2004) differentiates the potential with respect to productivity. Other potential analyses obtain results in the same order of magnitude, ranging from 372 to 469 PWh/a (cf. Krewitt et al., 2009; REN21, 2008; Hofman et al., 2002; Goldemberg, 2000). However, it needs to be noted, that both the assessment in Krewitt et al. (2009) and REN21 (2008) are based on the data from Hoogwijk (2004). Only de Vries et al. (2007) arrive at a far higher potential of 4,105 PWh/a in 2050 based on the same general methodology used in Hoogwijk (2004) but using more optimistic assumptions.

Within the global technical potential, there exist vast differences in annual power yields, expressed in terms of annual FLH (see Fig. A.12). Sites with more than 2,050 FLH/a only exist in Africa, Asia, Oceania and the Middle East. The overall potential for Africa is vast, accounting for approximately 40% of global potential, with virtually no suitable sites yielding less than 1,500 FLH/a. In contrast, Western Europe is characterized by a comparatively small overall technical potential of 4 PWh/a, with less than 1 PWh/a from sites yielding more than 1,500 FLH/a, concentrated in Southern Europe.

Hoogwijk (2004) bases her potential analysis on a performance ratio of 0.75. Since performance ratios already reach values of 0.85 and more, correspondingly higher FLH/a may be realistic. Based on this higher performance ratio, yields of above 2,400 FLH/a could be reached at some sites with very favorable conditions, primarily located in Africa and the Middle East.

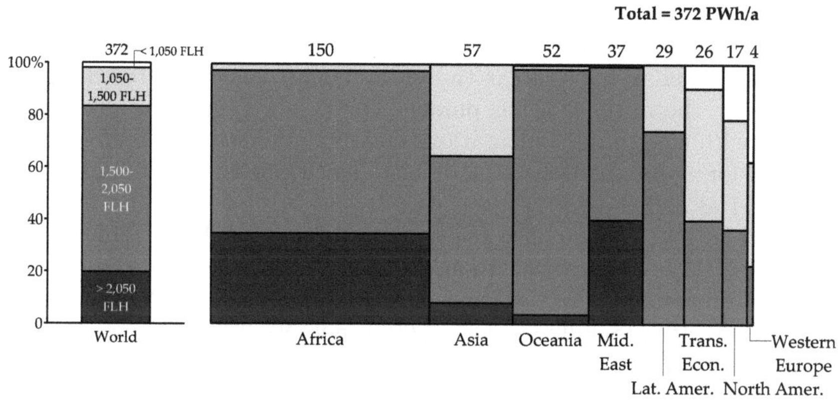

Fig. A.12 Global technical potential for PV power
Source: Hoogwijk (2004)

A.4 Solarthermal Power

A.4.1 Overview

Already by the middle of the 19th century solarthermal technology was employed to power steam engines. However, it took until 1968 for the first concentrating solar power (CSP) plant with a rated capacity of one MW to be installed. The first wave of grid-connected CSP deployment then took place from 1985 to 1990, where a series of nine plants, SEGS[19] I–IX, with a total rated capacity of 354 MW went on line in the Mojave desert. In 2007, after a break of almost 20 years, the next large-scale plants were commissioned in the USA and Spain. Since then, more than 600 MW have been commissioned or are still under construction, with most plants being located in Spain. At the end of 2009 global capacity amounted to 610 MW (REN21, 2010).

CSP plants use reflectors to concentrate the direct solar radiation onto a point or line, thereby increasing the temperature of a transportation medium. The thermal energy is then used to drive steam or combined cycle turbines, converting the thermal into mechanical and then, via a generator, into electrical energy. This process of converting thermal into electrical energy is comparable to the one found in conventional power plants, for example in coal-fired steam or gas-fired combined cycles.

CSP technologies can be grouped into *point concentrators* and *line concentrators*. Point concentrators achieve higher transport medium temperatures, approximately 600–1,200 °C, and can therefore be combined with gas or combined cycle turbines. Central receiver solar tower and dish/ engine technologies fall into this category. Line concentrators, on the other hand, operate

[19] Solar Energy Generating System.

at lower transport medium temperatures and are therefore limited to combination with steam turbines. Examples of this category are parabolic trough and linear fresnel systems. Solar trough and central receiver tower systems are currently the options being pursued for power generation at commercial scale, with trough implementations clearly dominating the market. They accounted for approximately 97% of installed capacity at the end of 2008 (cf. Trieb et al., 2009, p. 69).

Other than PV, which can convert both diffuse and direct radiation into electricity, CSP can only draw upon the direct share of global radiation. Of all factors impacting beam radiance at ground, clouds have the greatest impact on production (Schillings et al., 2004), potentially even causing output to drop to zero. For this reason, meteorological conditions at the site of implementation play an even more important role for CSP than for PV and stochastic output fluctuations are more pronounced. Two options exist to mitigate negative impacts from fluctuating energy carrier availability: plant hybridization and thermal storage. Since the power block of CSP plants is similar to the ones found in conventional plants, the turbine and grid connection can be shared with a fossil-fuel based solution, such as a combined cycle gas turbine. The latter feeds power into the grid during those times, when solar-based generation is not possible. Plants with integrated thermal storage produce more thermal energy during times of sufficient radiation than can be converted directly into electricity by the power block. This excess energy is stored and released to drive the power block during periods characterized by insufficient radiation. Storage capacity in current implementations, such as Andasol, is equivalent to approximately 6–8 hours of full load power production but larger storages are technically feasible. To a certain extent dispatchability of storage-enhanced CSP plants is thus given and, provided a sufficiently favorable solar regime at the site of implementation, even continuous power generation would technically be feasible.

A.4.2 Generation Costs

Specific investment costs for CSP plants strongly depend on the amount of thermal storage capacity. The thermal storage system itself, however, is not the main driver of higher capital costs but rather the enlarged collector field required to charge the storage. The addition of 6–8h storage capacity to a contemporary solar trough plant requires a rough doubling of the collector field's size. Yet, storage capacity addition does not require additional investment into power block capacity.

Considerable cost reductions are expected for CSP technology in the future. It is therefore not surprising to find capital forecasts at the lower end of the range of the data points shown in Fig. A.13. With less than one GW installed to date, CSP can still be classified as relatively immature. Albeit

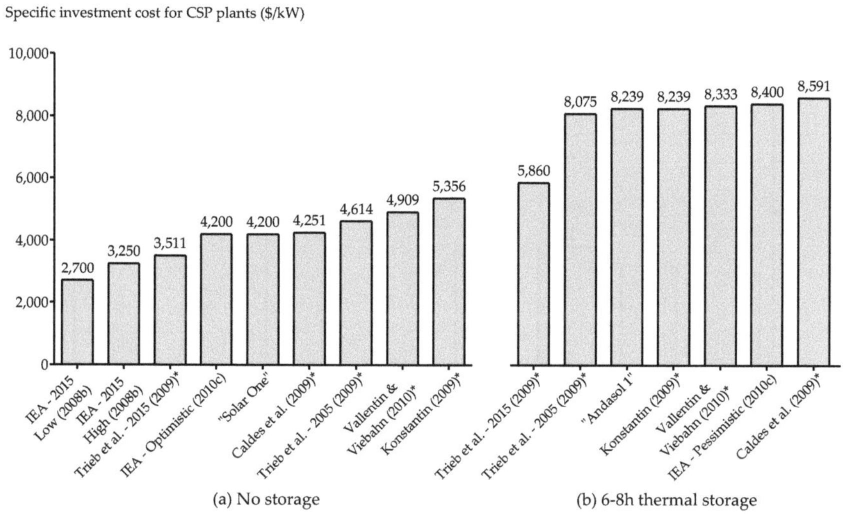

Fig. A.13 Overnight capital costs for CSP plants[20]

most research expects lower relative cost reductions for CSP than for PV,[21] the absolute potential for cost reductions is still considerable due to the early development stage. Current investment costs of 4,200 $/kW for a solar trough plant without storage are considered realistic, as this figure has been reported for the US-based project Solar One and several literature sources state values in this range.[22] The data for plants with a 6–8h storage is remarkably homogeneous. Likely most sources base their estimate on Andasol 1–3 figures, reported at 6,000 €/kW and minor differences may primarily be attributable to currency conversion. 8,100 $/kW will further be used as reference value. A stylized plant with 18 hours of thermal storage and a collector field four times the size of a plant without storage, would require investment of 15,029 $/kW[23] but would also be able to deliver 7,900 FLH/a and thus almost continuous production in a location with direct normal irradiance around 2,800 kWh/m^{2}a (Trieb et al., 2009).

[20] Calculation methodology for non-storage plant values for Caldés et al. (2009) and Vallentin and Viebahn (2010): investment for plant with storage - (50% collector field costs) - (100% storage costs) - (25% professional fees and contingencies). All data points have also been adjusted to exclude costs related to hybridization, if applicable and possible.

[21] Cost reductions of about 10% per doubling of capacity, see chap. 4.3.

[22] The value of converted 4,614 $/kW, presented by Trieb et al. (2009), is based on the year 2005, when total CSP capacity was at about half of today's. A corresponding cost reduction of 10% over this doubling of capacity, would also bring specific investment costs into the range of 4,200 $/kW.

[23] 10,920 €/kW; this is the reported 2005 value, 2015 forecast amounts to 7,688 €/kW (10,533 $/kW).

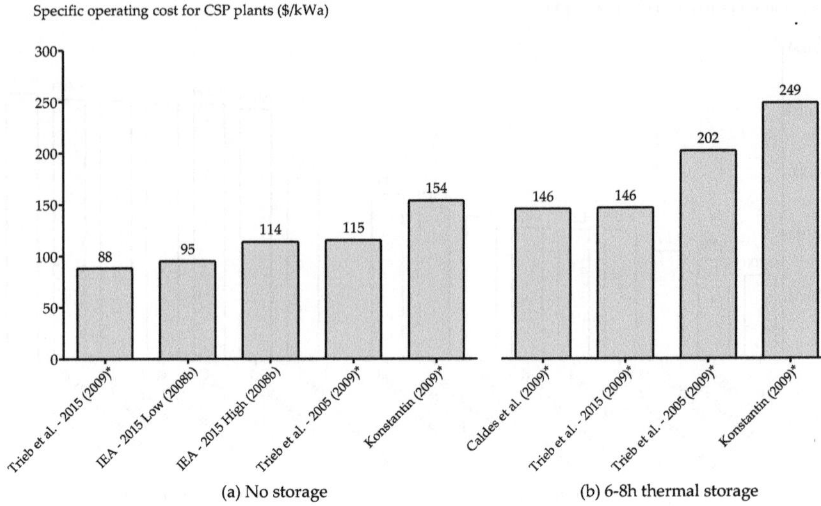

Fig. A.14 Annual operating costs for CSP plants

Operating cost estimates, presented in Fig. A.14 differ considerably, especially for plants with thermal storage. For further reference specific operating costs of 110 $/kWa for plants without and 200 $/kWa for plants with storage are used. In comparison to PV technology with reference operating costs of 40 $/kWa, operating costs for CSP are high. Power block maintenance and efforts required to maintain collector functionality are among the main drivers of these comparatively high values.

Economic lifetime assumptions in literature mostly fall into the range of 25 to 30 years. As the first commercial plant, SEGS I, has been commissioned in 1986 and is still in operation, a useful life longer than 25 years appears reasonable and 30 years are used for further reference. Construction of Andasol 1 required two years, which will also be used as further reference for the calculation of construction financing charges.

Generation costs are presented for two sample sites. First, a site in North Africa with a very high direct normal irradiance of 2,800 kWh/m^{2}a, yielding approximately 2,600 FLH/a at a plant without and 4,900 FLH/a at a plant with a 6-hour-storage.[24] Second, a site in Southern Spain with direct normal irradiance of 2,200 kWh/m^{2}a, corresponding to 1,800 and 3,700 FLH/a for plants without and with storage, respectively.

Calculated generation costs for these stylized plants lie in the range of 18 to 26 ct/kWh (see Tab. A.5), implying similar generation costs for CSP and PV technologies at sites with high radiation. The key advantage of CSP over PV power is the option to store energy or hybridize a plant at comparatively

[24] Annual full load hours calculated based on Trieb et al. (2009).

Table A.5 Generation costs for stylized CSP plants

CSP without storage			**CSP with 6–8h storage**		
		Cost assumptions			
	\$/kW	\$/kWa		\$/kW	\$/kWa
Investment	4,200	339	Investment	8,100	653
Construction fin.	299	24	Construction fin.	577	46
Operation		110	Operation		200
		Other specifications			
Useful life (a)		30	Useful life (a)		30
Constr. period (a)		2	Constr. period (a)		2
FLH: Africa (h/a)		2,600	FLH: Africa (h/a)		4,900
FLH: Spain (h/a)		1,800	FLH: Spain (h/a)		3,700
		Generation cost			
Africa:		18.2 ct/kWh	Africa:		18.3 ct/kWh
Spain:		26.3 ct/kWh	Spain:		24.3 ct/kWh

low additional costs. Technically, CSP could be implemented at large scale without incurring the penalties for primary energy carrier fluctuations, delivering dispatchable power for peak- or midload. In principle, even continuous production would be feasible. Unfortunately sites characterized by high direct radiation levels are often located far away from the points of consumption, so that power would have to be transported over long distances in case of large-scale implementation.

A.4.3 Potential

In comparison to other renewable technologies, the number of technical potential analyses available for CSP is low. Trieb et al. (2009) and Hofman et al. (2002) carry out bottom-up analyses with vastly different results. Whereas Trieb et al. identify a global technical potential of 2,946 PWh/a, Hofman et al. state 69 PWh/a. Krewitt et al. (2009) attempt to reconcile both analyses and find that the key driver of this difference is the assumption that only 5% of suitable land for CSP installations identified by Hofman et al. is actually accounted for as technical potential. The modification of this assumption, along with some others of lower impact, results in a global potential of 2,234 PWh/a (Krewitt et al., 2009). This figure lies in the same order of magnitude as Trieb et al. (2009). The latter analysis will be used for further reference due to its more up-to-date assumptions and the differentiation of

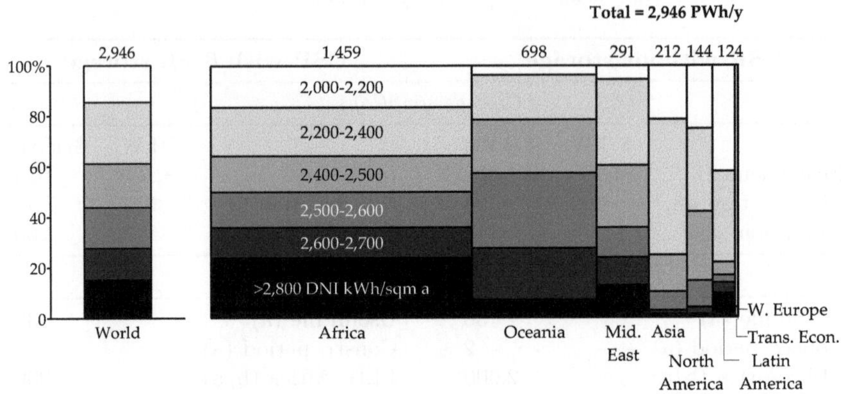

Fig. A.15 Global technical potential for CSP
Source: Trieb et al. (2009)

technical potential by irradiance. The global potential figures for both CSP and PV, despite their deviation by a factor of eight,[25] illustrate that solar power has by far the largest global technical potential of all renewable technologies. There is certainly a high degree of overlap among sites suitable for CSP and PV power production.

Figure A.15 illustrates the vast regional differences in CSP potentials. Half of global potential is located in Africa, with Oceania contributing another quarter. Africa, Oceania and the Middle East also possess a large number of locations with direct normal irradiance above 2,800 kWh/m^{2}a per year, translating into more than 2,600 FLH/a for a CSP plant without storage. The total technical potential of Western Europe is, at 2 PWh/a, comparatively small. Moreover, 90% of this potential can be attributed to sites within the lowest two classes of direct normal irradiance shown. Full load hours in this lowest reported category, however, are still around 1,700 FLH per year, depending on latitude. Plants at locations with less than 2,000 kWh/m^{2}a would only yield marginal power output and are therefore not included in the technical potential.

[25] The reference technical potential for PV, 372 PWh/a and CSP, 2,946 PWh/a, are from different sources, namely Hoogwijk (2004) and Trieb et al. (2009). They are based on different methodologies and assumptions. de Vries et al. (2007) derive a considerably higher potential for Solar PV, 4,105 PWh/a, based on the methodology applied in Hoogwijk (2004). This figure is of the same order of magnitude as the CSP reference potential.

A.5 Power from Biomass

A.5.1 Overview

Humans have already used biomass for heating and cooking in prehistoric times, making it the most important energy carrier well into modern age. Today domestic cooking and heating in developing countries is still the most widespread use of energy from biomass and accounts for 7% of global primary energy consumption (IEA, 2006). The term biomass includes a large number of substances with organic origin. Kaltschmitt et al. (2009) group the ones serving as primary energy carriers for potential power conversion into four categories:[26]

i. *Energy crops*, e.g. Corn, Willow, Poplar, Miscanthus, Triticale, Brassica
ii. *Crop residues*, e.g. straw, forest wood residues
iii. *Organic byproducts*, e.g. manure, wood scraps
iv. *Organic waste*, e.g. organic municipal and industrial waste

Globally around 200 TWh[27] of electricity were generated from biomass in 2007, representing about 1% of global power production (IEA, 2010a). As biomass can also be used as transport fuel, for centralized heat production or domestic heating and cooking, the electricity sector is in competition with other sectors for low-cost feedstock. A vast number of different biomass substances, possible energy conversion pathways and end use options exist. For a detailed coverage of this topic the interested reader is referred to Kaltschmitt et al. (2009).

A major issue when using biomass for energy production is the potential replacement of food through energy crops. In light of rising prices for agricultural products, use of edible crops for energy production or replacement of food crops by dedicated energy crops have received criticism. A solution to this problem is seen in the cultivation of dedicated, non-edible crops in areas unsuitable or not required for food production. Biomass feedstock produced in this manner is often referred to as *second generation* biofuel (cf. IEA, 2010b).

The use of biomass for power generation requires its combustion using technologies similar to the ones used for power production from coal or gas. Although CO_2 is emitted during the combustion process, power from biomass is considered CO_2-neutral, as the emitted CO_2 has previously been removed from the atmosphere during the crop's growth period. Different technological options exist for power conversion. The most cost-efficient technology to date is co-firing, where a share of the fossil feedstock in coal plants, usually around 10%, is replaced by biomass. Few additional investments are required for such low shares of biomass addition, whereas higher biomass shares generally require plant modifications resulting in uncompetitive economics. The

[26] Translation by the author.
[27] This figure includes some power from non-organic waste.

maximum penetration of co-firing is consequently linked to the system penetration of coal power with its comparatively high specific CO_2 emissions. Co-firing may therefore be appealing to reduce CO_2 emissions of existing coal plants in the short-term at comparatively low cost but is no serious option for a future, carbon-reduced power generation system. Most of today's large-scale biomass power plants operate in a similar manner to coal ones. Thermal energy is obtained through combustion, converted into mechanical energy via a steam turbine and then transformed into electricity via a generator. Most of the installations in operation today have been constructed opportunistically in locations with ample supply of low-cost biomass, e.g. close to a pulp-and-paper facility or sugar refinery. Some dedicated biomass plants also operate in CHP mode, which are characterized by substantially higher overall efficiencies than power-only units. Other technologies available for power generation include combustion of biogas and integrated gasification combined cycle (IGCC) stations, where solid biomass can be used as feedstock. In fact, a substantial portion of plants in operation today, for example in Germany, are rather small-scale, often only with a capacity of several kW, and based on the combustion of biogas from agricultural products or landfills. The key constraint for large-scale biomass technology deployment is local supply of biomass. Transportation of unprocessed biomass is generally characterized by prohibitively high costs attributable to the comparatively low energy density.

A.5.2 Generation Costs

Specific capital costs for dedicated steam cycle (DSC) or IGCC biomass power plants heavily depend on plant size. Since competitive local biomass feedstock is usually not sufficient to support plants with rated capacities beyond 100 MW, current implementations are of small size when compared to their coal-fired counterparts. Figure A.16 displays a selection of overnight capital cost data points for DSC and IGCC plants from literature. Since IGCC technology deployment has not yet entered the commercial phase, DSC plants are used as reference for generation cost calculation with overnight capital costs of 2,700 $/kW.

Identified literature data for non-fuel operating costs range from 50 $/kWa to more than 300 $/kWa and it is often not obvious whether and how much fuel costs are included in a figure. Therefore, non-fuel operating costs are assumed at 90 $/kWa, representing a markup of roughly 30% over a corresponding coal plant with non-fuel operating costs in the 60–80 $/kWa range. The two key reasons for the markup over the cost for a coal-fired plant are the smaller biomass plant scale and the greater material handling efforts assuming local, unprocessed biofuel supply. Economic life of 30 years is assumed for generation cost calculation. Due to their comparatively small size, DSC biomass plants can be constructed within a shorter time frame than their coal-fired counterparts and a construction duration of two years is used for

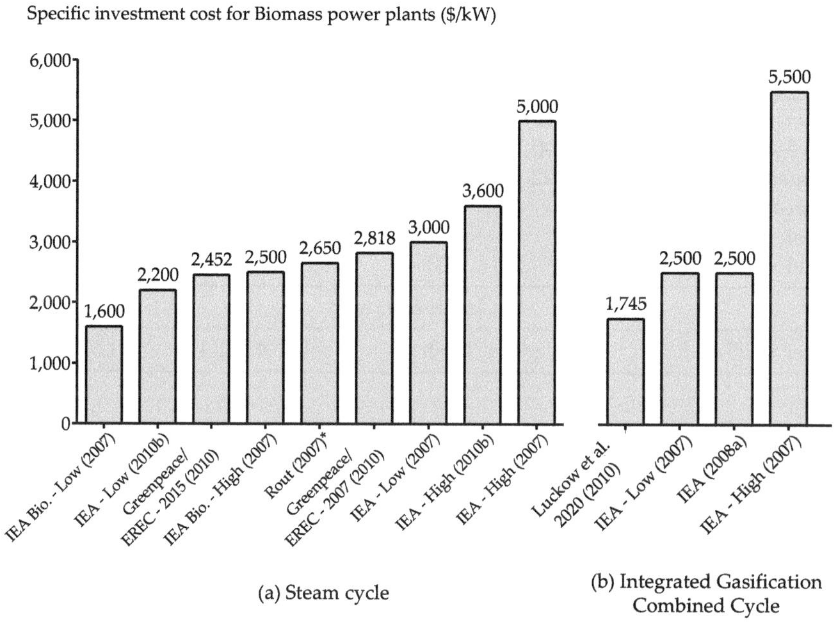

Fig. A.16 Specific overnight capital costs for biomass power plants

further reference. Current generation costs of approximately 8–11 $/kWh (see tab A.6) compare to the costs of wind power and are above current wholesale prices, although with a significantly smaller gap to competitiveness than power from solar technologies.

Given adequate supply of feedstock, solid biomass combustion plants can be operated in a similar manner as their conventional counterparts, i.e. coal steam cycle or IGCC plants, and thus produce mid- or even base-load power. Load factors in the range of 0.4 to 0.8 can be considered realistic, certainly with deviations depending on plant type, scale and feedstock supply. Plant efficiencies depend on the type of technology. Whereas dedicated steam-turbine based conversion is characterized by efficiencies in the 30–35% range, IGCC conversion may reach efficiencies upwards 40%. Since the vast majority of current biomass-fueled power plants are steam-turbine based, a conversion efficiency of 33% is used for generation cost calculation. A capacity factor of 0.6 (ca. 5,300 FLH/a) is used, corresponding to mid-load power production.

A.5.3 Potential

In the future, the largest share of biomass feedstock is expected to come from energy crops on surplus land, whereas today residues still play a more important role (IEA, 2010b). In a recent review of biomass potential literature by

Table A.6 Generation costs for stylized biomass power plants

Cost assumptions			*Other specifications*	
	\$/kW	\$/kWa		
Investment	2,700	218	Useful life (a)	30
Construction fin.	192	15	Constr. period (a)	2
Non-fuel operation		90	FLH (h/a)	5,300
Fuel @ 2\$/GJ		116	Conversion efficiency	33%
Fuel @ 4\$/GJ		231		
		Generation cost		
Fuel @ 2\$/GJ:		8.3 ct/kWh	Fuel @ 4\$/GJ	10.5 ct/kWh

Offermann et al. (2011), the majority of the analyses identify a geographical potential in the 200 to 600 EJ/a range for biomass from energy crops and between 62 and 325 EJ/a for residues.[28] Yet there also exist a number of potential analyses outside these ranges. Wolf et al. (2003), for example, identify a geographical potential of zero[29] for biomass from energy crops, whereas Smeets et al. (2007) obtain a value of 1,550 EJ/a.[30] These large variations are primarily due to different assumptions regarding the amount of available surplus land. Land availability in turn, is determined by a number of highly uncertain factors, such as population growth, future nutritional habits, agricultural yields or total farming land availability.

In light of high prices for agricultural products, some recent publications have argued that truly sustainable biomass potentials from dedicated crops are much lower than previous potential analyses may suggest. Krewitt et al. (2009), for example, only find a global potential from crops of 97 EJ/a realistic.[31] Moreover, most studies state potentials for the year 2050 and assume considerable crop yield increases in developing countries until then due to efficiency improvements. So far, however, there is no empirical evidence that these assumed yield increases are realistic (IEA, 2010b). Based on these recent discussions on sustainability and feasibility of yield improvements Offermann et al. (2011) conclude that a sustainable potential would likely be located towards the lower range of the stated 200 to 600 EJ/a. For further reference the geographical potential derived by Hoogwijk (2004) under assumption of global trade and orientation towards ecological and social values[32] will be used (see Fig. A.17). In addition to regional differentiation, this

[28] For additional biomass potential literature reviews the interested reader is referred to Lysen et al. (2008), IEA (2010b), Krewitt et al. (2009) or Berndes et al. (2003).

[29] Low scenario.

[30] High scenario.

[31] Their analysis is based on Seidenberger et al. (2008).

[32] SRES scenario B1.

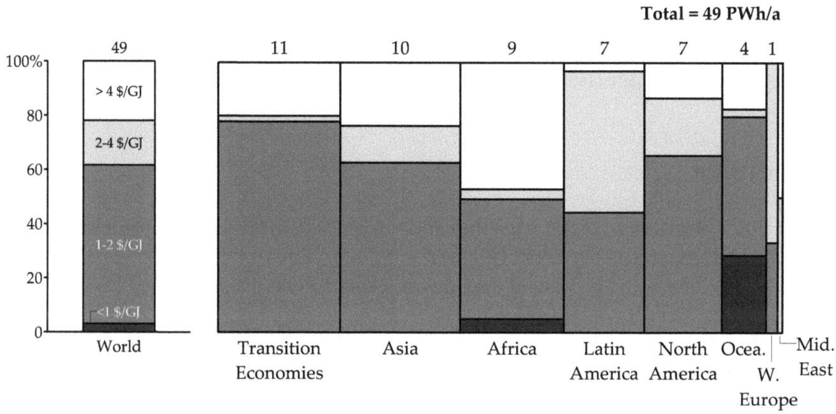

Fig. A.17 Global technical potential for power from biomass (2050)
Source: Adapted from Hoogwijk (2004)[34]

bottom-up analysis has the benefit of also providing information on biomass feedstock costs. With a global potential of 443 EJ/a it stills falls within the 200 to 600 EJ/a range, although somewhat above the sustainable potential suggested by Offermann et al. (2011). A clear benefit of using this data is that the assessment methodology is consistent with the ones for the reference onshore wind and PV potential.

The regional technical potentials for biomass are more evenly distributed than the ones for the other discussed renewable technologies. Regions with large, non-desert surfaces are characterized by higher potentials than small regions. Biomass with production costs in the 1–2 $/GJ range represent the largest share of technical potential. The fact that almost two thirds of technical potential is available at costs below 2 $/GJ is not fully in line with feedstock costs found in other literature. Often feedstock costs in the 2–5 $/GJ range are reported (cf., for example, IEA 2007; IEA Bioenergy 2007; IEA 2010b). The reasons for this difference is difficult to assess. A contributing factor may be biomass processing and transportation costs to the power plant. Whereas Hoogwijk (2004) assumes values between 0.26 and 0.45 $/GJ for transportation to local plants without pellet production,

[34] Scenario B1 (global trade and orientation towards social and ecological values) has been used as reference, since it is compatible with the goal of a carbon-reduced power sector. Moreover, the scenario's potential values lie in between the extremes of scenarios A1 and A2. The data shown here is based on the hypothetical case that all available biomass were used for power generation. Geographic feedstock potentials have been multiplied with a power conversion efficiency of 0.4. However, only 20% of the potential displayed here is considered available to the power sector for the model presented in chapter 6.

van Vliet et al. (2009), for example, arrive at a value of 1.87 $/GJ for local transportation with pellet production. Hoogwijk (2004) also incorporates cost reductions from expected innovations until 2050 into her analysis.

References

Archer, C.L., Jacobson, M.Z.: Evaluation of global wind power. Journal of Geophysical Research 110, D12110 (2005)

Bartle, A.: Hydropower potential and development activities. Energy Policy 30(14), 1231–1239 (2002)

Berndes, G., Hoogwijk, M.M., van den Broek, R.: The contribution of biomass in the future global energy supply: a review of 17 studies. Biomass and Bioenergy 25(1), 1–28 (2003)

Betz, A.: Windenergie und ihre Ausnutzung durch Windmühlen (Reprint 1994 ed.). Ökobuch Verlag, Staufen (1926)

BMU. Erfahrungsbericht 2007 zum Erneuerbare-Energien-Gesetz, Berlin (2007)

Caldés, N., Varela, M., Sanatmaría, R.S.: Economic impact of solar thermal electricity deployment in Spain. Energy Policy 37(5), 1628–1636 (2009)

de Vries, B., van Vuuren, D.P., Hoogwijk, M.M.: Renewable energy sources: their global potential for the first-half of the 21st century at a global level: an integrated approach. Energy Policy 35(4), 2590–2610 (2007)

Denholm, P., Margolis, R.M.: Evaluating the limits of solar photovoltaics (PV) in traditional electric power systems. Energy Policy 35(5), 2852–2861 (2007)

Deutsche Offshore-Testfeld und Infrastruktur GmbH & Co. KG. Ein Offhore-Windpark ensteht: Alpha Ventus, Oldenburg (2010)

European Wind Energy Association. The economics of wind energy, Brussels (2009)

Fichtner. Die Wettbewerbsfähigkeit von großen Laufwasserkraftwerken im liberalisierten deutschen Strommarkt: Endbericht. Stuttgart (2003)

Frey, G.W., Linke, D.M.: Hydropower as a renewable and sustainable energy resource meeting global energy challenges in a reasonable way. Energy Policy 30(14), 1261–1265 (2002)

Giesecke, J., Mosonyi, E., Heimer, S.: Wasserkraftanlagen: Planung, Bau und Betrieb, 5th revised and enlarged edn. Springer, Heidelberg (2009)

Goldemberg, J. (ed.): World energy assessment: Energy and the challenge of sustainability. UNDP, New York (2000)

Greenpeace International and EREC. energy [r]evolution: A Sustainable World Energy Outlook, Amsterdam (2010)

Grubb, M., Meyer, N.: Wind energy: resources, systems, and regional strategies. In: Johansson, T., Kelly, H., Reddy, A., Williams, R. (eds.) Renewable Energy: Sources for Fuels and Electricity, pp. 157–212. Island Press, Washington, DC (1993)

GWEC. Global Wind 2009 Report, Brussels (2010)

Hofman, Y., Jager, D., Molenbroek, E., Schilig, F., Voogt, M.: The potential of solar electricity to reduce CO2 emissions, Utrecht (2002)

Hoogwijk, M.: On the global and regional potential of renewable energy sources: Doctoral thesis. Doctoral thesis, Utrecht University, Utrecht (2004)

Hoogwijk, M., de Vries, B., Turkenburg, W.: Assessment of the global and regional geographical, technical and economic potential of onshore wind energy. Energy Economics 26(5), 889–919 (2004)

IEA. Renewables for power generation: Status and prospects. IEA Publications, Paris (2003)
IEA. World Energy Outlook 2006. IEA Publications, Paris (2006)
IEA. IEA Energy Technology Essentials: Biomass for Power Generation and CHP, Paris (2007)
IEA. Energy Technology Perspectives: Scenarios & Strategies to 2050. IEA Publications, Paris (2008a)
IEA. World Energy Outlook 2008. IEA Publications, Paris (2008b)
IEA. Prospects for Large-Scale Energy Storage in Decarbonised Power Grids, Paris (2009a)
IEA. Technology Roadmap: Wind energy, Paris (2009b)
IEA. World Energy Outlook 2009. IEA Publications, Paris (2009c)
IEA. Electricity/ Heat in World in 2007 (2010a)
IEA. Sustainable Production of Second-Generation Biofuels: Potential and perspectives in major economies and developing countries, Paris (2010b)
IEA. Technology Roadmap: Concentrating Solar Power, Paris (2010c)
IEA. Technology Roadmap: Solar photovoltaic energy, Paris (2010d)
IEA Bioenergy. Potential Contribution of Bioenergy to the World's Future Energy Demand, Whakarewarewa (2007)
Jarass, L., Obermair, G.M.: Windenergie: Zuverlässige Integration in die Energieversorgung, 2nd revised edn. Springer, Heidelberg (2009)
Junginger, M.: Learning in Renewable Energy Technology Development: Doctoral thesis. Doctoral thesis, Utrecht University, Utrecht (2005)
Kaltschmitt, M., Hartmann, H., Hofbauer, H. (eds.): Energie aus Biomasse: Grundlagen, Techniken und Verfahren, 2nd revised edn. Springer, Heidelberg (2009)
Kaltschmitt, M., Streicher, W., Wiese, A. (eds.): Erneuerbare Energien: Systemtechnik, Wirtschaftlichkeit, Umweltaspekte. Springer, Heidelberg (2006)
Konstantin, P.: Praxisbuch Energiewirtschaft: Energieumwandlung, -transport und -bescha ung im liberalisierten Markt, 2nd revised edn. 46. Springer, Heidelberg (2009)
Krewitt, W., Nienhaus, K., Kleßmann, C., Capone, C., Stricker, E., Graus, W., Hoogwijk, M., Supersberger, N., Winterfeld, U., Samadi, S.: Role and Potential of Renewable Energy and Energy Efficiency for Global Energy Supply, Dessau-Roßlau (2009)
Luckow, P., Wise, M.A., Dooleey, J.J., Kim, S.H.: Biomass Energy for Transport and Electricity: Large Scale Utilization Under Low CO2 Concentration Scenarios. Springfield (2010)
Lysen, E., van Egmond, S., Dornburg, V., Faaij, A., Verweij, P., Langeveld, H., van de Ven, G., Wester, F., van Keulen, H., van Diepen, K., Meeusen, M., Banse, M., Ros, J., van Vuuren, D.P., van den Born, G.J., van Oorschot, M., Smout, F., van Vliet, J., Aiking, H., Londo, M., Mozaffarian, H., Smekens, K.: Biomass Assessment: Assessment of global biomass potentials and their links to food, water, biodiversity, energy demand and economy, Bilthoven (2008)
Oanda.com. Historical Exchange Rates (2010)
Offermann, R., Seidenberger, T., Thrän, D., Kaltschmitt, M., Zinoviev, S., Miertus, S.: Assessment of global bioenergy potentials. Mitigation and Adaptation Strategies for Global Change 16(1), 103–115 (2011)
pvXchange. PV price index (2012)
REN21. Global Potential of Renewable Energy Sources: A Literature Assessment, Paris (2008)

REN21. Renewables 2010 — Global Status Report, Paris (2010)

Rout, U.K.: Modelling of Endogenous Technological Learning of Energy Technologies — an Analysis with a Global Multi-regional Energy System Model: Dissertation. Stuttgart University, Stuttgart (2007)

Schillings, C., Mannstein, H., Meyer, R.: Operational method for deriving high resolution direct normal irradiance from satellite data. Solar Energy 76(4), 475–484 (2004)

Seidenberger, T., Thrän, D., Offermann, R., Seyfert, U., Buchhorn, M., Zeddis, J.: Global Biomass Potentials, Leipzig (2008)

Smeets, E.M.W., Faaij, A.P.C., Lewandowski, I.M., Turkenburg, W.C.: A bottom-up assessment and review of global bio-energy potentials to 2050. Progress in Energy and Combustion Science 33(1), 56–106 (2007)

Solarserver. Photovoltaik Preisindex (2010)

Stodola, N., Modi, V.: Penetration of solar power without storage. Energy Policy 37(11), 4730–4736 (2009)

Swider, D.J., Weber, C.: The costs of wind's intermittency in Germany: application of a stochastic electricity market model. European Transactions on Electrical Power 17(2), 151–172 (2007)

Trieb, F., O'Sullivan, M., Pregger, T., Schillings, C., Krewitt, W.: Characterisation of Solar Electricity Import Corridors from MENA to Europe: Potential, Infrastructure and Cost, Stuttgart (2009)

Trieb, F., Schillings, C., O'Sullivan, M., Pregger, T., Hoyer-Klick, C.: Global Potential of Concentrating Solar Power. Presented at SolarPaces Conference, Berlin (2009)

Vallentin, D., Viebahn, P.: Economic opportunities resulting from a global deployment of concentrated solar power (CSP) technologies — The example of German technology providers. Energy Policy 38(8), 4467–4478 (2010)

van Vliet, O.P.R., Faaij, A.P.C., Turkenburg, W.C.: Fischer-Tropsch diesel production in a well-to-wheel perspective: A carbon, energy flow and cost analysis. Energy Conversion and Management 50(4), 855–876 (2009)

Wagner, A.: Photovoltaik Engineering: Handbuch für Planung, Entwicklung und Anwendung, 3rd edn. Springer, Heidelberg (2010)

Wolf, J., Bindraban, P.S., Luijten, J.C., Vleeshouwers, L.M.: Exploratory study on the land area required for global food supply and the potential global production of bioenergy. Agricultural Systems 76(3), 841–861 (2003)

World Energy Council. 2007 Survey of Energy Resources, London (2007)

World Energy Council. Survey of Energy Resources: Interim Update 2009, London (2009)

Appendix B
Analytical Model Backup

B.1 Basic Model with Perfect Foresight

B.1.1 Extensive Model Formulation

For the solution process and to enable application of the solution to a real world problem, the basic deterministic model described in Eqs. (5.1) through (5.4) is transformed into an extensive form.

Minimize

$$
\begin{aligned}
z = {} & \alpha_A \theta_1 \psi_1 w_{A1} + \alpha_A \theta_2 \psi_2 w_{A2} + \alpha_B \theta_1 \psi_1 w_{B1} + \alpha_B \theta_2 \psi_2 w_{B2} \\
& + \beta_A \psi_1 y_{A1} + \beta_A \psi_2 y_{A2} + \beta_B \psi_1 y_{B1} + \beta_B \psi_2 y_{B2} \\
& + \frac{\phi_A \psi_1}{1 - \gamma_A} \left[(\kappa_{A0} + x_{A1})^{(1-\gamma_A)} - \kappa_{A0}^{(1-\gamma_A)} \right] \\
& + \frac{\phi_A \psi_2 \theta_2}{\pi_A (1 - \gamma_A)} \left[(\kappa_{A0} + x_{A1} + x_{A2})^{(1-\gamma_A)} - (\kappa_{A0} + x_{A1})^{(1-\gamma_A)} \right] \\
& + \frac{\phi_B \psi_1}{1 - \gamma_B} \left[(\kappa_{B0} + x_{B1})^{(1-\gamma_B)} - \kappa_{B0}^{(1-\gamma_B)} \right] \\
& + \frac{\phi_B \psi_2 \theta_2}{\pi_B (1 - \gamma_B)} \left[(\kappa_{B0} + x_{B1} + x_{B2})^{(1-\gamma_B)} - (\kappa_{B0} + x_{B1})^{(1-\gamma_B)} \right]
\end{aligned}
\tag{B.1}
$$

Subject to

$$
\begin{aligned}
y_{A1} + y_{B1} &\geq \theta_1 \delta_1 & \theta_2 \sigma_B w_{B2} &\geq y_{B2} \\
y_{A2} + y_{B2} &\geq \theta_2 \delta_2 & x_{A1} &= w_{A1} \\
\theta_1 \sigma_A w_{A1} &\geq y_{A1} & w_{A1} + x_{A2} - \mu_A x_{A1} &= w_{A2} \\
\theta_2 \sigma_A w_{A2} &\geq y_{A2} & x_{B1} &= w_{B1} \\
\theta_1 \sigma_B w_{B1} &\geq y_{B1} & w_{B1} + x_{B2} - \mu_B x_{B1} &= w_{B2}
\end{aligned}
\tag{B.2}
$$

B.1.2 First-Order Derivatives with Respect to Decision Variables

$$Z_{w_{A1}} = \alpha_A\theta_1\psi_1 - \theta_1\sigma_A\lambda_3 + \lambda_7 - \lambda_8 \tag{B.3}$$

$$Z_{w_{A2}} = \alpha_A\theta_2\psi_2 - \theta_2\sigma_A\lambda_4 + \lambda_8 \tag{B.4}$$

$$Z_{w_{B1}} = \alpha_B\theta_1\psi_1 - \theta_1\sigma_B\lambda_5 + \lambda_9 - \lambda_{10} \tag{B.5}$$

$$Z_{w_{B2}} = \alpha_B\theta_2\psi_2 - \theta_2\sigma_B\lambda_6 + \lambda_{10} \tag{B.6}$$

$$\begin{aligned} Z_{x_{A1}} &= \phi_A\psi_1(k_{A0} + x_{A1})^{-\gamma_A} + \frac{\phi_A\psi_2\theta_2}{\pi_A}(k_{A0} + x_{A1} + x_{A2})^{-\gamma_A} \\ &\quad - \frac{\phi_A\psi_2\theta_2}{\pi_A}(k_{A0} + x_{A1})^{-\gamma_A} - \lambda_7 + \mu_A\lambda_8 \end{aligned} \tag{B.7}$$

$$Z_{x_{A2}} = \frac{\phi_A\psi_2\theta_2}{\pi_A}(k_{A0} + x_{A1} + x_{A2})^{-\gamma_A} - \lambda_8 \tag{B.8}$$

$$\begin{aligned} Z_{x_{B1}} &= \phi_B\psi_1(k_{B0} + x_{B1})^{-\gamma_B} + \frac{\phi_B\psi_2\theta_2}{\pi_B}(k_{B0} + x_{B1} + x_{B2})^{-\gamma_B} \\ &\quad - \frac{\phi_B\psi_2\theta_2}{\pi_B}(k_{B0} + x_{B1})^{-\gamma_B} - \lambda_9 + \mu_B\lambda_{10} \end{aligned} \tag{B.9}$$

$$Z_{x_{B2}} = \frac{\phi_B\psi_2\theta_2}{\pi_B}(k_{B0} + x_{B1} + x_{B2})^{-\gamma_B} - \lambda_{10} \tag{B.10}$$

$$Z_{y_{A1}} = \beta_A\psi_1 - \lambda_1 + \lambda_3 \tag{B.11}$$

$$Z_{y_{A2}} = \beta_A\psi_2 - \lambda_2 + \lambda_4 \tag{B.12}$$

$$Z_{y_{B1}} = \beta_B\psi_1 - \lambda_1 + \lambda_5 \tag{B.13}$$

$$Z_{y_{B2}} = \beta_B\psi_2 - \lambda_2 + \lambda_6 \tag{B.14}$$

B.2 Deterministic Model with Capacity Endowment

B.2.1 Lagrangian Function

$$\begin{aligned} Z &= \alpha_A\theta_1\psi_1 w_{A1} + \alpha_A\theta_2\psi_2 w_{A2} + \alpha_B\theta_1\psi_1 w_{B1} + \alpha_B\theta_2\psi_2 w_{B2} \\ &\quad + \beta_A\psi_1 y_{A1} + \beta_A\psi_2 y_{A2} + \beta_B\psi_1 y_{B1} + \beta_B\psi_2 y_{B2} \\ &\quad + \frac{\phi_A\psi_1}{1-\gamma_A}\left[(k_{A0} + x_{A1})^{(1-\gamma_A)} - k_{A0}^{(1-\gamma_A)}\right] \\ &\quad + \frac{\phi_A\psi_2\theta_2}{\pi_A(1-\gamma_A)}\left[(k_{A0} + x_{A1} + x_{A2})^{(1-\gamma_A)} - (k_{A0} + x_{A1})^{(1-\gamma_A)}\right] \\ &\quad + \frac{\phi_B\psi_1}{1-\gamma_B}\left[(k_{B0} + x_{B1})^{(1-\gamma_B)} - k_{B0}^{(1-\gamma_B)}\right] \\ &\quad + \frac{\phi_B\psi_2\theta_2}{\pi_B(1-\gamma_B)}\left[(k_{B0} + x_{B1} + x_{B2})^{(1-\gamma_B)} - (k_{B0} + x_{B1})^{(1-\gamma_B)}\right] \\ &\quad + \lambda_1(\theta_1\delta_1 - y_{A1} - y_{B1}) + \lambda_2(\theta_2\delta_2 - y_{A2} - y_{B2}) \\ &\quad + \lambda_3(y_{A1} - \theta_1\sigma_A w_{A1}) + \lambda_4(y_{A2} - \theta_2\sigma_A w_{A2}) \\ &\quad + \lambda_5(y_{B1} - \theta_1\sigma_B w_{B1}) + \lambda_6(y_{B2} - \theta_2\sigma_B w_{B2}) \\ &\quad + \lambda_7(w_{A1} - x_{A1} - \boldsymbol{\epsilon_A}) + \lambda_8(w_{A2} - w_{A1} - x_{A2} + \mu_A x_{A1} + \boldsymbol{\nu_A}) \\ &\quad + \lambda_9(w_{B1} - x_{B1} - \boldsymbol{\epsilon_B}) + \lambda_{10}(w_{B2} - w_{B1} - x_{B2} + \mu_B x_{B1} + \boldsymbol{\nu_B}) \end{aligned} \tag{B.15}$$

B.3 Dynamic Model with Uncertainty

B.3.1 Lagrangian Function

$$\begin{aligned}
Z &= \alpha_A\psi_1\theta_1 w_{A1} + \alpha_B\psi_1\theta_1 w_{B1} + \beta_{A1}\psi_1 y_{A1} + \psi_1\beta_{B1}y_{B1} \\
&+ \frac{\phi_A\psi_1}{1-\gamma_{A1}}(k_{A0}+x_{A1})^{1-\gamma_{A1}} - \frac{\phi_A\psi_1}{1-\gamma_{A1}}k_{A0}^{1-\gamma_{A1}} \\
&+ \frac{\phi_B\psi_1}{1-\gamma_{B1}}(k_{B0}+x_{B1})^{1-\gamma_{B1}} - \frac{\phi_B\psi_1}{1-\gamma_{B1}}k_{B0}^{1-\gamma_{B1}} \\
&+ p_a\alpha_A\psi_2\theta_2 w_{A2a} + p_a\alpha_B\psi_2\theta_2 w_{B2a} + p_a\beta_{A2a}\psi_2 y_{A2a} + p_a\beta_{B2a}\psi_2 y_{B2a} \\
&+ \frac{p_a\phi_A\psi_2\theta_2}{\pi_A(1-\gamma_{A2a})}(k_{A0}+x_{A1}+x_{A2a})^{1-\gamma_{A2a}} - \frac{p_a\phi_A\psi_2\theta_2}{\pi_A(1-\gamma_{A2a})}(k_{A0}+x_{A1})^{1-\gamma_{A2a}} \\
&+ \frac{p_a\phi_A\psi_2\theta_2}{\pi_A}x_{A2a}(k_{A0}+x_{A1})^{-\gamma_{A1}} - \frac{p_a\phi_A\psi_2\theta_2}{\pi_A}x_{A2a}(k_{A0}+x_{A1})^{-\gamma_{A2a}} \\
&+ \frac{p_a\phi_B\psi_2\theta_2}{\pi_B(1-\gamma_{B2a})}(k_{B0}+x_{B1}+x_{B2a})^{1-\gamma_{B2a}} - \frac{p_a\phi_B\psi_2\theta_2}{\pi_B(1-\gamma_{B2a})}(k_{B0}+x_{B1})^{1-\gamma_{B2a}} \\
&+ \frac{p_a\phi_B\psi_2\theta_2}{\pi_B}x_{B2a}(k_{B0}+x_{B1})^{-\gamma_{B1}} - \frac{p_a\phi_B\psi_2\theta_2}{\pi_B}x_{B2a}(k_{B0}+x_{B1})^{-\gamma_{B2a}} \\
&+ p_b\alpha_A\psi_2\theta_2 w_{A2b} + p_b\alpha_B\psi_2\theta_2 w_{B2b} + p_b\beta_{A2b}\psi_2 y_{A2b} + p_b\beta_{B2b}\psi_2 y_{B2b} \\
&+ \frac{p_b\phi_A\psi_2\theta_2}{\pi_A(1-\gamma_{A2b})}(k_{A0}+x_{A1}+x_{A2b})^{1-\gamma_{A2b}} - \frac{p_b\phi_A\psi_2\theta_2}{\pi_A(1-\gamma_{A2b})}(k_{A0}+x_{A1})^{1-\gamma_{A2b}} \\
&+ \frac{p_b\phi_A\psi_2\theta_2}{\pi_A}x_{A2b}(k_{A0}+x_{A1})^{-\gamma_{A1}} - \frac{p_b\phi_A\psi_2\theta_2}{\pi_A}x_{A2b}(k_{A0}+x_{A1})^{-\gamma_{A2b}} \\
&+ \frac{p_b\phi_B\psi_2\theta_2}{\pi_B(1-\gamma_{B2b})}(k_{B0}+x_{B1}+x_{B2b})^{1-\gamma_{B2b}} - \frac{p_b\phi_B\psi_2\theta_2}{\pi_B(1-\gamma_{B2b})}(k_{B0}+x_{B1})^{1-\gamma_{B2b}} \\
&+ \frac{p_b\phi_B\psi_2\theta_2}{\pi_B}x_{B2b}(k_{B0}+x_{B1})^{-\gamma_{B1}} - \frac{p_b\phi_B\psi_2\theta_2}{\pi_B}x_{B2b}(k_{B0}+x_{B1})^{-\gamma_{B2b}} \\
&+ \lambda_1(\delta_1\theta_1 - y_{A1} - y_{B1}) + \lambda_2(\delta_2\theta_2 - y_{A2a} - y_{B2a}) + \lambda_3(\delta_2\theta_2 - y_{A2b} - y_{B2b}) \\
&+ \lambda_4(y_{A1} - \sigma_A\theta_1 w_{A1}) + \lambda_5(y_{B1} - \sigma_B\theta_1 w_{B1}) + \lambda_6(y_{A2a} - \sigma_A\theta_2 w_{A2a}) \\
&+ \lambda_7(y_{B2a} - \sigma_B\theta_2 w_{B2a}) + \lambda_8(y_{A2b} - \sigma_A\theta_2 w_{A2b}) + \lambda_9(y_{B2b} - \sigma_B\theta_2 w_{B2b}) \\
&+ \lambda_{10}(w_{A1} - x_{A1}) + \lambda_{11}(w_{B1} - x_{B1}) \\
&+ \lambda_{12}(w_{A2a} - w_{A1} - x_{A2a} + \mu_A x_{A1}) + \lambda_{13}(w_{B2a} - w_{B1} - x_{B2a} + \mu_B x_{B1}) \\
&+ \lambda_{14}(w_{A2b} - w_{A1} - x_{A2b} + \mu_A x_{A1}) + \lambda_{15}(w_{B2b} - w_{B1} - x_{B2b} + \mu_B x_{B1})
\end{aligned} \tag{B.16}$$

B.3.2 First-Order Derivatives with Respect to Decision Variables

$$Z_{w_{A1}} = \alpha_A\psi_1\theta_1 - \sigma_A\theta_1\lambda_4 + \lambda_{10} - \lambda_{12} - \lambda_{14} \tag{B.17}$$

$$Z_{w_{B1}} = \alpha_B\psi_1\theta_1 - \sigma_B\theta_1\lambda_5 + \lambda_{11} - \lambda_{13} - \lambda_{15} \tag{B.18}$$

$$Z_{w_{A2a}} = p_a\alpha_A\psi_2\theta_2 - \sigma_A\theta_2\lambda_6 + \lambda_{12} \tag{B.19}$$

$$Z_{w_{A2b}} = p_b\alpha_A\psi_2\theta_2 - \sigma_A\theta_2\lambda_8 + \lambda_{14} \tag{B.20}$$

$$Z_{w_{B2a}} = p_a\alpha_B\psi_2\theta_2 - \sigma_B\theta_2\lambda_7 + \lambda_{13} \tag{B.21}$$

$$Z_{w_{B2b}} = p_b\alpha_B\psi_2\theta_2 - \sigma_B\theta_2\lambda_9 + \lambda_{15} \tag{B.22}$$

$$\begin{aligned}
Z_{x_{A1}} &= \psi_1\phi_A(k_{A0}+x_{A1})^{-\gamma_{A1}} - \lambda_{10} + \mu_A\lambda_{12} + \mu_A\lambda_{14} \\
&+ \frac{p_a\phi_A\psi_2\theta_2}{\pi_A}(k_{A0}+x_{A1}+x_{A2a})^{-\gamma_{A2a}} - \frac{p_a\phi_A\psi_2\theta_2}{\pi_A}(k_{A0}+x_{A1})^{-\gamma_{A2a}} \\
&+ \frac{p_a\phi_A\psi_2\theta_2\gamma_{A2a}}{\pi_A}x_{A2a}(k_{A0}+x_{A1})^{-(\gamma_{A2a}+1)} \\
&- \frac{p_a\phi_A\psi_2\theta_2\gamma_{A1}}{\pi_A}x_{A2a}(k_{A0}+x_{A1})^{-(\gamma_{A1}+1)} \\
&+ \frac{p_b\phi_A\psi_2\theta_2}{\pi_A}(k_{A0}+x_{A1}+x_{A2b})^{-\gamma_{A2b}} - \frac{p_b\phi_A\psi_2\theta_2}{\pi_A}(k_{A0}+x_{A1})^{-\gamma_{A2b}} \\
&+ \frac{p_b\phi_A\psi_2\theta_2\gamma_{A2b}}{\pi_A}x_{A2b}(k_{A0}+x_{A1})^{-(\gamma_{A2b}+1)} \\
&- \frac{p_b\phi_A\psi_2\theta_2\gamma_{A1}}{\pi_A}x_{A2b}(k_{A0}+x_{A1})^{-(\gamma_{A1}+1)}
\end{aligned} \tag{B.23}$$

$$\begin{aligned}
Z_{x_{B1}} &= \psi_1\phi_B(k_{B0}+x_{B1})^{-\gamma_{B1}} - \lambda_{11} + \mu_B\lambda_{13} + \mu_B\lambda_{15} \\
&+ \frac{p_a\phi_B\psi_2\theta_2}{\pi_B}(k_{B0}+x_{B1}+x_{B2a})^{-\gamma_{B2a}} - \frac{p_a\phi_B\psi_2\theta_2}{\pi_B}(k_{B0}+x_{B1})^{-\gamma_{B2a}} \\
&+ \frac{p_a\phi_B\psi_2\theta_2\gamma_{B2a}}{\pi_B}x_{B2a}(k_{B0}+x_{B1})^{-(\gamma_{B2a}+1)} \\
&- \frac{p_a\phi_B\psi_2\theta_2\gamma_{B1}}{\pi_B}x_{B2a}(k_{B0}+x_{B1})^{-(\gamma_{B1}+1)} \\
&+ \frac{p_b\phi_B\psi_2\theta_2}{\pi_B}(k_{B0}+x_{B1}+x_{B2b})^{-\gamma_{B2b}} - \frac{p_b\phi_B\psi_2\theta_2}{\pi_B}(k_{B0}+x_{B1})^{-\gamma_{B2b}} \\
&+ \frac{p_b\phi_B\psi_2\theta_2\gamma_{B2b}}{\pi_B}x_{B2b}(k_{B0}+x_{B1})^{-(\gamma_{B2b}+1)} \\
&- \frac{p_b\phi_B\psi_2\theta_2\gamma_{B1}}{\pi_B}x_{B2b}(k_{B0}+x_{B1})^{-(\gamma_{B1}+1)}
\end{aligned} \tag{B.24}$$

$$\begin{aligned}
Z_{x_{A2a}} &= \frac{p_a\phi_A\psi_2\theta_2}{\pi_A}(k_{A0}+x_{A1}+x_{A2a})^{-\gamma_{A2a}} - \lambda_{12} \\
&+ \frac{p_a\phi_A\psi_2\theta_2}{\pi_A}(k_{A0}+x_{A1})^{-\gamma_{A1}} - \frac{p_a\phi_A\psi_2\theta_2}{\pi_A}(k_{A0}+x_{A1})^{-\gamma_{A2a}}
\end{aligned} \tag{B.25}$$

$$\begin{aligned}
Z_{x_{B2a}} &= \frac{p_a\phi_B\psi_2\theta_2}{\pi_B}(k_{B0}+x_{B1}+x_{B2a})^{-\gamma_{B2a}} - \lambda_{13} \\
&+ \frac{p_a\phi_B\psi_2\theta_2}{\pi_B}(k_{B0}+x_{B1})^{-\gamma_{B1}} - \frac{p_a\phi_B\psi_2\theta_2}{\pi_B}(k_{B0}+x_{B1})^{-\gamma_{B2a}}
\end{aligned} \tag{B.26}$$

$$\begin{aligned}
Z_{x_{A2b}} &= \frac{p_b\phi_A\psi_2\theta_2}{\pi_A}(k_{A0}+x_{A1}+x_{A2b})^{-\gamma_{A2b}} - \lambda_{14} \\
&+ \frac{p_b\phi_A\psi_2\theta_2}{\pi_A}(k_{A0}+x_{A1})^{-\gamma_{A1}} - \frac{p_b\phi_A\psi_2\theta_2}{\pi_A}(k_{A0}+x_{A1})^{-\gamma_{A2b}}
\end{aligned} \tag{B.27}$$

$$\begin{aligned}
Z_{x_{B2b}} &= \frac{p_b\phi_B\psi_2\theta_2}{\pi_B}(k_{B0}+x_{B1}+x_{B2b})^{-\gamma_{B2b}} - \lambda_{15} \\
&+ \frac{p_b\phi_B\psi_2\theta_2}{\pi_B}(k_{B0}+x_{B1})^{-\gamma_{B1}} - \frac{p_b\phi_B\psi_2\theta_2}{\pi_B}(k_{B0}+x_{B1})^{-\gamma_{B2b}}
\end{aligned} \tag{B.28}$$

$$Z_{y_{A1}} = \psi_1 \beta_{A1} - \lambda_1 + \lambda_4 \tag{B.29}$$
$$Z_{y_{B1}} = \psi_1 \beta_{B1} - \lambda_1 + \lambda_5 \tag{B.30}$$
$$Z_{y_{A2a}} = p_a \beta_{A2a} \psi_2 - \lambda_2 + \lambda_6 \tag{B.31}$$
$$Z_{y_{A2b}} = p_b \beta_{A2b} \psi_2 - \lambda_3 + \lambda_8 \tag{B.32}$$
$$Z_{y_{B2a}} = p_a \beta_{B2a} \psi_2 - \lambda_2 + \lambda_7 \tag{B.33}$$
$$Z_{y_{B2b}} = p_b \beta_{B2b} \psi_2 - \lambda_3 + \lambda_9 \tag{B.34}$$

B.3.3 Solutions for Even-Numbered Cases

Solution for case P2 — Dominance of technology A

$$\begin{array}{lll}
w^*_{A1} = \dfrac{\delta_1}{\sigma_A} & w^*_{A2a} = \dfrac{\delta_2}{\sigma_A} & w^*_{A2b} = \dfrac{\delta_2}{\sigma_A} \\
w^*_{B1} = 0 & w^*_{B2a} = 0 & w^*_{B2b} = 0 \\
x^*_{A1} = \dfrac{\delta_1}{\sigma_A} & x^*_{A2a} = (\mu_A - 1)\dfrac{\delta_1}{\sigma_A} + \dfrac{\delta_2}{\sigma_A} & x^*_{A2b} = (\mu_A - 1)\dfrac{\delta_1}{\sigma_A} + \dfrac{\delta_2}{\sigma_A} \\
x^*_{B1} = 0 & x^*_{B2a} = 0 & x^*_{B2b} = 0 \\
y^*_{A1} = \delta_1\theta_1 & y^*_{A2a} = \delta_2\theta_2 & y^*_{A2b} = \delta_2\theta_2 \\
y^*_{B1} = 0 & y^*_{B2a} = 0 & y^*_{B2b} = 0
\end{array}$$

Solution for case P4 — Advantage of technology A

$$\begin{array}{lll}
w^*_{A1} = \dfrac{\delta_1}{\sigma_A} & w^*_{A2a} = \dfrac{\delta_2}{\sigma_A} & w^*_{A2b} = (1 - \mu_A)\dfrac{\delta_1}{\sigma_A} \\
w^*_{B1} = 0 & w^*_{B2a} = 0 & w^*_{B2b} = \dfrac{\delta_2 - \delta_1(1 - \mu_A)}{\sigma_B} \\
x^*_{A1} = \dfrac{\delta_1}{\sigma_A} & x^*_{A2a} = (\mu_A - 1)\dfrac{\delta_1}{\sigma_A} + \dfrac{\delta_2}{\sigma_A} & x^*_{A2b} = 0 \\
x^*_{B1} = 0 & x^*_{B2a} = 0 & x^*_{B2b} = \dfrac{\delta_2 - \delta_1(1 - \mu_A)}{\sigma_B} \\
y^*_{A1} = \delta_1\theta_1 & y^*_{A2a} = \delta_2\theta_2 & y^*_{A2b} = (1 - \mu_A)\delta_1\theta_2 \\
y^*_{B1} = 0 & y^*_{B2a} = 0 & y^*_{B2b} = \theta_2[\delta_2 - \delta_1(1 - \mu_A)]
\end{array}$$

Solution for case P6 — Investment shift from A to B

$$
\begin{aligned}
w^*_{A1} &= \frac{\delta_1}{\sigma_A} & w^*_{A2a} &= (1-\mu_A)\frac{\delta_1}{\sigma_A} & w^*_{A2b} &= (1-\mu_A)\frac{\delta_1}{\sigma_A} \\
w^*_{B1} &= 0 & w^*_{B2a} &= \frac{\delta_2-\delta_1(1-\mu_A)}{\sigma_B} & w^*_{B2b} &= \frac{\delta_2-\delta_1(1-\mu_A)}{\sigma_B} \\
x^*_{A1} &= \frac{\delta_1}{\sigma_A} & x^*_{A2a} &= 0 & x^*_{A2b} &= 0 \\
x^*_{B1} &= 0 & x^*_{B2a} &= \frac{\delta_2-\delta_1(1-\mu_A)}{\sigma_B} & x^*_{B2b} &= \frac{\delta_2-\delta_1(1-\mu_A)}{\sigma_B} \\
y^*_{A1} &= \delta_1\theta_1 & y^*_{A2a} &= (1-\mu_A)\delta_1\theta_2 & y^*_{A2b} &= (1-\mu_A)\delta_1\theta_2 \\
y^*_{B1} &= 0 & y^*_{B2a} &= \theta_2[\delta_2-\delta_1(1-\mu_A)] & y^*_{B2b} &= \theta_2[\delta_2-\delta_1(1-\mu_A)]
\end{aligned}
$$

Solution for case P8 — Advantage technology A with potential full shift to B

$$
\begin{aligned}
w^*_{A1} &= \frac{\delta_1}{\sigma_A} & w^*_{A2a} &= \frac{\delta_2}{\sigma_A} & w^*_{A2b} &= (1-\mu_A)\frac{\delta_1}{\sigma_A} \\
w^*_{B1} &= 0 & w^*_{B2a} &= 0 & w^*_{B2b} &= \frac{\delta_2}{\sigma_B} \\
x^*_{A1} &= \frac{\delta_1}{\sigma_A} & x^*_{A2a} &= (\mu_A-1)\frac{\delta_1}{\sigma_A}+\frac{\delta_2}{\sigma_A} & x^*_{A2b} &= 0 \\
x^*_{B1} &= 0 & x^*_{B2a} &= 0 & x^*_{B2b} &= \frac{\delta_2}{\sigma_B} \\
y^*_{A1} &= \delta_1\theta_1 & y^*_{A2a} &= \delta_2\theta_2 & y^*_{A2b} &= 0 \\
y^*_{B1} &= 0 & y^*_{B2a} &= 0 & y^*_{B2b} &= \delta_2\theta_2
\end{aligned}
$$

Solution for case P10 — Potential full shift from A to B

$$
\begin{aligned}
w^*_{A1} &= \frac{\delta_1}{\sigma_A} & w^*_{A2a} &= (1-\mu_A)\frac{\delta_1}{\sigma_A} & w^*_{A2b} &= (1-\mu_A)\frac{\delta_1}{\sigma_A} \\
w^*_{B1} &= 0 & w^*_{B2a} &= \frac{\delta_2+(\mu_A-1)\delta_1}{\sigma_B} & w^*_{B2b} &= \frac{\delta_2}{\sigma_B} \\
x^*_{A1} &= \frac{\delta_1}{\sigma_A} & x^*_{A2a} &= 0 & x^*_{A2b} &= 0 \\
x^*_{B1} &= 0 & x^*_{B2a} &= \frac{\delta_2+(\mu_A-1)\delta_1}{\sigma_B} & x^*_{B2b} &= \frac{\delta_2}{\sigma_B} \\
y^*_{A1} &= \delta_1\theta_1 & y^*_{A2a} &= (1-\mu_A)\delta_1\theta_2 & y^*_{A2b} &= 0 \\
y^*_{B1} &= 0 & y^*_{B2a} &= \theta_2[\delta_2+(\mu_A-1)\delta_1] & y^*_{B2b} &= \delta_2\theta_2
\end{aligned}
$$

Solution for case P12 — Full shift from A to B

$$
\begin{array}{lll}
w^*_{A1} = \dfrac{\delta_1}{\sigma_A} & w^*_{A2a} = (1-\mu_A)\dfrac{\delta_1}{\sigma_A} & w^*_{A2b} = (1-\mu_A)\dfrac{\delta_1}{\sigma_A} \\
w^*_{B1} = 0 & w^*_{B2a} = \dfrac{\delta_2}{\sigma_B} & w^*_{B2b} = \dfrac{\delta_2}{\sigma_B} \\
x^*_{A1} = \dfrac{\delta_1}{\sigma_A} & x^*_{A2a} = 0 & x^*_{A2b} = 0 \\
x^*_{B1} = 0 & x^*_{B2a} = \dfrac{\delta_2}{\sigma_B} & x^*_{B2b} = \dfrac{\delta_2}{\sigma_A} \\
y^*_{A1} = \delta_1\theta_1 & y^*_{A2a} = 0 & y^*_{A2b} = 0 \\
y^*_{B1} = 0 & y^*_{B2a} = \delta_2\theta_2 & y^*_{B2b} = \delta_2\theta_2
\end{array}
$$

B.3.4 Decision Rules for Even-Numbered Cases

$$
\begin{aligned}
z^{P2} = {} & \psi_1\theta_1\delta_1\left(\frac{\alpha_A}{\sigma_A} + \beta_A\right) + \frac{\phi_A\psi_1}{1-\gamma_{A1}}\left(\kappa_{A1}^{1-\gamma_{A1}} - \kappa_{A0}^{1-\gamma_{A1}}\right) \\
& + \psi_2\theta_2\delta_2\left(\frac{\alpha_A}{\sigma_A} + (p_a\beta_{A2a} + p_b\beta_{A2b})\right) \\
& + \frac{\phi_A\psi_2\theta_2}{\pi_A}\left[\frac{p_a}{1-\gamma_{A2a}}\left(\kappa_{A2s}^{1-\gamma_{A2a}} - \kappa_{A1}^{1-\gamma_{A2a}}\right) + \frac{p_b}{1-\gamma_{A2b}}\left(\kappa_{A2s}^{1-\gamma_{A2b}} - \kappa_{A1}^{1-\gamma_{A2b}}\right)\right] \\
& + \frac{\phi_A\psi_2\theta_2}{\pi_A}\left[(\mu_A - 1)\frac{\delta_1}{\sigma_A} + \frac{\delta_2}{\sigma_A}\right]\left[\kappa_{A1}^{-\gamma_{A1}} - \left(p_a\kappa_{A1}^{-\gamma_{A2a}} + p_b\kappa_{A1}^{-\gamma_{A2b}}\right)\right]
\end{aligned}
\tag{B.35}
$$

with $\kappa_{A1} = \kappa_{A0} + \dfrac{\delta_1}{\sigma_A}$ and $\kappa_{A2s} = \kappa_{A0} + \mu_A\dfrac{\delta_1}{\sigma_A} + \dfrac{\delta_2}{\sigma_A}$

$$
\begin{aligned}
z^{P4} = {} & \psi_1\theta_1\delta_1\left(\frac{\alpha_A}{\sigma_A} + \beta_{A1}\right) + \frac{\phi_A\psi_1}{1-\gamma_{A1}}\left(\kappa_{A1}^{1-\gamma_{A1}} - \kappa_{A0}^{1-\gamma_{A1}}\right) \\
& + p_b\psi_2\theta_2\left[(\delta_2 - \delta_1(1-\mu_A))\left(\frac{\alpha_B}{\sigma_B} + \beta_{B2b}\right) + \delta_1(1-\mu_A)\left(\frac{\alpha_A}{\sigma_A} + \beta_{A2b}\right)\right] \\
& + p_a\psi_2\theta_2\delta_2\left(\frac{\alpha_A}{\sigma_A} + \beta_{A2a}\right) \\
& + \frac{p_b\phi_B\psi_2\theta_2}{\pi_B(1-\gamma_{B2b})}\left(\kappa_{B2b}^{1-\gamma_{B2b}} - \kappa_{B0}^{1-\gamma_{B2b}}\right) \\
& + \frac{p_a\phi_A\psi_2\theta_2}{\pi_A(1-\gamma_{A2a})}\left(\kappa_{A2a}^{1-\gamma_{A2a}} - \kappa_{A1}^{1-\gamma_{A2a}}\right) \\
& + \frac{p_b\phi_B\psi_2\theta_2[\delta_2 - \delta_1(1-\mu_A)]}{\pi_B\sigma_B}\left(\kappa_{B1}^{-\gamma_{B1}} - \kappa_{B1}^{-\gamma_{B2b}}\right) \\
& + \frac{p_a\phi_A\psi_2\theta_2}{\pi_A}\left[(\mu_A - 1)\frac{\delta_1}{\sigma_A} + \frac{\delta_2}{\sigma_A}\right]\left(\kappa_{A1}^{-\gamma_{A1}} - \kappa_{A1}^{-\gamma_{A2a}}\right)
\end{aligned}
\tag{B.36}
$$

$$\text{with}\quad \kappa_{A1} = \kappa_{A0} + \frac{\delta_1}{\sigma_A} \qquad \kappa_{B2b} = \kappa_{B0} + \frac{\delta_2 - \delta_1(1-\mu_A)}{\sigma_B}$$
$$\kappa_{A2a} = \kappa_{A0} + \frac{\mu_A\delta_1 + \delta_2}{\sigma_A}$$

$$\begin{aligned} z^{P6} = {} & \psi_1 \left[\theta_1\delta_1\left(\frac{\alpha_A}{\sigma_A} + \beta_{A1}\right) + \frac{\phi_A}{1-\gamma_{A1}}\left(\kappa_{A1}^{1-\gamma_{A1}} - \kappa_{A0}^{1-\gamma_{A1}}\right)\right] \\ & + \psi_2\theta_2[\delta_2 - \delta_1(1-\mu_A)]\left[\frac{\alpha_B}{\sigma_B} + (p_b\beta_{B2b} + p_a\beta_{B2a})\right] \\ & + \psi_2\theta_2\delta_1(1-\mu_A)\left[\frac{\alpha_A}{\sigma_A} + (p_b\beta_{A2b} + p_a\beta_{A2a})\right] \\ & + \frac{\phi_B\psi_2\theta_2}{\pi_B}\left[\frac{p_b}{1-\gamma_{B2b}}\left(\kappa_{B2s}^{1-\gamma_{B2b}} - \kappa_{B0}^{1-\gamma_{B2b}}\right) + \frac{p_a}{1-\gamma_{B2a}}\left(\kappa_{B2s}^{1-\gamma_{B2a}} - \kappa_{B0}^{1-\gamma_{B2a}}\right)\right] \\ & + \frac{\phi_B\psi_2\theta_2[\delta_2 - \delta_1(1-\mu_A)]}{\pi_B\sigma_B}\left[\kappa_{B0}^{-\gamma_{B1}} - \left(p_b\kappa_{B0}^{-\gamma_{B2b}} + p_a\kappa_{B0}^{-\gamma_{B2a}}\right)\right] \end{aligned} \tag{B.37}$$

$$\text{with}\quad \kappa_{A1} = \kappa_{A0} + \frac{\delta_1}{\sigma_A} \quad\text{and}\quad \kappa_{B2s} = \kappa_{B0} + \frac{\delta_2 - \delta_1(1-\mu_A)}{\sigma_B}$$

$$\begin{aligned} z^{P8} = {} & \psi_1 \left[\theta_1\delta_1\left(\frac{\alpha_A}{\sigma_A} + \beta_{A1}\right) + \frac{\phi_A}{1-\gamma_{A1}}\left(\kappa_{A1}^{1-\gamma_{A1}} - \kappa_{A0}^{1-\gamma_{A1}}\right)\right] \\ & + p_b\psi_2\theta_2\left[\delta_2\left(\frac{\alpha_B}{\sigma_B} + \beta_{B2b}\right) + \delta_1(1-\mu_A)\frac{\alpha_A}{\sigma_A}\right] + p_a\psi_2\theta_2\delta_2\left(\frac{\alpha_A}{\sigma_A} + \beta_{A2a}\right) \\ & + \frac{p_b\phi_B\psi_2\theta_2}{\pi_B}\left[\frac{1}{1-\gamma_{B2b}}\left(\kappa_{B2b}^{1-\gamma_{B2b}} - \kappa_{B1}^{1-\gamma_{B2b}}\right) + \frac{\delta_2}{\sigma_B}\left(\kappa_{B1}^{-\gamma_{B1}} - \kappa_{B1}^{-\gamma_{B2b}}\right)\right] \\ & + \frac{p_a\phi_A\psi_2\theta_2}{\pi_A(1-\gamma_{A2a})}\left(\kappa_{A2a}^{1-\gamma_{A2a}} - \kappa_{A1}^{1-\gamma_{A2a}}\right) \\ & + \frac{p_a\phi_A\psi_2\theta_2}{\pi_A}\left((\mu_A - 1)\frac{\delta_1}{\sigma_A} + \frac{\delta_2}{\sigma_A}\right)\left(\kappa_{A1}^{-\gamma_{A1}} - \kappa_{A1}^{-\gamma_{A2a}}\right) \end{aligned} \tag{B.38}$$

$$\text{with}\quad \kappa_{B1} = \kappa_{B0} \qquad \kappa_{A1} = \kappa_{A0} + \frac{\delta_1}{\sigma_A}$$
$$\kappa_{B2b} = \kappa_{B0} + \frac{\delta_2}{\sigma_B} \qquad \kappa_{A2a} = \kappa_{A0} + \mu_A\frac{\delta_1}{\sigma_A} + \frac{\delta_2}{\sigma_A}$$

$$
\begin{aligned}
z^{P10} = {} & \psi_1 \left[\theta_1 \delta_1 \left(\frac{\alpha_A}{\sigma_A} + \beta_{A1}\right) + \frac{\phi_A}{1-\gamma_{A1}} \left(\kappa_{A1}^{1-\gamma_{A1}} - \kappa_{A0}^{1-\gamma_{A1}}\right)\right] \\
& + \psi_2 \theta_2 \left[p_b \delta_2 \left(\frac{\alpha_B}{\sigma_B} + \beta_{B2b}\right) + (1-\mu_A)\delta_1 \left(\frac{\alpha_A}{\sigma_A} + p_a \beta_{A2a}\right)\right] \\
& + \psi_2 \theta_2 p_a [\delta_2 + (\mu_A - 1)\delta_1] \left(\frac{\alpha_B}{\sigma_B} + \beta_{B2a}\right) \\
& + \frac{\phi_B \psi_2 \theta_2}{\pi_B} \left[\frac{p_b}{1-\gamma_{B2b}} \left(\kappa_{B2b}^{1-\gamma_{B2b}} - \kappa_{B1}^{1-\gamma_{B2b}}\right) + \frac{p_a}{1-\gamma_{B2a}} \left(\kappa_{B2a}^{1-\gamma_{B2a}} - \kappa_{B1}^{1-\gamma_{B2a}}\right)\right] \\
& + \frac{\phi_B \psi_2 \theta_2}{\pi_B \sigma_B} \left[p_b \delta_2 \left(\kappa_{B1}^{-\gamma_{B1}} - \kappa_{B1}^{-\gamma_{B2b}}\right) + p_a [\delta_2 + (\mu_A - 1)\delta_1] \left(\kappa_{B1}^{-\gamma_{B1}} - \kappa_{B1}^{-\gamma_{B2a}}\right)\right]
\end{aligned}
\tag{B.39}
$$

$$
\text{with} \quad \kappa_{A1} = \kappa_{A0} + \frac{\delta_1}{\sigma_A} \qquad \kappa_{B1} = \kappa_{B0}
$$

$$
\kappa_{B2b} = \kappa_{B0} + \frac{\delta_2}{\sigma_B} \qquad \kappa_{B2a} = \kappa_{B0} + \frac{\delta_2 + (\mu_A - 1)\delta_1}{\sigma_B}
$$

$$
\begin{aligned}
z^{P12} = {} & \psi_1 \left[\theta_1 \delta_1 \left(\frac{\alpha_A}{\sigma_A} + \beta_{A1}\right) + \frac{\phi_A}{1-\gamma_{A1}} \left(\kappa_{A1}^{1-\gamma_{A1}} - \kappa_{A0}^{1-\gamma_{A1}}\right)\right] \\
& + \psi_2 \theta_2 \left[\delta_2 \left(\frac{\alpha_B}{\sigma_B} + p_b \beta_{B2b} + p_a \beta_{B2a}\right) + (1-\mu_A)\delta_1 \frac{\alpha_A}{\sigma_A}\right] \\
& + \frac{\phi_B \psi_2 \theta_2}{\pi_B} \left[\frac{p_b}{1-\gamma_{B2b}} \left(\kappa_{B2b}^{1-\gamma_{B2b}} - \kappa_{B1}^{1-\gamma_{B2b}}\right) + \frac{p_a}{1-\gamma_{B2a}} \left(\kappa_{B2a}^{1-\gamma_{B2a}} - \kappa_{B1}^{1-\gamma_{B2a}}\right)\right] \\
& + \frac{\phi_B \psi_2 \theta_2 \delta_2}{\pi_B \sigma_B} \left[\kappa_{B1}^{-\gamma_{B1}} - \left(p_b \kappa_{B1}^{-\gamma_{B2b}} + p_a \kappa_{B1}^{-\gamma_{B2a}}\right)\right]
\end{aligned}
\tag{B.40}
$$

$$
\text{with} \quad \kappa_{A1} = \kappa_{A0} + \frac{\delta_1}{\sigma_A} \qquad \kappa_{B1} = \kappa_{B0}
$$

$$
\kappa_{B2b} = \kappa_{B0} + \frac{\delta_2}{\sigma_B} \qquad \kappa_{B2a} = \kappa_{B0} + \frac{\delta_2}{\sigma_B}
$$

B.3.5 First-Order Derivative with Respect to Learning Index

The following transformations illustrate the derivation of the first-oder derivative of the expression $h = (x_{T1}+\kappa_{i0})^{-\gamma}+\kappa_{i0}^{-\gamma}-(x_{T1}-x_{B1}+\kappa_{i0})^{-\gamma}-(x_{B1}+\kappa_{i0})^{-\gamma}$, which is required to show the superiority of pure strategy P3 over M7 for a positive vertical shift on pp. 121ff. and leading to Equation 5.193. Starting point is

$$h(\gamma) = (x_{T1} + \kappa_{i0})^{-\gamma} + \kappa_{i0}^{-\gamma} - (x_{T1} - x_{B1} + \kappa_{i0})^{-\gamma} - (x_{B1} + \kappa_{i0})^{-\gamma} \quad \text{(B.41)}$$

To enable derivation the equation is transformed into

$$\begin{aligned} h(\gamma) &= \mathrm{e}^{[\ln(x_{T1}+\kappa_{i0})]^{-\gamma}} + \mathrm{e}^{(\ln \kappa_{i0})^{-\gamma}} - \mathrm{e}^{[\ln(x_{T1}-x_{B1}+\kappa_{i0})]^{-\gamma}} - \mathrm{e}^{[\ln(x_{B1}+\kappa_{i0})]^{-\gamma}} \\ &= \mathrm{e}^{-\gamma\ln(x_{T1}+\kappa_{i0})} + \mathrm{e}^{-\gamma\ln \kappa_{i0}} - \mathrm{e}^{-\gamma\ln(x_{T1}-x_{B1}+\kappa_{i0})} - \mathrm{e}^{[\ln(x_{B1}+\kappa_{i0})]^{-\gamma}} \end{aligned} \quad \text{(B.42)}$$

Then, the first-order derivative with respect to γ is

$$\begin{aligned} h'(\gamma) = &-\ln(x_{T1} + \kappa_{i0})\mathrm{e}^{[\ln(x_{T1}+\kappa_{i0})]^{-\gamma}} - (\ln \kappa_{i0})\mathrm{e}^{(\ln \kappa_{i0})^{-\gamma}} \\ &+\ln(x_{T1} - x_{B1} + \kappa_{i0})\mathrm{e}^{[\ln(x_{T1}-x_{B1}+\kappa_{i0})]^{-\gamma}} + \ln(x_{B1} + \kappa_{i0})\mathrm{e}^{[\ln(x_{B1}+\kappa_{i0})]^{-\gamma}} \end{aligned} \quad \text{(B.43)}$$

Simplification and rearranging then yields the expression in Eq. B.44.

$$h'(\gamma) = \frac{\ln(x_{T1} - x_{B1} + \kappa_{i0})}{(x_{T1} - x_{B1} + \kappa_{i0})^{\gamma}} + \frac{\ln(x_{B1} + \kappa_{i0})}{(x_{B1} + \kappa_{i0})^{\gamma}} - \frac{\ln(x_{T1} + \kappa_{i0})}{(x_{T1} + \kappa_{i0})^{\gamma}} - \frac{\ln(\kappa_{i0})}{(\kappa_{i0})^{\gamma}} \quad \text{(B.44)}$$

B.4 Pure versus Mixed Investment Strategies: Sample Solutions

B.4.1 No Change in Learning Rates

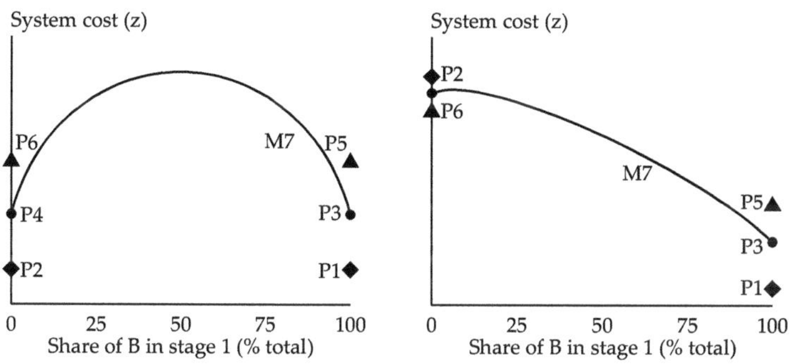

a) $\gamma_{A1} = \gamma_{B1} = \gamma_{A2a} = \gamma_{B2b}$ b) $\gamma_{A1} = \gamma_{A2a} < \gamma_{B1} = \gamma_{B2b}$

B.4.2 Learning Rate Increase

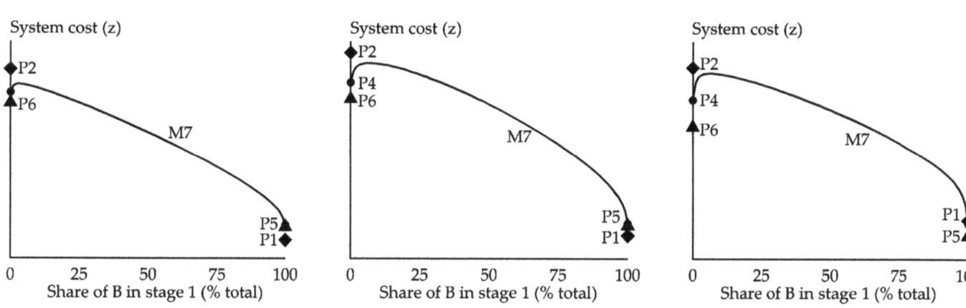

a) $\gamma_{A1} < \gamma_{A2a} < \gamma_{B1} < \gamma_{B2b}$ b) $\gamma_{A1} < \gamma_{B1} = \gamma_{A2a} < \gamma_{B2b}$ c) $\gamma_{A1} < \gamma_{B1} < \gamma_{A2a} < \gamma_{B2b}$

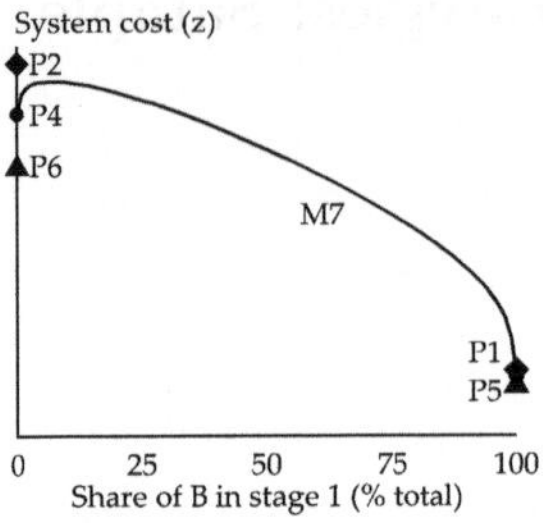

d) $\gamma_{A1} < \gamma_{B1} < \gamma_{A2a} = \gamma_{B2b}$

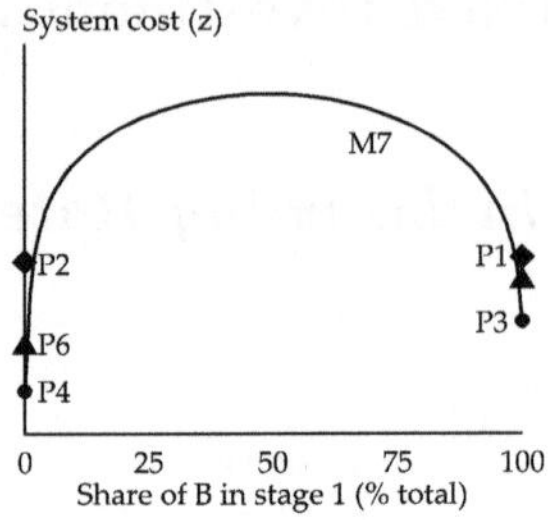

e) $\gamma_{A1} = \gamma_{B1} < \gamma_{A2a} < \gamma_{B2b}$

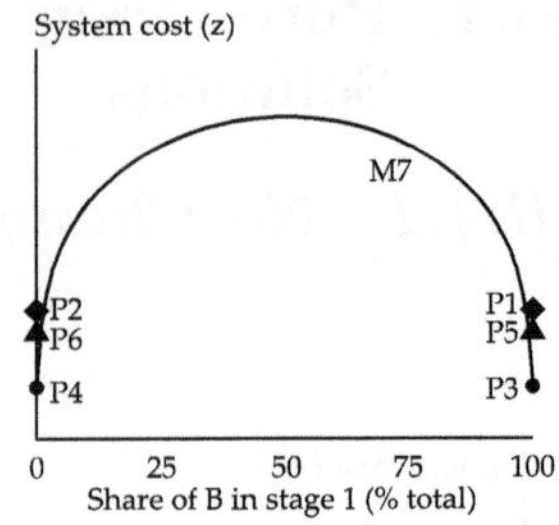

f) $\gamma_{A1} = \gamma_{B1} < \gamma_{A2a} = \gamma_{B2b}$

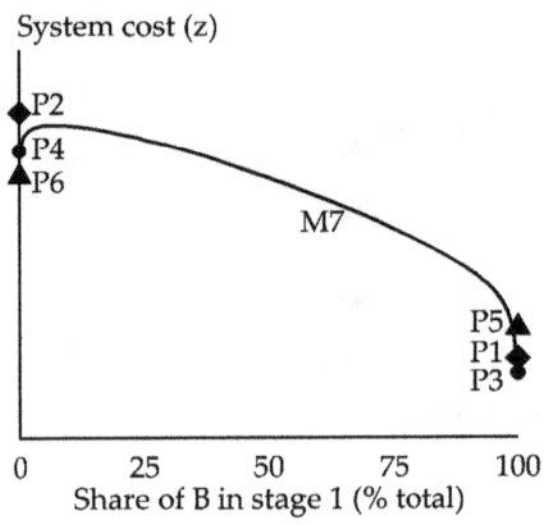

g) $\gamma_{A1} < \gamma_{B1} < \gamma_{B2b} < \gamma_{A2a}$

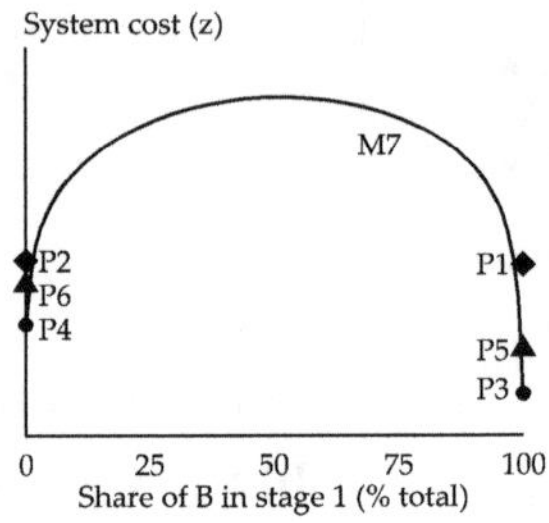

h) $\gamma_{A1} = \gamma_{B1} < \gamma_{B2b} < \gamma_{A2a}$

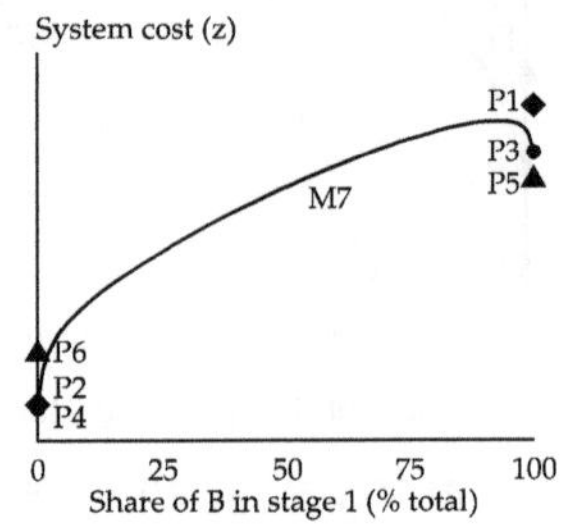

i) $\gamma_{B1} < \gamma_{A1} < \gamma_{A2a} = \gamma_{B2b}$

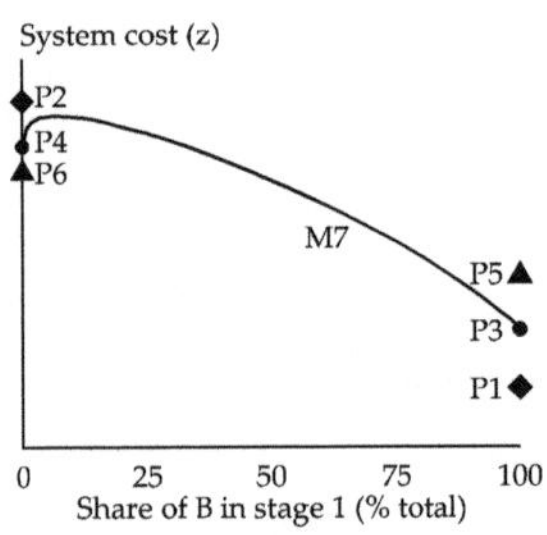

j) $\gamma_{A1} = \gamma_{A2a} < \gamma_{B1} < \gamma_{B2b}$

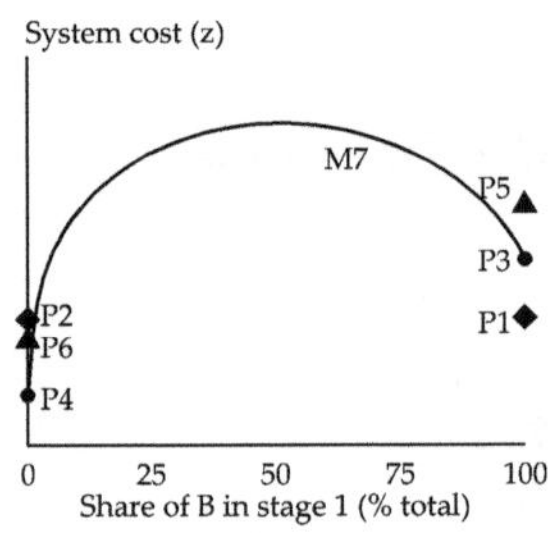

k) $\gamma_{A1} = \gamma_{B1} = \gamma_{A2a} < \gamma_{B2b}$

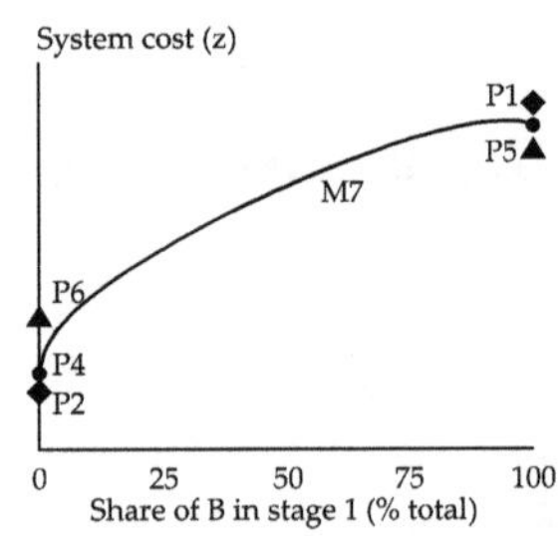

l) $\gamma_{B1} < \gamma_{A1} = \gamma_{A2a} = \gamma_{B2b}$

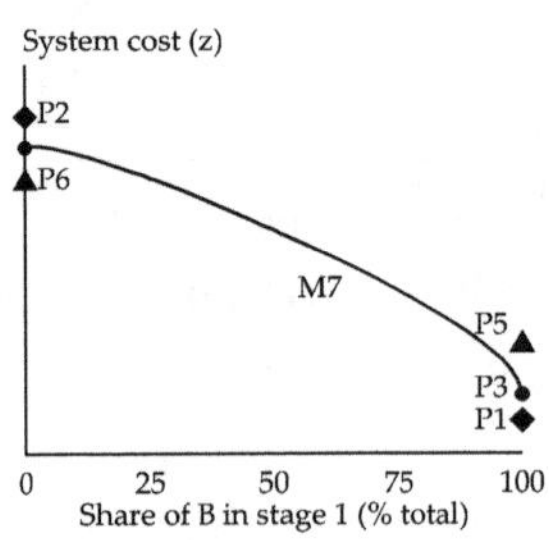

m) $\gamma_{A1} < \gamma_{A2a} < \gamma_{B1} = \gamma_{B2b}$

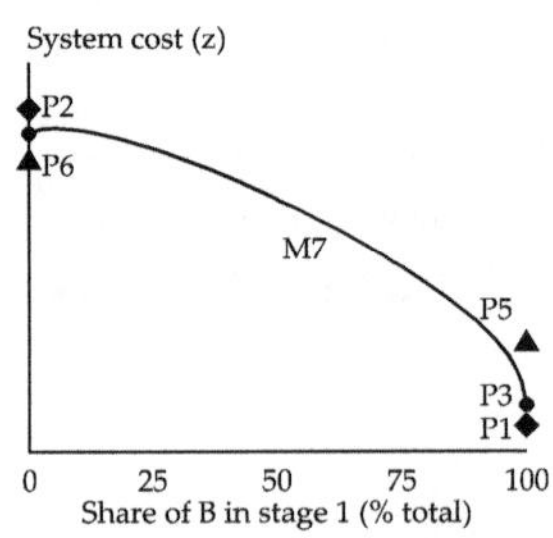

n) $\gamma_{A1} < \gamma_{B1} = \gamma_{A2a} = \gamma_{B2b}$

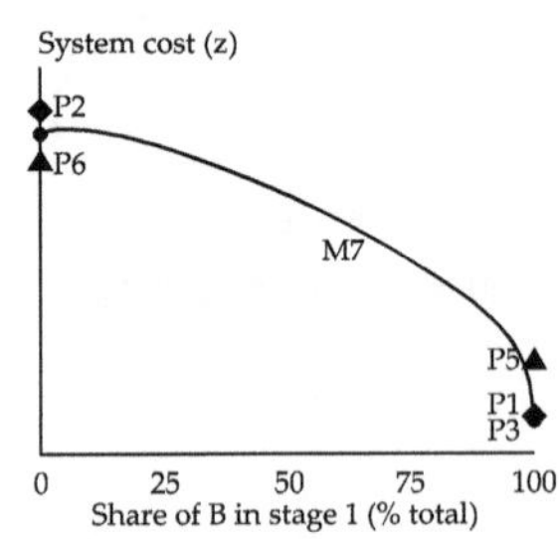

o) $\gamma_{A1} < \gamma_{B1} = \gamma_{B2b} < \gamma_{A2a}$

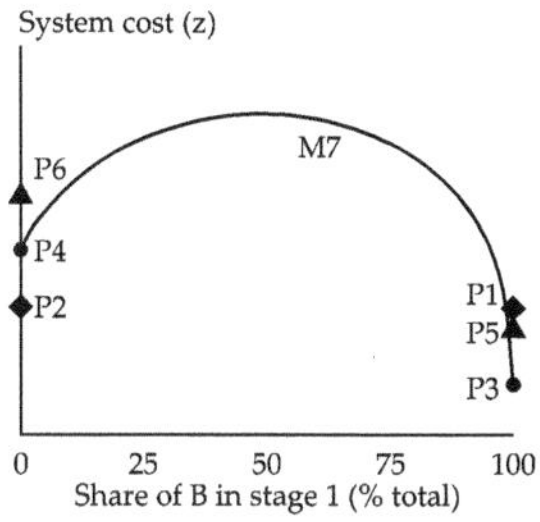

p) $\gamma_{A1} = \gamma_{B1} = \gamma_{B2b} < \gamma_{A2a}$

B.4.3 Learning Rate Decrease

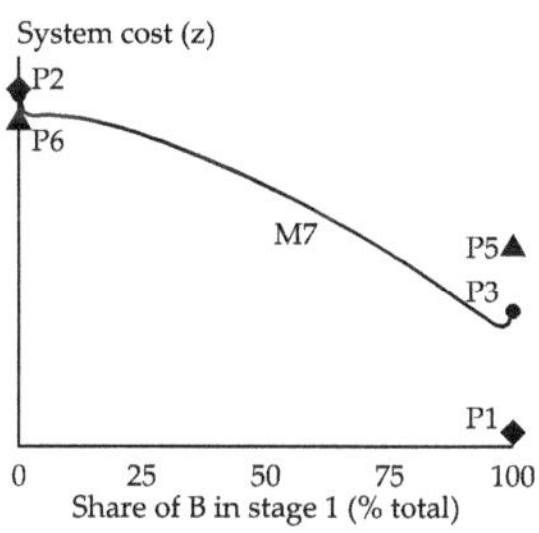

a) $\gamma_{A2a} < \gamma_{A1} < \gamma_{B2b} < \gamma_{B1}$

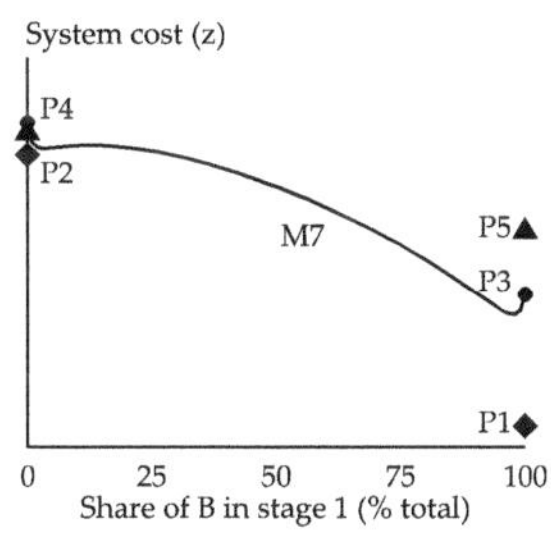

b) $\gamma_{A2a} < \gamma_{A1} = \gamma_{B2b} < \gamma_{B1}$

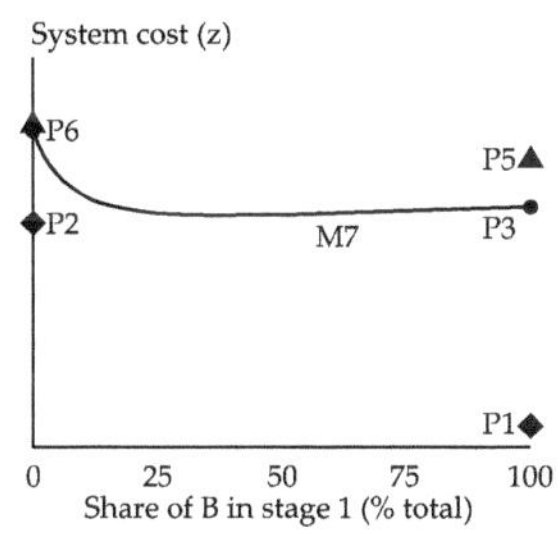

c) $\gamma_{A2a} < \gamma_{B2b} < \gamma_{A1} < \gamma_{B1}$

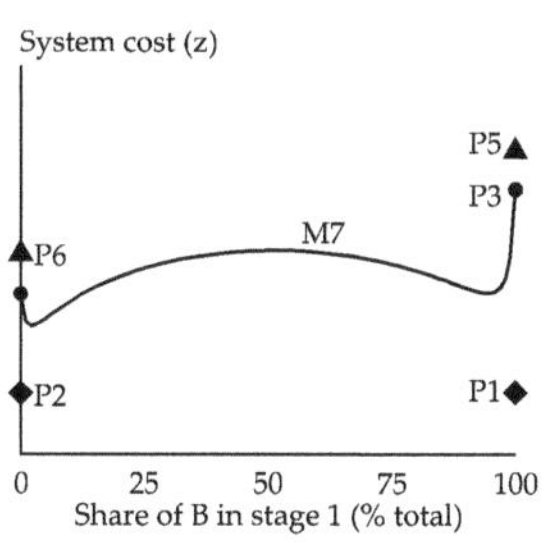

d) $\gamma_{A2a} < \gamma_{B2b} < \gamma_{A1} = \gamma_{B1}$

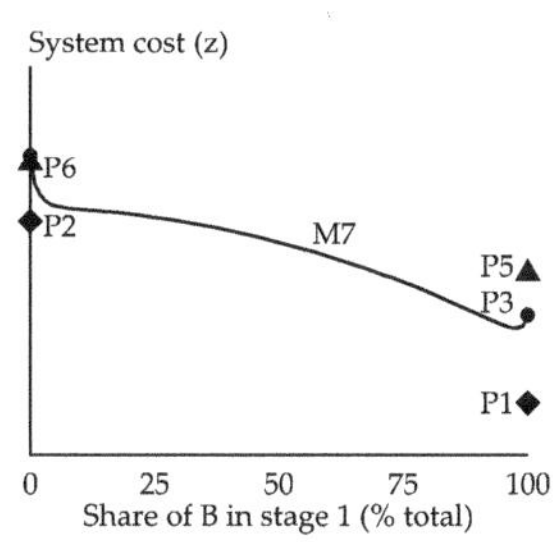

e) $\gamma_{A2a} = \gamma_{B2b} < \gamma_{A1} < \gamma_{B1}$

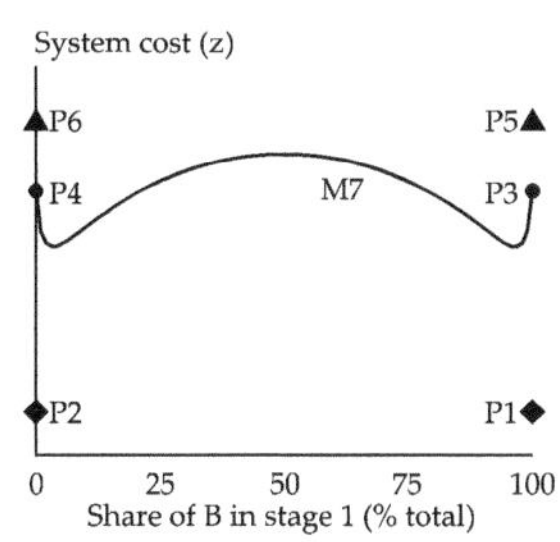

f) $\gamma_{A2a} = \gamma_{B2b} < \gamma_{A1} = \gamma_{B1}$

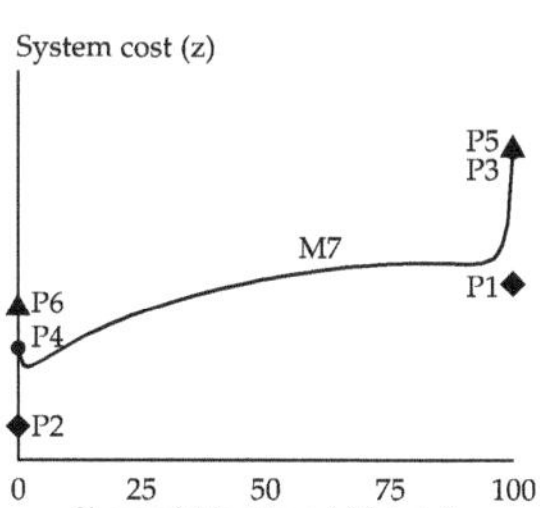

g) $\gamma_{A2a} < \gamma_{B2b} < \gamma_{B1} < \gamma_{A1}$

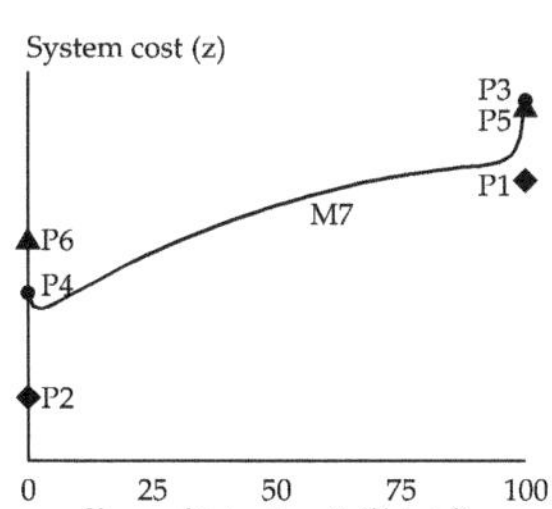

h) $\gamma_{A2a} = \gamma_{B2b} < \gamma_{B1} < \gamma_{A1}$

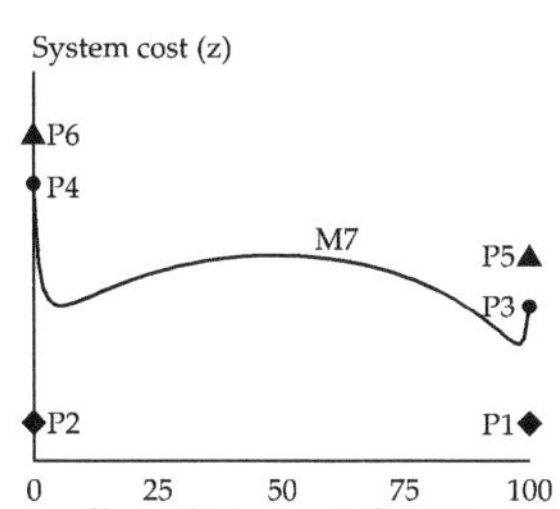

i) $\gamma_{B2b} < \gamma_{A2a} < \gamma_{A1} = \gamma_{B1}$

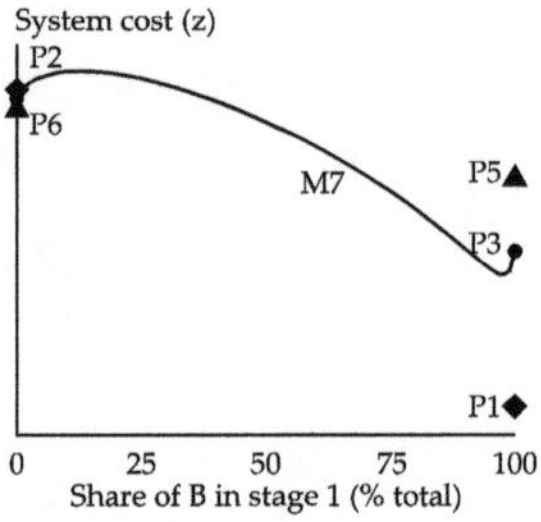

j) $\gamma_{A2a} < \gamma_{A1} < \gamma_{B1} = \gamma_{B2b}$

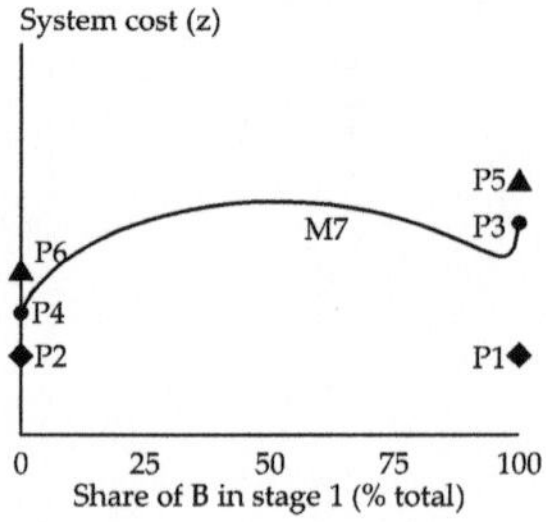

k) $\gamma_{A2a} < \gamma_{A1} = \gamma_{B1} = \gamma_{B2b}$

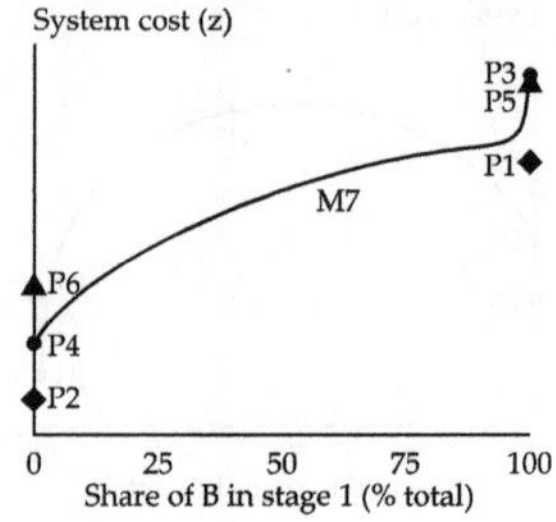

l) $\gamma_{A2a} < \gamma_{B1} = \gamma_{B2b} < \gamma_{A1}$

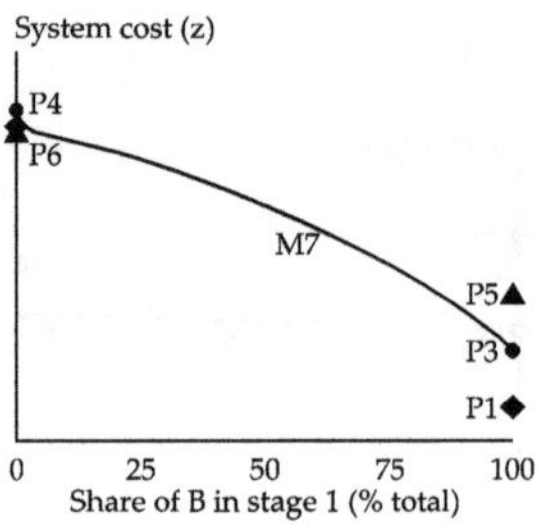

m) $\gamma_{B1} = \gamma_{A2a} = \gamma_{B2b} < \gamma_{A1}$

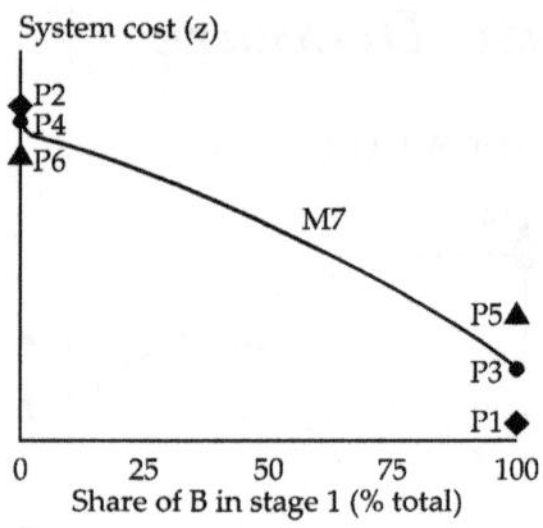

n) $\gamma_{A1} = \gamma_{A2a} < \gamma_{B2b} < \gamma_{B1}$

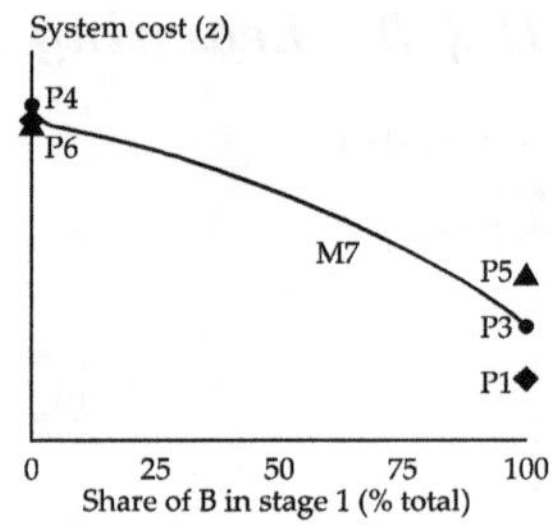

o) $\gamma_{A1} = \gamma_{A2a} = \gamma_{B2b} < \gamma_{B1}$

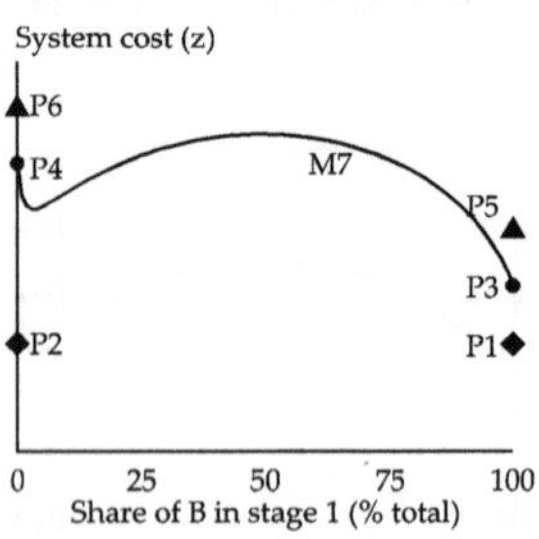

p) $\gamma_{B2b} < \gamma_{A1} = \gamma_{B1} = \gamma_{A2a}$

B.4.4 Learning Rate Increase and Decrease

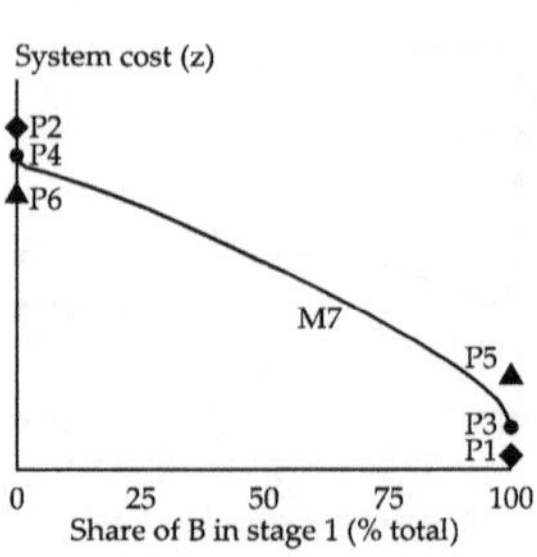

a) $\gamma_{A1} < \gamma_{A2a} < \gamma_{B2b} < \gamma_{B1}$

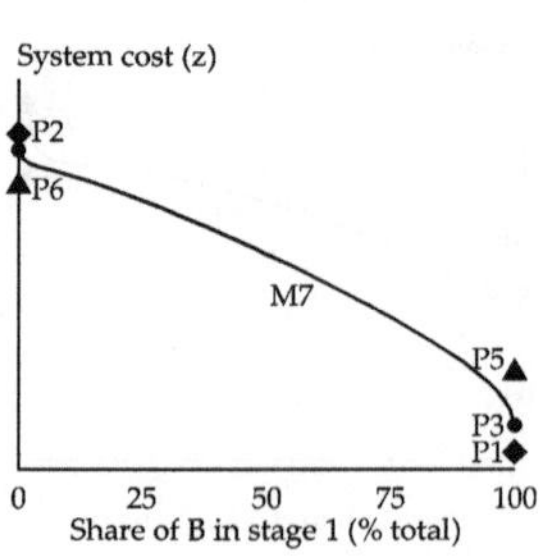

b) $\gamma_{A1} < \gamma_{A2a} = \gamma_{B2b} < \gamma_{B1}$

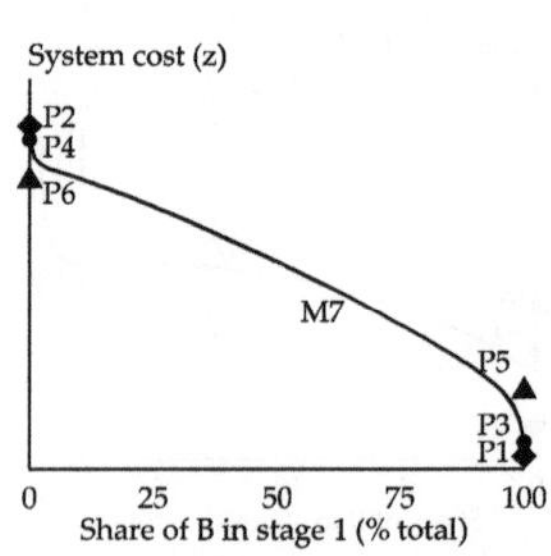

c) $\gamma_{A1} < \gamma_{B2b} < \gamma_{A2a} < \gamma_{B1}$

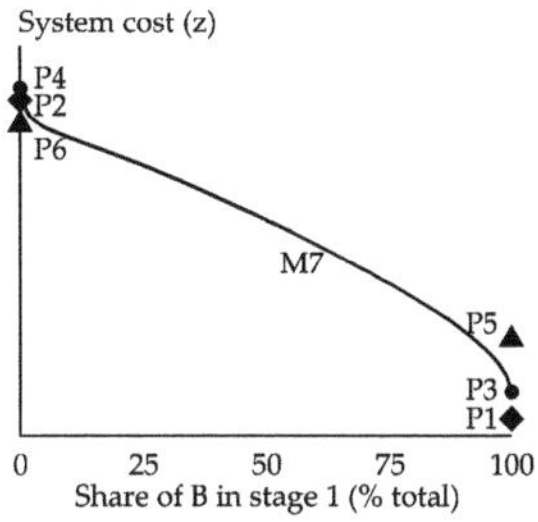

d) $\gamma_{A1} = \gamma_{B2b} < \gamma_{A2a} < \gamma_{B1}$

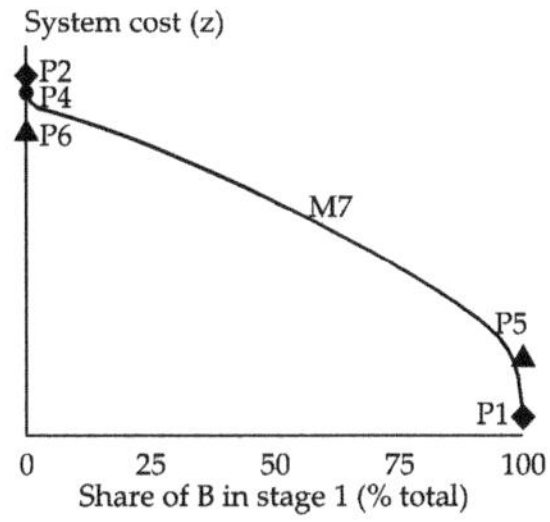

e) $\gamma_{A1} < \gamma_{B2b} < \gamma_{B1} < \gamma_{A2a}$

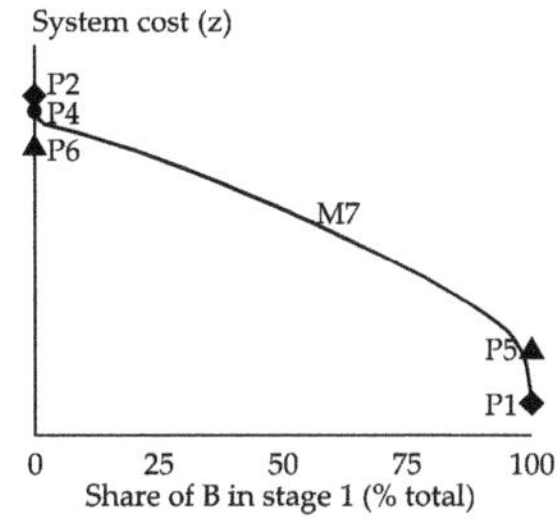

f) $\gamma_{A1} < \gamma_{B2b} < \gamma_{B1} < \gamma_{A2a}$

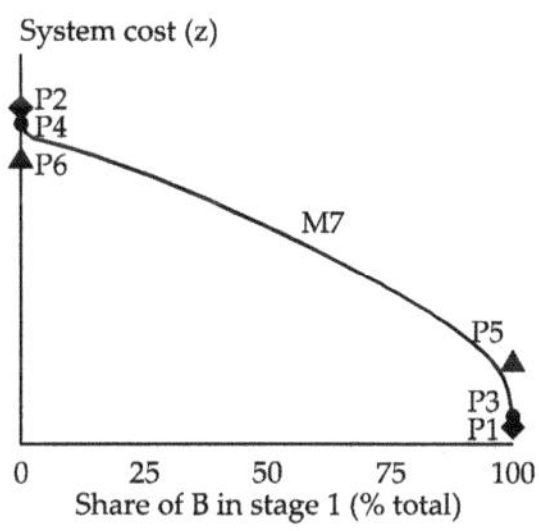

g) $\gamma_{A1} < \gamma_{B2b} < \gamma_{B1} = \gamma_{A2a}$

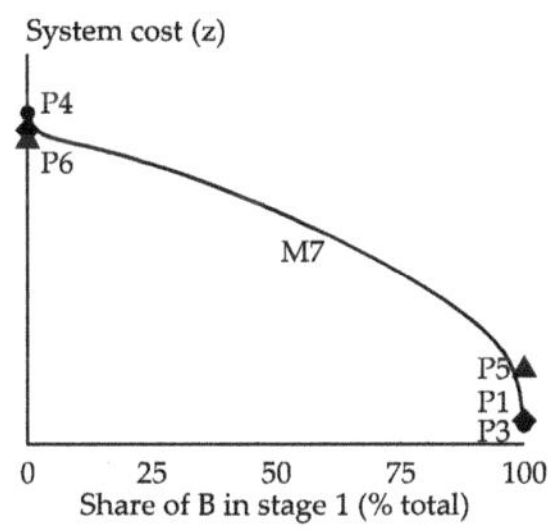

h) $\gamma_{A1} = \gamma_{B2b} < \gamma_{B1} < \gamma_{A2a}$

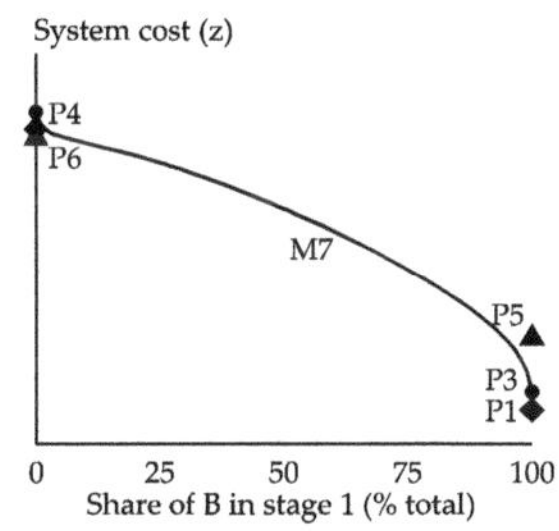

i) $\gamma_{A1} = \gamma_{B2b} < \gamma_{B1} = \gamma_{A2a}$

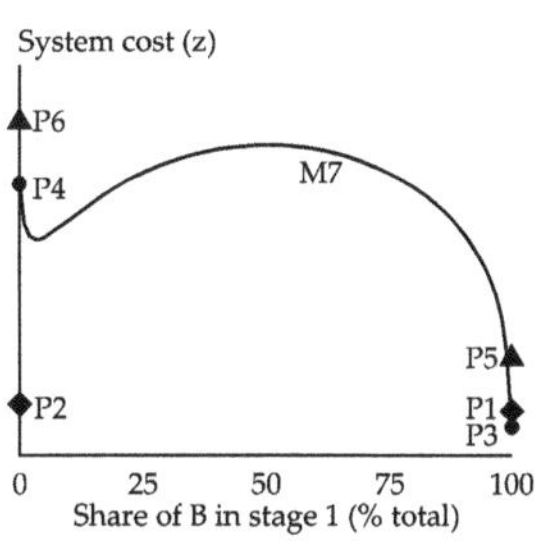

j) $\gamma_{B2b} < \gamma_{A1} = \gamma_{B1} < \gamma_{A2a}$

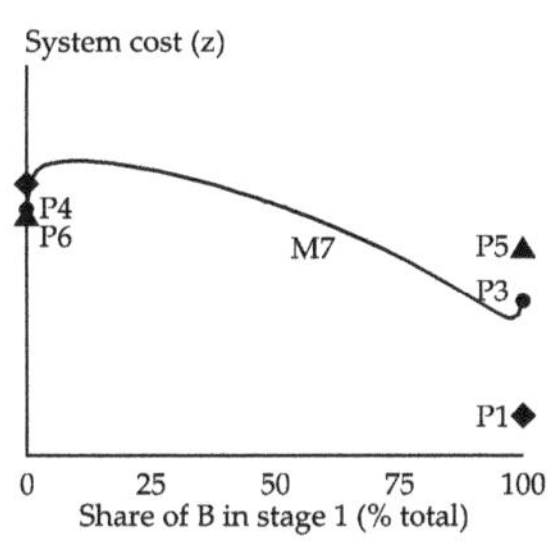

k) $\gamma_{A2a} < \gamma_{A1} < \gamma_{B1} < \gamma_{B2b}$

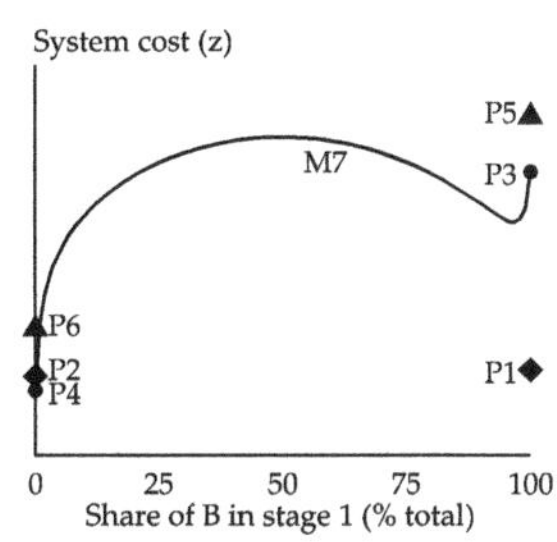

l) $\gamma_{A2a} < \gamma_{A1} = \gamma_{B1} < \gamma_{B2b}$

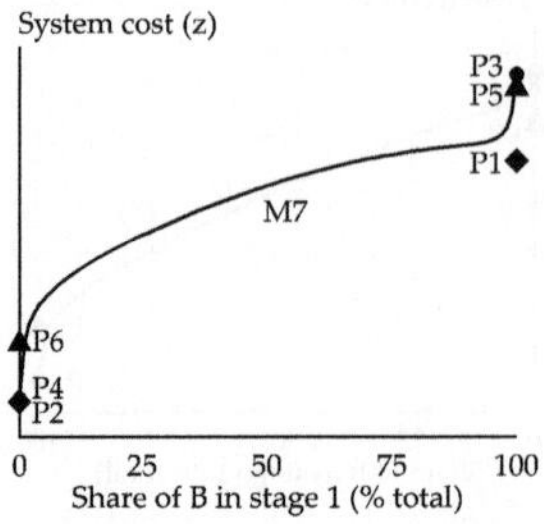

m) $\gamma_{A2a} < \gamma_{B1} < \gamma_{A1} < \gamma_{B2b}$

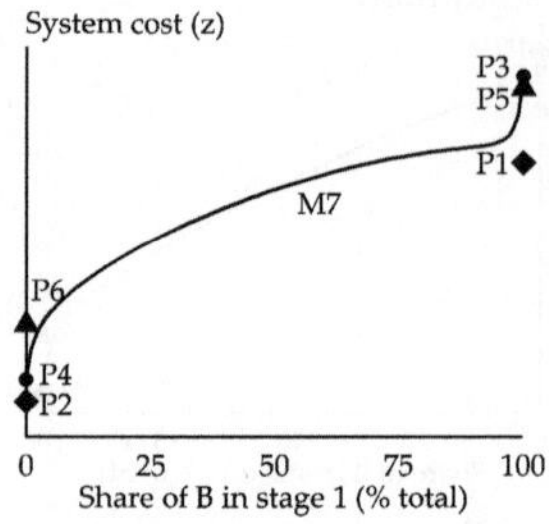

n) $\gamma_{A2a} < \gamma_{B1} < \gamma_{A1} = \gamma_{B2b}$

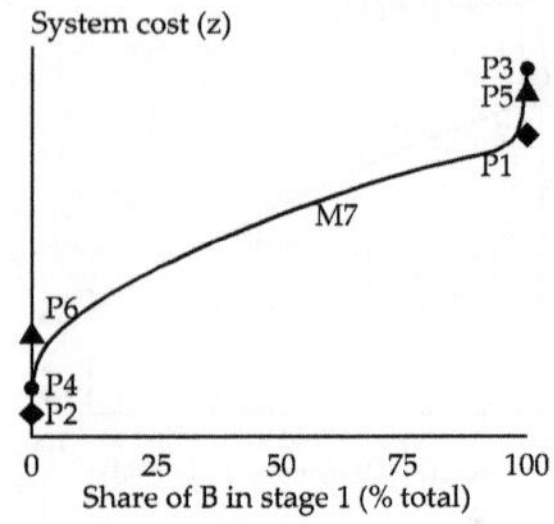

o) $\gamma_{A2a} < \gamma_{B1} < \gamma_{B2b} < \gamma_{A1}$

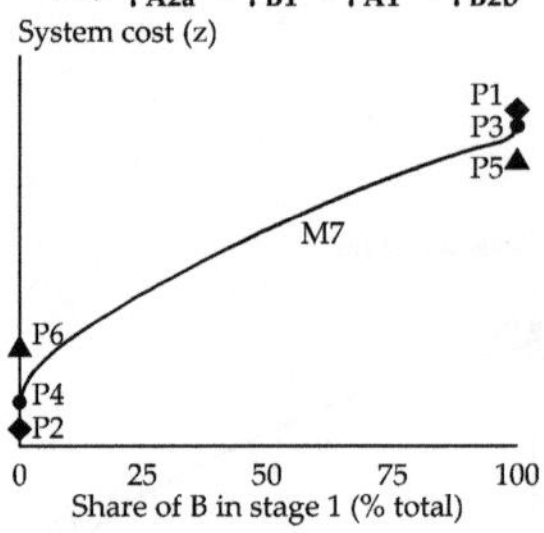

p) $\gamma_{B1} < \gamma_{A2a} = \gamma_{B2b} < \gamma_{A1}$

Appendix C
Numerical Model Results

This appendix contains the main results of the numerical CORE and SCORE models in tabular form for all model years and technologies.[1] For the deterministic model, base case results, specifically power production and operational capacity stock, are shown in section C.1. Globally aggregated solution values are followed by all eight regional ones.

Section C.2 then contains the globally aggregated results from the stochastic model for different types of parameter uncertainty. The results can be classified into three major groups:

- Deterministic reference case (indicated by subscript D)
- Stochastic setup (indicated by subscript S)
- Scenario setup (indicated by subscript C)

In the deterministic reference case expected values for the uncertain parameters have been used. In the scenario setup, by contrast, the existence of different states of the world is acknowledged, yet the outcome is known ex-ante. In sum, there are four different scenarios (aa, ab, ba, bb). Uncertain parameters considered are:

- PV learning rate
- Learning rates for all renewable technologies
- CSP learning rates
- Carbon bound
- Fossil fuel prices (gas and coal)
- Natural gas price
- Renewable potential
- "Multiple" uncertainties[2]
- PV learning rate without vertical shift

[1] New and replacement hydro plants are aggregated for the stochastic results — yet they are considered separately in the corresponding model runs.
[2] Learning rates, carbon bound, fossil fuel prices, renewable potential.

C.1 Deterministic Model Results

C.1.1 Base Case — Global Results

Table C.1 CORE base case — GLOBAL

	Power generation (TWh)							
	2008	*2010*	*2020*	*2030*	*2050*	*2075*	*2100*	*Mean**
Hard Coal	9,290	10,152	11,113	10,088	6,597	1,777	-	7,003
Lignite	941	887	681	369	-	146	-	432
Gas OC	624	91	88	41	45	30	23	135
Gas CC	869	1,874	5,204	6,336	3,462	3,124	1,561	3,204
Oil	-	-	-	-	-	-	-	-
Nuclear	2,967	2,848	2,255	2,077	9,803	12,513	7,636	5,728
Wind Ons	302	272	337	3,392	5,415	6,469	5,993	3,169
Wind Offs	6	6	3	127	551	805	503	286
Biomass	100	174	384	1,471	4,452	5,340	5,963	2,555
PV	24	22	12	3	-	5,429	15,272	2,966
CSP Base	1	1	1	-	-	-	-	1
CSP Stor	-	-	-	-	-	-	-	-
Hydro Repl	-	67	491	1,200	2,320	3,128	3,983	1,598
Hydro New	3,501	3,743	3,087	2,682	1,858	1,350	2,420	2,663
Total	**18,626**	**20,137**	**23,655**	**27,786**	**34,503**	**40,112**	**43,352**	**29,739**

	Capacity (GW)							
	2008	*2010*	*2020*	*2030*	*2050*	*2075*	*2100*	*Mean**
Hard Coal	1,345	1,407	1,511	1,435	1,088	374	-	1,023
Lignite	139	128	91	52	-	23	-	62
Gas OC	835	221	186	193	216	242	355	321
Gas CC	365	354	963	1,391	1,281	1,611	1,032	1,000
Oil	442	-	-	-	-	-	-	63
Nuclear	377	361	286	264	1,244	1,603	1,199	762
Wind Ons	119	107	126	1,249	2,023	2,414	2,234	1,182
Wind Offs	2	1	1	31	142	216	129	74
Biomass	47	25	49	187	633	840	1,395	454
PV	14	13	7	2	-	2,500	7,578	1,445
CSP Base	1	-	-	-	-	-	-	-
CSP Stor	-	-	-	-	-	-	-	-
Hydro Repl	-	22	132	340	656	949	1,035	448
Hydro New	941	917	802	690	530	415	719	716
Total	**4,626**	**3,557**	**4,154**	**5,832**	**7,812**	**11,186**	**15,678**	**7,549**

* Unweighted mean across reference periods.

C.1.2 Base Case — Regional Results

Table C.2 CORE base case — Model region AFRICA

	Power generation (TWh)							
	2008	*2010*	*2020*	*2030*	*2050*	*2075*	*2100*	*Mean**
Hard Coal	311	522	500	368	117	48	-	267
Lignite	-	-	-	-	-	-	-	-
Gas OC	104	30	-	-	-	-	-	19
Gas CC	34	20	33	32	92	99	1	44
Oil	-	-	-	-	-	-	-	-
Nuclear	16	15	12	9	3	-	-	8
Wind Ons	1	1	-	-	-	-	-	-
Wind Offs	-	-	-	18	34	187	-	34
Biomass	-	-	24	287	733	733	733	359
PV	-	-	-	-	-	500	1,003	215
CSP Base	-	-	-	-	-	-	-	-
CSP Stor	-	-	-	-	-	-	-	-
Hydro Repl	-	3	21	40	73	105	361	86
Hydro New	131	112	286	322	499	417	264	290
Total	**596**	**703**	**875**	**1,077**	**1,550**	**2,089**	**2,362**	**1,322**

	Capacity (GW)							
	2008	*2010*	*2020*	*2030*	*2050*	*2075*	*2100*	*Mean**
Hard Coal	42	71	68	56	25	16	-	40
Lignite	-	-	-	-	-	-	-	-
Gas OC	36	13	-	-	-	-	-	7
Gas CC	4	4	11	10	33	62	11	19
Oil	23	-	-	-	-	-	-	3
Nuclear	2	2	2	1	-	-	-	1
Wind Ons	1	1	-	-	-	-	-	-
Wind Offs	-	-	-	5	9	55	-	10
Biomass	-	-	3	36	94	98	154	55
PV	-	-	-	-	-	230	462	99
CSP Base	-	-	-	-	-	-	-	-
CSP Stor	-	-	-	-	-	-	-	-
Hydro Repl	-	1	5	9	17	31	73	19
Hydro New	24	31	68	64	121	107	65	69
Total	**133**	**121**	**156**	**180**	**299**	**600**	**765**	**322**

* Unweighted mean across reference periods.

Table C.3 CORE base case — Model region Asia

	Power generation (TWh)							
	2008	*2010*	*2020*	*2030*	*2050*	*2075*	*2100*	*Mean**
Hard Coal	4,819	5,666	7,242	7,997	5,562	857	-	4,592
Lignite	174	164	114	65	-	146	-	95
Gas OC	-	17	9	5	35	14	3	12
Gas CC	-	125	1,444	2,037	1,250	1,548	872	1,040
Oil	-	-	-	-	-	-	-	-
Nuclear	604	579	459	753	6,894	8,887	4,503	3,240
Wind Ons	43	39	17	-	-	-	-	14
Wind Offs	-	-	-	108	396	401	396	186
Biomass	-	32	30	592	1,082	1,320	1,438	642
PV	5	5	3	1	-	4,486	10,668	2,167
CSP Base	-	-	-	-	-	-	-	-
CSP Stor	-	-	-	-	-	-	-	-
Hydro Repl	-	24	188	367	668	973	1,055	468
Hydro New	920	1,145	912	713	364	309	1,063	775
Total	**6,565**	**7,796**	**10,418**	**12,639**	**16,252**	**18,941**	**19,999**	**13,230**

	Capacity (GW)							
	2008	*2010*	*2020*	*2030*	*2050*	*2075*	*2100*	*Mean**
Hard Coal	720	795	985	1,087	845	149	-	654
Lignite	23	22	15	9	-	23	-	13
Gas OC	176	114	93	92	160	123	124	126
Gas CC	44	41	337	487	424	773	548	379
Oil	128	-	-	-	-	-	-	18
Nuclear	77	74	58	96	875	1,128	682	427
Wind Ons	24	22	10	-	-	-	-	8
Wind Offs	-	-	-	26	103	105	103	48
Biomass	6	6	4	75	137	200	417	121
PV	2	2	1	-	-	2,066	5,356	1,061
CSP Base	-	-	-	-	-	-	-	-
CSP Stor	-	-	-	-	-	-	-	-
Hydro Repl	-	10	59	108	207	296	313	142
Hydro New	296	286	236	204	106	118	368	230
Total	**1,496**	**1,371**	**1,798**	**2,185**	**2,857**	**4,980**	**7,911**	**3,228**

* Unweighted mean across reference periods.

Table C.4 CORE base case — Model region EASTERN EUROPE/ EURASIA

	Power generation (TWh)							
	2008	*2010*	*2020*	*2030*	*2050*	*2075*	*2100*	*Mean**
Hard Coal	542	511	406	230	-	-	-	241
Lignite	204	192	134	76	-	-	-	87
Gas OC	-	-	5	3	-	-	-	1
Gas CC	-	56	625	958	253	132	19	292
Oil	-	-	-	-	-	-	-	-
Nuclear	327	314	249	183	52	-	-	161
Wind Ons	-	-	-	108	458	668	642	268
Wind Offs	-	-	-	-	-	-	-	-
Biomass	9	5	5	2	1,097	1,265	1,375	537
PV	-	-	-	-	-	-	-	-
CSP Base	-	-	-	-	-	-	-	-
CSP Stor	-	-	-	-	-	-	-	-
Hydro Repl	-	-	2	111	214	269	356	136
Hydro New	292	379	258	223	95	-	-	178
Total	**1,374**	**1,458**	**1,684**	**1,894**	**2,169**	**2,333**	**2,392**	**1,901**

	Capacity (GW)							
	2008	*2010*	*2020*	*2030*	*2050*	*2075*	*2100*	*Mean**
Hard Coal	84	70	55	31	-	-	-	34
Lignite	27	26	18	10	-	-	-	12
Gas OC	112	2	2	2	-	-	-	17
Gas CC	28	17	96	145	99	85	45	74
Oil	31	-	-	-	-	-	-	4
Nuclear	42	40	32	23	7	-	-	20
Wind Ons	-	-	-	39	166	243	234	98
Wind Offs	-	-	-	-	-	-	-	-
Biomass	1	1	1	-	146	183	237	81
PV	-	-	-	-	-	-	-	-
CSP Base	-	-	-	-	-	-	-	-
CSP Stor	-	-	-	-	-	-	-	-
Hydro Repl	-	-	1	33	64	91	91	40
Hydro New	91	88	73	58	27	-	-	48
Total	**416**	**243**	**277**	**343**	**509**	**602**	**606**	**428**

* Unweighted mean across reference periods.

Table C.5 CORE base case — Model region LATIN AMERICA

	Power generation (TWh)							
	2008	*2010*	*2020*	*2030*	*2050*	*2075*	*2100*	*Mean**
Hard Coal	44	42	29	14	-	-	-	18
Lignite	6	6	4	2	-	-	-	3
Gas OC	60	-	-	-	-	-	-	9
Gas CC	66	253	421	186	93	93	-	159
Oil	-	-	-	-	-	-	-	-
Nuclear	24	23	19	14	4	-	-	12
Wind Ons	2	2	1	437	609	847	708	372
Wind Offs	-	-	-	-	-	-	-	-
Biomass	40	38	24	57	513	513	546	247
PV	-	-	-	-	-	-	-	-
CSP Base	-	-	-	-	-	-	-	-
CSP Stor	-	-	-	-	-	-	-	-
Hydro Repl	-	18	133	220	472	577	751	310
Hydro New	743	658	531	459	145	294	740	510
Total	**986**	**1,040**	**1,161**	**1,389**	**1,836**	**2,324**	**2,745**	**1,640**

	Capacity (GW)							
	2008	*2010*	*2020*	*2030*	*2050*	*2075*	*2100*	*Mean**
Hard Coal	6	6	4	2	-	-	-	3
Lignite	1	1	1	-	-	-	-	-
Gas OC	32	-	-	-	-	-	-	5
Gas CC	8	30	57	55	58	58	-	38
Oil	32	-	-	-	-	-	-	5
Nuclear	3	3	2	2	-	-	-	2
Wind Ons	1	1	-	173	242	336	281	148
Wind Offs	-	-	-	-	-	-	-	-
Biomass	5	5	3	7	79	74	102	39
PV	-	-	-	-	-	-	-	-
CSP Base	-	-	-	-	-	-	-	-
CSP Stor	-	-	-	-	-	-	-	-
Hydro Repl	-	5	28	51	98	140	140	66
Hydro New	140	135	112	89	42	74	172	109
Total	**228**	**185**	**207**	**380**	**519**	**682**	**695**	**414**

* Unweighted mean across reference periods.

Table C.6 CORE base case — Model region MIDDLE EAST

	Power generation (TWh)							
	2008	*2010*	*2020*	*2030*	*2050*	*2075*	*2100*	*Mean**
Hard Coal	38	35	25	14	-	-	-	16
Lignite	1	1	-	-	-	-	-	-
Gas OC	460	38	38	8	8	11	11	82
Gas CC	191	654	831	1,249	374	403	219	560
Oil	-	-	-	-	-	-	-	-
Nuclear	-	-	-	-	1,758	2,405	1,675	834
Wind Ons	-	-	-	-	-	-	-	-
Wind Offs	-	-	-	-	15	15	-	4
Biomass	-	-	-	-	-	37	73	16
PV	-	-	-	-	-	443	2,202	378
CSP Base	-	-	-	-	-	-	-	-
CSP Stor	-	-	-	-	-	-	-	-
Hydro Repl	-	-	-	-	15	21	22	8
Hydro New	20	34	18	14	6	-	48	20
Total	**709**	**762**	**913**	**1,285**	**2,177**	**3,335**	**4,250**	**1,919**

	Capacity (GW)							
	2008	*2010*	*2020*	*2030*	*2050*	*2075*	*2100*	*Mean**
Hard Coal	5	5	3	2	-	-	-	2
Lignite	-	-	-	-	-	-	-	-
Gas OC	91	26	26	24	39	71	96	53
Gas CC	23	82	109	174	90	133	123	105
Oil	65	-	-	-	-	-	-	9
Nuclear	-	-	-	-	223	305	271	114
Wind Ons	-	-	-	-	-	-	-	-
Wind Offs	-	-	-	-	4	4	-	1
Biomass	-	-	-	-	-	6	26	5
PV	-	-	-	-	-	204	1,014	174
CSP Base	-	-	-	-	-	-	-	-
CSP Stor	-	-	-	-	-	-	-	-
Hydro Repl	-	-	-	-	7	10	10	4
Hydro New	10	10	8	7	3	-	26	9
Total	**194**	**123**	**147**	**207**	**367**	**733**	**1,566**	**477**

* Unweighted mean across reference periods.

Table C.7 CORE base case — Model region NORTH AMERICA

	Power generation (TWh)							
	2008	*2010*	*2020*	*2030*	*2050*	*2075*	*2100*	*Mean**
Hard Coal	2,509	2,366	2,176	1,040	874	529	-	1,356
Lignite	-	-	63	24	-	-	-	12
Gas OC	-	5	31	3	1	4	10	8
Gas CC	532	563	589	463	400	429	252	461
Oil	-	-	-	-	-	-	-	-
Nuclear	911	874	692	510	146	474	478	584
Wind Ons	75	68	246	1,988	2,944	3,545	3,351	1,745
Wind Offs	-	-	-	-	-	-	-	-
Biomass	49	93	224	402	733	972	1,118	513
PV	1	1	1	-	-	-	1,398	200
CSP Base	1	1	1	-	-	-	-	-
CSP Stor	-	-	-	-	-	-	-	-
Hydro Repl	-	22	140	261	495	650	825	342
Hydro New	774	785	574	564	573	331	305	558
Total	**4,853**	**4,778**	**4,737**	**5,255**	**6,167**	**6,934**	**7,736**	**5,780**

	Capacity (GW)							
	2008	*2010*	*2020*	*2030*	*2050*	*2075*	*2100*	*Mean**
Hard Coal	341	322	296	198	209	140	-	215
Lignite	13	8	8	5	-	-	-	5
Gas OC	264	46	46	56	11	39	130	84
Gas CC	176	129	146	234	267	290	170	202
Oil	94	-	-	-	-	-	-	13
Nuclear	116	111	88	65	18	72	90	80
Wind Ons	27	25	90	723	1,071	1,290	1,219	635
Wind Offs	-	-	-	-	-	-	-	-
Biomass	14	12	28	51	135	201	341	112
PV	1	1	-	-	-	-	746	107
CSP Base	-	-	-	-	-	-	-	-
CSP Stor	-	-	-	-	-	-	-	-
Hydro Repl	-	6	37	67	128	183	211	90
Hydro New	183	177	147	144	172	117	88	147
Total	**1,229**	**836**	**886**	**1,544**	**2,012**	**2,331**	**2,996**	**1,691**

* Unweighted mean across reference periods.

Table C.8 CORE base case — Model region OCEANIA

	Power generation (TWh)							
	2008	*2010*	*2020*	*2030*	*2050*	*2075*	*2100*	*Mean**
Hard Coal	97	103	95	63	11	-	-	53
Lignite	77	72	50	24	-	-	-	32
Gas OC	-	1	-	-	1	-	-	-
Gas CC	16	29	52	46	23	24	1	27
Oil	-	-	-	-	-	-	-	-
Nuclear	69	66	52	38	11	-	-	34
Wind Ons	4	4	2	59	103	109	22	43
Wind Offs	-	-	-	-	-	-	-	-
Biomass	1	1	59	100	239	335	513	178
PV	1	1	-	-	-	-	-	-
CSP Base	-	-	-	-	-	-	-	-
CSP Stor	-	-	-	-	-	-	-	-
Hydro Repl	-	1	5	11	20	24	36	14
Hydro New	28	36	28	19	9	-	-	17
Total	**292**	**313**	**343**	**360**	**417**	**492**	**572**	**399**

	Capacity (GW)							
	2008	*2010*	*2020*	*2030*	*2050*	*2075*	*2100*	*Mean**
Hard Coal	14	14	13	9	3	-	-	8
Lignite	10	10	7	4	-	-	-	4
Gas OC	11	6	4	4	6	8	5	6
Gas CC	7	7	12	16	14	15	7	11
Oil	3	-	-	-	-	-	-	-
Nuclear	9	8	7	5	1	-	-	4
Wind Ons	2	1	1	21	38	40	8	16
Wind Offs	-	-	-	-	-	-	-	-
Biomass	-	-	7	13	34	48	76	25
PV	-	-	-	-	-	-	-	-
CSP Base	-	-	-	-	-	-	-	-
CSP Stor	-	-	-	-	-	-	-	-
Hydro Repl	-	-	1	3	6	9	9	4
Hydro New	9	9	7	6	3	-	-	5
Total	**65**	**56**	**60**	**81**	**105**	**120**	**105**	**85**

* Unweighted mean across reference periods.

Table C.9 CORE base case — Model region WESTERN EUROPE

	Power generation (TWh)							
	2008	*2010*	*2020*	*2030*	*2050*	*2075*	*2100*	*Mean**
Hard Coal	930	906	640	362	33	343	-	459
Lignite	479	452	315	178	-	-	-	203
Gas OC	-	-	5	23	-	-	-	4
Gas CC	30	174	1,209	1,365	977	397	196	621
Oil	-	-	-	-	-	-	-	-
Nuclear	1,016	976	773	569	935	747	980	857
Wind Ons	176	159	71	800	1,300	1,300	1,270	725
Wind Offs	6	6	3	1	107	202	107	62
Biomass	3	6	18	30	55	165	165	63
PV	16	15	8	2	-	-	-	6
CSP Base	-	-	-	-	-	-	-	-
CSP Stor	-	-	-	-	-	-	-	-
Hydro Repl	-	-	3	189	363	509	577	234
Hydro New	593	593	480	368	166	-	-	314
Total	**3,251**	**3,286**	**3,524**	**3,887**	**3,936**	**3,664**	**3,296**	**3,549**

	Capacity (GW)							
	2008	*2010*	*2020*	*2030*	*2050*	*2075*	*2100*	*Mean**
Hard Coal	133	125	87	49	5	69	-	67
Lignite	64	61	42	24	-	-	-	27
Gas OC	113	15	15	15	-	-	-	23
Gas CC	75	45	194	270	295	194	128	172
Oil	67	-	-	-	-	-	-	10
Nuclear	129	124	98	72	119	98	157	114
Wind Ons	64	58	26	291	506	506	493	278
Wind Offs	2	1	1	-	26	52	26	15
Biomass	21	2	2	4	8	31	42	16
PV	10	9	5	1	-	-	-	4
CSP Base	-	-	-	-	-	-	-	-
CSP Stor	-	-	-	-	-	-	-	-
Hydro Repl	-	-	1	67	128	188	188	82
Hydro New	188	181	150	119	56	-	-	99
Total	**865**	**622**	**622**	**913**	**1,144**	**1,137**	**1,034**	**905**

* Unweighted mean across reference periods.

C.2 Stochastic Model Results

C.2.1 *Uncertain Learning Rates for Renewable Technologies*

The table in this section displays the stochastic results for uncertainty in the learning rates of all renewable technologies modeled with endogenous technological learning. Additional information on this topic can be found in chapter 7, in particular within section 7.3.1.

Table C.10 Results for uncertain renewable learning rates (Investment in GW)

	2015			2030						2050					
	0_D	0_S	M_C	0_D	a_S	b_S	M_S	M_{C_a}	M_{C_b}	0_D	a_S	b_S	M_S	M_{C_a}	M_{C_b}
H Coal	485	493	508	452	521	354	437	490	317	604	623	670	646	659	747
Lignite	-	-	-	-	-	-	-	-	-	-	23	-	12	23	-
Gas OC	-	-	-	91	84	76	80	2	36	114	-	139	69	2	118
Gas CC	183	176	182	1,050	1,052	1,043	1,048	1,127	1,107	181	372	207	289	332	153
Oil	-	-	-	-	-	-	-	-	-	-	-	-	-	-	-
Nuclear	-	-	-	119	-	316	158	-	349	1,172	223	1,030	626	280	1,078
Win Ons	26	95	4	906	1,719	149	934	1,526	235	1,565	2,239	1,346	1,793	2,233	1,181
Win Offs	3	-	-	48	159	-	79	240	-	120	538	90	314	402	57
Biomass	7	7	2	226	56	210	133	137	209	417	572	444	508	492	429
PV	-	-	-	-	-	-	-	-	-	-	2,009	-	1,005	2,009	-
CSP Bas	-	-	-	-	67	-	34	42	-	-	447	-	223	440	-
CSP Sto	-	-	-	-	121	-	60	96	-	1	542	-	271	597	-
Hydro	93	83	76	290	296	330	313	293	297	614	336	624	480	328	616

	2075							2100						
	0_D	aa_S	ab_S	ba_S	bb_S	M_S	M_C	0_D	aa_S	ab_S	ba_S	bb_S	M_S	M_C
H Coal	-	-	-	-	-	-	-	-	-	-	-	-	-	-
Lignite	-	-	-	-	-	-	-	-	-	-	-	-	-	-
Gas OC	-	10	55	83	28	44	50	241	-	166	262	45	118	35
Gas CC	1,030	989	909	968	962	957	997	33	94	159	146	180	145	186
Oil	-	-	-	-	-	-	-	-	-	-	-	-	-	-
Nuclear	255	1,057	1,226	133	137	638	575	903	886	632	975	976	867	882
Win Ons	1,745	2,325	1,943	2,013	2,101	2,096	2,090	1,592	2,024	2,064	2,002	2,284	2,094	2,881
Win Offs	160	243	201	643	894	495	467	871	228	74	227	219	187	335
Biomass	492	268	245	441	472	357	418	1,013	1,051	885	982	923	960	798
PV	2,316	3,171	3,196	2,316	1,295	2,495	2,615	6,915	3,423	4,952	7,886	5,782	5,511	4,025
CSP Bas	155	66	-	4	-	18	19	1	505	74	52	-	158	814
CSP Sto	98	224	51	11	-	71	45	23	556	23	48	-	157	404
Hydro	956	490	543	791	1,170	749	812	290	384	869	352	1,019	656	981

Legend

Node	Nodes of stochastic program (0, a, b, aa, ab, ba, bb)
M	Probability weighted mean
Subscript D	for deterministic reference case
Subscript S	for stochastic setup
Subscript C	for/ across all four scenarios (aa, ab, ba, bb)
Subscript C_a	for/ across scenarios aa, ab
Subscript C_b	for/ across scenarios ba, bb

C.2.2 Uncertain Learning Rate for PV

The table in this section displays the stochastic results for uncertainty in the learning rate of PV. Additional information on this topic can be found in chapter 7, in particular within section 7.3.1.

Table C.11 Stochastic results for uncertain PV learning rate (Investment in GW)

	2015			2030						2050					
	0_D	0_S	M_C	0_D	a_S	b_S	M_S	M_{C_a}	M_{C_b}	0_D	a_S	b_S	M_S	M_{C_a}	M_{C_b}
H Coal	485	486	507	452	454	455	454	415	418	604	509	630	570	598	651
Lignite	-	-	-	-	-	-	-	-	-	-	23	-	12	-	-
Gas OC	-	-	-	91	94	104	99	38	38	114	166	113	140	223	165
Gas CC	183	182	182	1,050	1,059	1,055	1,057	1,135	1,133	181	393	208	301	320	163
Oil	-	-	-	-	-	-	-	-	-	-	-	-	-	-	-
Nuclear	-	-	-	119	113	122	117	153	153	1,172	714	1,143	928	658	1,119
Win Ons	26	27	-	906	905	906	905	921	922	1,565	1,670	1,948	1,809	1,658	1,846
Win Offs	3	3	-	48	74	74	74	51	51	120	100	118	109	116	151
Biomass	7	247	63	226	221	221	221	213	215	417	395	391	393	390	397
PV	-	-	-	-	-	-	-	-	-	-	2,009	-	1,005	2,009	-
CSP Bas	-	-	-	-	-	-	-	-	-	-	-	-	-	-	-
CSP Sto	-	-	-	-	-	-	-	-	-	1	-	1	-	-	-
Hydro	93	92	77	290	293	279	286	291	288	614	450	501	475	469	502

	2075							2100						
	0_D	aa_S	ab_S	ba_S	bb_S	M_S	M_C	0_D	aa_S	ab_S	ba_S	bb_S	M_S	M_C
H Coal	-	-	-	-	-	-	-	-	-	-	-	-	-	-
Lignite	-	-	-	-	-	-	-	-	-	-	-	-	-	-
Gas OC	-	263	257	189	14	181	157	241	88	14	30	14	37	59
Gas CC	1,030	829	817	1,039	1,115	950	1,050	33	248	254	48	64	154	118
Oil	-	-	-	-	-	-	-	-	-	-	-	-	-	-
Nuclear	255	923	923	310	305	615	615	903	385	386	878	863	628	623
Win Ons	1,745	1,644	1,641	2,057	2,120	1,865	1,981	1,592	1,984	1,716	1,989	2,277	1,992	2,054
Win Offs	160	152	155	587	587	370	372	871	228	150	198	420	249	190
Biomass	492	448	454	456	459	454	464	1,013	870	688	690	1,085	833	915
PV	2,316	3,316	3,338	2,500	-	2,289	2,247	6,915	7,179	7,765	8,076	-	5,755	5,658
CSP Bas	155	1	3	5	1,925	483	503	1	51	62	95	2,917	781	808
CSP Sto	98	2	8	15	84	27	25	23	50	57	87	616	203	189
Hydro	956	495	498	697	776	617	557	290	745	940	714	463	716	700

Legend	
Node	Nodes of stochastic program (0, a, b, aa, ab, ba, bb)
M	Probability weighted mean
Subscript D	for deterministic reference case
Subscript S	for stochastic setup
Subscript C	for/ across all four scenarios (aa, ab, ba, bb)
Subscript C_a	for/ across scenarios aa, ab
Subscript C_b	for/ across scenarios ba, bb

C.2.3 Uncertain Learning Rate for CSP

The table in this section displays the stochastic results for an uncertain CSP learning rate. Additional information on this topic can be found in chapter 7, in particular within section 7.3.1.

Table C.12 Results for uncertain CSP learning rate (Investment in GW)

	2015			2030						2050					
	0_D	0_S	M_C	0_D	a_S	b_S	M_S	M_{C_a}	M_{C_b}	0_D	a_S	b_S	M_S	M_{C_a}	M_{C_b}
H Coal	485	481	507	452	468	456	462	430	417	604	520	508	514	653	637
Lignite	-	-	-	-	-	-	-	-	-	-	23	23	23	12	-
Gas OC	-	-	-	91	59	94	76	-	38	114	-	166	83	-	149
Gas CC	183	182	182	1,050	1,017	1,058	1,037	1,085	1,132	181	519	396	457	402	165
Oil	-	-	-	-	-	-	-	-	-	-	-	-	-	-	-
Nuclear	-	-	-	119	73	117	95	102	154	1,172	563	710	637	590	1,159
Win Ons	26	46	3	906	949	870	909	977	921	1,565	1,677	1,681	1,679	1,665	1,710
Win Offs	3	3	-	48	55	74	64	51	50	120	103	96	99	106	146
Biomass	7	247	63	226	221	221	221	213	215	417	389	391	390	401	398
PV	-	-	-	-	-	-	-	-	-	-	-	2,009	1,005	-	-
CSP Bas	-	-	-	-	74	-	37	70	-	-	1,888	-	944	1,791	-
CSP Sto	-	-	-	-	124	-	62	121	-	1	444	-	222	456	-
Hydro	93	92	77	290	274	293	284	291	288	614	330	450	390	331	500

	2075							2100						
	0_D	aa_S	ab_S	ba_S	bb_S	M_S	M_C	0_D	aa_S	ab_S	ba_S	bb_S	M_S	M_C
H Coal	-	-	9	-	-	2	-	-	-	-	-	-	-	-
Lignite	-	-	-	-	-	-	-	-	-	-	-	-	-	-
Gas OC	-	-	105	263	264	158	70	241	-	8	86	111	51	129
Gas CC	1,030	921	829	829	829	852	1,013	33	174	275	248	227	231	112
Oil	-	-	-	-	-	-	-	-	-	-	-	-	-	-
Nuclear	255	1,034	1,198	921	921	1,018	636	903	364	171	386	387	327	615
Win Ons	1,745	1,698	1,554	1,646	1,653	1,638	1,665	1,592	2,027	2,173	1,984	1,986	2,042	2,045
Win Offs	160	43	187	158	152	135	353	871	127	240	228	227	205	238
Biomass	492	361	416	452	452	420	459	1,013	1,289	1,060	869	889	1,027	1,052
PV	2,316	-	-	3,300	3,302	1,651	1,204	6,915	-	2,316	7,187	7,263	4,192	4,422
CSP Bas	155	557	215	1	-	193	256	1	2,862	2,776	51	-	1,422	1,346
CSP Sto	98	836	43	2	-	220	201	23	1,913	199	50	-	541	595
Hydro	956	523	646	496	496	540	766	290	387	772	745	784	672	472

Legend

Node	Nodes of stochastic program (0, a, b, aa, ab, ba, bb)
M	Probability weighted mean
Subscript D	for deterministic reference case
Subscript S	for stochastic setup
Subscript C	for/ across all four scenarios (aa, ab, ba, bb)
Subscript C_a	for/ across scenarios aa, ab
Subscript C_b	for/ across scenarios ba, bb

C.2.4 Uncertain Carbon Bound

The table in this section displays the stochastic results for an uncertain carbon bound. Additional information on this topic can be found in chapter 7, in particular within section 7.3.2.

Table C.13 Results for uncertain carbon bound (Investment in GW)

	2015			2030						2050					
	0_D	0_S	M_C	0_D	a_S	b_S	M_S	M_{C_a}	M_{C_b}	0_D	a_S	b_S	M_S	M_{C_a}	M_{C_b}
H Coal	485	450	501	452	-	1,487	744	-	1,410	604	474	378	426	481	446
Lignite	-	-	-	-	-	-	-	-	-	-	-	-	-	-	-
Gas OC	-	-	-	91	126	94	110	45	36	114	143	225	184	221	273
Gas CC	183	213	181	1,050	634	632	633	676	742	181	790	373	581	721	273
Oil	-	-	-	-	-	-	-	-	-	-	-	-	-	-	-
Nuclear	-	-	-	119	796	-	398	924	-	1,172	450	867	658	358	864
Win Ons	26	60	5	906	1,720	25	873	1,505	114	1,565	1,933	1,679	1,806	1,815	1,785
Win Offs	3	-	-	48	216	-	108	240	-	120	3	243	123	3	167
Biomass	7	197	51	226	240	4	122	228	3	417	407	507	457	432	513
PV	-	-	-	-	-	-	-	-	-	-	2,009	-	1,005	2,009	-
CSP Bas	-	-	-	-	-	-	-	-	-	-	-	-	-	-	-
CSP Sto	-	-	-	-	-	-	-	-	-	1	-	-	-	-	-
Hydro	93	83	77	290	361	193	277	311	182	614	409	505	457	452	529

	2075							2100						
	0_D	aa_S	ab_S	ba_S	bb_S	M_S	M_C	0_D	aa_S	ab_S	ba_S	bb_S	M_S	M_C
H Coal	-	-	-	24	291	79	49	-	-	-	-	-	-	-
Lignite	-	-	-	-	23	6	6	-	-	-	-	-	-	-
Gas OC	-	84	271	190	186	183	148	241	75	75	15	97	65	93
Gas CC	1,030	351	586	1,066	1,095	774	895	33	187	470	152	230	260	182
Oil	-	-	-	-	-	-	-	-	-	-	-	-	-	-
Nuclear	255	183	141	893	786	501	453	903	885	905	390	591	693	738
Win Ons	1,745	1,790	2,036	1,554	1,731	1,778	1,860	1,592	2,011	1,970	488	2,094	1,641	2,088
Win Offs	160	356	234	426	286	325	281	871	228	165	98	207	174	222
Biomass	492	466	442	357	206	368	371	1,013	1,418	674	892	859	961	1,034
PV	2,316	5,296	4,153	2,500	2,500	3,612	3,906	6,915	7,465	8,872	9,097	6,504	7,984	7,386
CSP Bas	155	7	5	4	1	4	2	1	317	91	81	48	134	69
CSP Sto	98	20	16	11	2	12	7	23	481	81	89	47	175	73
Hydro	956	753	542	671	566	633	566	290	593	925	603	577	674	740

Legend	
Node	Nodes of stochastic program (0, a, b, aa, ab, ba, bb)
M	Probability weighted mean
Subscript D	for deterministic reference case
Subscript S	for stochastic setup
Subscript C	for/ across all four scenarios (aa, ab, ba, bb)
Subscript C_a	for/ across scenarios aa, ab
Subscript C_b	for/ across scenarios ba, bb

C.2.5 Uncertain Price Development for Fossil Fuels

The table in this section displays the stochastic results for uncertain gas and coal price development. Additional information on this topic can be found in chapter 7, in particular within section 7.3.3.

Table C.14 Results for uncertain coal and natural gas prices (Investment in GW)

	2015			2030						2050					
	0_D	0_S	M_C	0_D	a_S	b_S	M_S	M_{C_a}	M_{C_b}	0_D	a_S	b_S	M_S	M_{C_a}	M_{C_b}
H Coal	485	399	420	452	837	-	418	733	-	604	550	-	275	671	-
Lignite	-	765	191	-	28	-	14	15	-	-	59	-	30	83	-
Gas OC	-	-	-	91	103	92	98	38	87	114	68	220	144	44	229
Gas CC	183	265	254	1,050	403	2,030	1,217	605	2,110	181	193	1,130	661	192	1,031
Oil	-	-	-	-	-	-	-	-	-	-	-	-	-	-	-
Nuclear	-	-	-	119	314	-	157	252	-	1,172	1,123	495	809	1,209	526
Win Ons	26	16	-	906	723	-	361	929	-	1,565	1,360	1,831	1,595	1,819	1,792
Win Offs	3	-	-	48	78	-	39	78	-	120	177	260	218	99	208
Biomass	7	205	53	226	324	-	162	322	-	417	396	435	416	223	443
PV	-	-	-	-	-	-	-	-	-	-	18	-	9	-	-
CSP Bas	-	-	-	-	-	-	-	-	-	-	1	-	-	1	-
CSP Sto	-	-	-	-	-	-	-	-	-	1	3	-	1	2	-
Hydro	93	76	76	290	392	171	282	394	95	614	562	450	506	508	514

	2075							2100						
	0_D	aa_S	ab_S	ba_S	bb_S	M_S	M_C	0_D	aa_S	ab_S	ba_S	bb_S	M_S	M_C
H Coal	-	28	-	317	-	86	90	-	-	-	-	-	-	-
Lignite	-	15	-	23	-	10	10	-	-	-	-	-	-	-
Gas OC	-	80	64	203	232	145	202	241	185	244	119	5	138	45
Gas CC	1,030	857	935	340	1,113	811	957	33	85	68	766	197	279	205
Oil	-	-	-	-	-	-	-	-	-	-	-	-	-	-
Nuclear	255	-	-	1,407	1,121	632	577	903	1,075	1,075	-	477	657	757
Win Ons	1,745	1,497	1,482	1,438	2,224	1,660	2,414	1,592	1,968	2,034	665	2,093	1,690	2,427
Win Offs	160	724	738	299	243	501	206	871	228	209	65	206	177	170
Biomass	492	527	477	382	110	374	339	1,013	980	969	811	1,251	1,003	765
PV	2,316	2,681	2,681	2,500	2,500	2,590	1,829	6,915	7,956	7,893	8,608	5,795	7,563	3,310
CSP Bas	155	34	59	1	1	24	691	1	44	20	48	253	91	1,835
CSP Sto	98	49	99	2	4	39	417	23	41	12	47	132	58	481
Hydro	956	848	822	658	535	716	762	290	401	401	671	542	504	899

Legend

Node	Nodes of stochastic program (0, a, b, aa, ab, ba, bb)
M	Probability weighted mean
Subscript D	for deterministic reference case
Subscript S	for stochastic setup
Subscript C	for/ across all four scenarios (aa, ab, ba, bb)
Subscript C_a	for/ across scenarios aa, ab
Subscript C_b	for/ across scenarios ba, bb

C.2.6 Uncertain Price Development for Natural Gas

The table in this section displays the stochastic results for uncertain natural gas price development. Additional information on this topic can be found in chapter 7, in particular within section 7.3.3.

Table C.15 Results for uncertain natural gas prices (Investment in GW)

	2015			2030						2050					
	0_D	0_S	M_C	0_D	a_S	b_S	M_S	M_{C_a}	M_{C_b}	0_D	a_S	b_S	M_S	M_{C_a}	M_{C_b}
H Coal	485	400	419	452	952	-	476	838	-	604	557	-	279	668	-
Lignite	-	-	-	-	-	-	-	-	-	-	-	-	-	-	-
Gas OC	-	-	-	91	101	174	138	36	116	114	113	142	127	71	200
Gas CC	183	270	256	1,050	261	2,058	1,160	415	2,133	181	291	1,067	679	275	996
Oil	-	-	-	-	-	-	-	-	-	-	-	-	-	-	-
Nuclear	-	-	-	119	375	-	187	414	-	1,172	1,039	527	783	1,011	525
Win Ons	26	-	-	906	786	-	393	735	-	1,565	1,458	1,785	1,621	1,359	1,792
Win Offs	3	-	-	48	78	-	39	52	-	120	208	215	211	251	208
Biomass	7	2	1	226	324	-	162	324	-	417	373	439	406	390	440
PV	-	-	-	-	-	-	-	-	-	-	-	-	-	-	-
CSP Bas	-	-	-	-	-	-	-	-	-	-	1	-	-	-	-
CSP Sto	-	-	-	-	-	-	-	-	-	1	3	-	1	1	-
Hydro	93	76	76	290	375	92	234	370	68	614	531	518	524	487	541

	2075							2100						
	0_D	aa_S	ab_S	ba_S	bb_S	M_S	M_C	0_D	aa_S	ab_S	ba_S	bb_S	M_S	M_C
H Coal	-	116	-	454	-	142	89	-	-	-	-	-	-	-
Lignite	-	-	-	23	-	6	6	-	-	-	-	-	-	-
Gas OC	-	105	5	141	254	126	263	241	170	259	1	-	108	54
Gas CC	1,030	769	902	376	1,069	779	848	33	99	92	831	174	299	293
Oil	-	-	-	-	-	-	-	-	-	-	-	-	-	-
Nuclear	255	25	-	1,373	1,352	688	577	903	1,053	1,059	14	44	543	724
Win Ons	1,745	2,080	1,186	1,860	2,487	1,903	1,926	1,592	1,922	1,995	2,348	2,269	2,134	1,918
Win Offs	160	759	858	243	657	629	457	871	228	227	655	257	342	186
Biomass	492	536	571	423	322	463	575	1,013	987	949	902	1,257	1,024	822
PV	2,316	2,316	2,316	-	-	1,158	1,875	6,915	8,124	7,903	695	2,500	4,805	5,590
CSP Bas	155	32	57	1,877	3	492	638	1	53	19	642	2,455	792	739
CSP Sto	98	47	106	26	10	47	31	23	54	2	300	464	205	269
Hydro	956	806	953	705	583	762	604	290	429	384	1,084	480	594	640

Legend

Node	Nodes of stochastic program (0, a, b, aa, ab, ba, bb)
M	Probability weighted mean
Subscript D	for deterministic reference case
Subscript S	for stochastic setup
Subscript C	for/ across all four scenarios (aa, ab, ba, bb)
Subscript C_a	for/ across scenarios aa, ab
Subscript C_b	for/ across scenarios ba, bb

C.2.7 Uncertain Renewable Potential

The table in this section displays the stochastic results for uncertainty in the size of the renewable potential segments. Additional information on this topic can be found in chapter 7, in particular within section 7.3.4.

Table C.16 Results for uncertain renewable potential (Investment in GW)

	2015			**2030**						**2050**					
	0_D	0_S	M_C	0_D	a_S	b_S	M_S	M_{C_a}	M_{C_b}	0_D	a_S	b_S	M_S	M_{C_a}	M_{C_b}
H Coal	485	491	509	452	426	373	399	434	386	604	642	507	575	606	561
Lignite	-	-	-	-	-	-	-	-	-	-	-	-	-	-	-
Gas OC	-	-	-	91	99	234	166	39	190	114	108	238	173	177	287
Gas CC	183	176	185	1,050	1,001	1,163	1,082	1,076	1,298	181	198	525	361	187	237
Oil	-	-	-	-	-	-	-	-	-	-	-	-	-	-	-
Nuclear	-	-	-	119	18	413	216	-	290	1,172	1,098	778	938	1,086	1,136
Win Ons	26	84	5	906	850	621	735	1,070	1,253	1,565	1,586	1,628	1,607	1,659	1,590
Win Offs	3	-	-	48	45	30	37	43	85	120	178	103	141	162	241
Biomass	7	247	64	226	230	111	170	240	105	417	574	246	410	567	245
PV	-	-	-	-	-	-	-	-	-	-	-	1,922	961	-	939
CSP Bas	-	-	-	-	-	-	-	-	-	-	-	-	-	-	-
CSP Sto	-	-	-	-	-	-	-	-	-	1	-	-	-	-	-
Hydro	93	56	66	290	510	122	316	483	56	614	394	247	320	388	308

	2075							**2100**						
	0_D	aa_S	ab_S	ba_S	bb_S	M_S	M_C	0_D	aa_S	ab_S	ba_S	bb_S	M_S	M_C
H Coal	-	-	-	-	-	-	-	-	-	-	-	-	-	-
Lignite	-	-	-	-	-	-	-	-	-	-	-	-	-	-
Gas OC	-	156	26	226	313	180	224	241	88	70	143	-	75	71
Gas CC	1,030	877	895	861	909	886	1,115	33	196	234	243	198	218	73
Oil	-	-	-	-	-	-	-	-	-	-	-	-	-	-
Nuclear	255	492	535	324	362	428	389	903	760	719	811	792	771	855
Win Ons	1,745	1,613	1,723	1,975	1,985	1,824	2,037	1,592	1,929	1,756	2,162	1,682	1,882	1,911
Win Offs	160	218	108	133	123	146	466	871	226	-	65	397	172	261
Biomass	492	467	336	535	262	400	438	1,013	1,062	619	836	1,114	907	999
PV	2,316	2,500	2,500	4,199	6,175	3,844	3,433	6,915	5,422	9,403	7,865	9,179	7,967	7,446
CSP Bas	155	-	21	3	18	10	4	1	27	7	46	789	217	379
CSP Sto	98	-	41	8	53	25	11	23	27	2	41	1,136	301	260
Hydro	956	597	929	841	639	752	607	290	917	761	666	179	631	633

Legend

Node	Nodes of stochastic program (0, a, b, aa, ab, ba, bb)
M	Probability weighted mean
Subscript D	for deterministic reference case
Subscript S	for stochastic setup
Subscript C	for/ across all four scenarios (aa, ab, ba, bb)
Subscript C_a	for/ across scenarios aa, ab
Subscript C_b	for/ across scenarios ba, bb

C.2.8 Multiple Uncertainties

The table in this section displays the stochastic results for a model setup with multiple uncertainties: learning rates, carbon bound, fossil fuel prices and renewable potential. Additional information on this topic can be found in chapter 7, in particular within section 7.3.5.

Table C.17 Results for multiple uncertain parameters (Investment in GW)

	2015			**2030**						**2050**					
	0_D	0_S	M_C	0_D	a_S	b_S	M_S	M_{C_a}	M_{C_b}	0_D	a_S	b_S	M_S	M_{C_a}	M_{C_b}
H Coal	485	506	507	452	-	-	-	-	-	604	426	-	213	427	-
Lignite	-	-	-	-	-	-	-	-	-	-	88	-	44	88	-
Gas OC	-	-	-	91	-	193	96	21	186	114	19	242	130	18	266
Gas CC	183	174	176	1,050	658	2,177	1,417	672	2,171	181	582	1,696	1,139	577	1,698
Oil	-	-	-	-	-	-	-	-	-	-	-	-	-	-	-
Nuclear	-	-	-	119	374	-	187	387	-	1,172	65	201	133	62	264
Win Ons	26	5	5	906	1,511	-	755	1,525	-	1,565	2,066	702	1,384	2,051	621
Win Offs	3	-	-	48	251	-	125	251	-	120	-	19	10	-	10
Biomass	7	2	1	226	353	-	177	355	-	417	645	262	453	659	247
PV	-	-	-	-	1,531	-	766	1,531	-	-	1,781	-	891	1,762	-
CSP Bas	-	-	-	-	72	-	36	58	-	-	313	-	157	329	-
CSP Sto	-	-	-	-	95	-	48	94	-	1	467	-	233	465	-
Hydro	93	70	75	290	482	6	244	430	7	614	361	344	353	358	304

	2075							**2100**						
	0_D	aa_S	ab_S	ba_S	bb_S	M_S	M_C	0_D	aa_S	ab_S	ba_S	bb_S	M_S	M_C
H Coal	-	-	-	331	-	83	80	-	-	-	-	-	-	-
Lignite	-	-	-	23	-	6	6	-	-	-	-	-	-	-
Gas OC	-	22	149	5	72	62	122	241	9	165	399	-	143	126
Gas CC	1,030	417	631	356	1,089	623	670	33	141	409	638	632	455	458
Oil	-	-	-	-	-	-	-	-	-	-	-	-	-	-
Nuclear	255	646	1,130	1,619	706	1,025	999	903	859	409	-	-	317	321
Win Ons	1,745	2,052	1,673	1,835	2,477	2,009	1,934	1,592	1,913	2,067	2,353	2,816	2,287	2,252
Win Offs	160	251	228	167	876	381	430	871	251	183	52	670	289	320
Biomass	492	611	198	528	316	413	446	1,013	1,432	1,065	831	703	1,008	963
PV	2,316	3,988	3,140	2,500	2,500	3,032	3,095	6,915	4,194	6,049	5,980	8,158	6,095	6,212
CSP Bas	155	168	18	-	1	47	38	1	252	57	8	325	160	176
CSP Sto	98	195	-	-	2	49	48	23	421	65	12	481	245	250
Hydro	956	452	422	979	1,117	743	724	290	661	939	280	1,226	777	795

Legend

Node	Nodes of stochastic program (0, a, b, aa, ab, ba, bb)
M	Probability weighted mean
Subscript D	for deterministic reference case
Subscript S	for stochastic setup
Subscript C	for/ across all four scenarios (aa, ab, ba, bb)
Subscript C_a	for/ across scenarios aa, ab
Subscript C_b	for/ across scenarios ba, bb

C.2.9 Setup Not Considering "Vertical Shift"

The table in this section displays the stochastic results for a model setup with uncertain PV learning rate but not considering the "vertical shift". Additional information on this topic can be found in chapter 7, in particular within section 7.3.6.

Table C.18 Stochastic results for uncertain PV learning rate without vertical shift (Investment in GW)

	2015			2030						2050					
	0_D	0_S	M_C	0_D	a_S	b_S	M_S	M_{C_a}	M_{C_b}	0_D	a_S	b_S	M_S	M_{C_a}	M_{C_b}
H Coal	485	488	507	452	398	453	426	415	418	604	534	634	584	599	651
Lignite	-	-	-	-	-	-	-	-	-	-	23	-	12	-	-
Gas OC	-	-	-	91	187	104	146	38	38	114	173	114	143	135	165
Gas CC	183	176	182	1,050	1,043	1,057	1,050	1,135	1,133	181	553	205	379	321	163
Oil	-	-	-	-	-	-	-	-	-	-	-	-	-	-	-
Nuclear	-	-	-	119	-	122	61	152	153	1,172	556	1,144	850	655	1,119
Win Ons	26	20	-	906	284	911	597	921	922	1,565	838	1,960	1,399	1,658	1,846
Win Offs	3	-	-	48	10	77	44	51	51	120	105	114	109	116	151
Biomass	7	247	63	226	166	219	193	213	215	417	349	389	369	390	397
PV	-	-	-	-	1,531	-	766	2	-	-	2,823	-	1,412	2,022	-
CSP Bas	-	-	-	-	-	-	-	-	-	-	-	-	-	-	-
CSP Sto	-	-	-	-	-	-	-	-	-	1	-	1	-	-	-
Hydro	93	105	77	290	241	270	255	291	288	614	380	498	439	469	502

	2075							2100						
	0_D	aa_S	ab_S	ba_S	bb_S	M_S	M_C	0_D	aa_S	ab_S	ba_S	bb_S	M_S	M_C
H Coal	-	-	5	-	-	1	-	-	-	-	-	-	-	-
Lignite	-	-	-	-	-	-	-	-	-	-	-	-	-	-
Gas OC	-	284	284	123	14	176	157	241	89	88	183	13	93	70
Gas CC	1,030	619	613	1,054	1,116	851	1,051	33	494	506	44	64	277	118
Oil	-	-	-	-	-	-	-	-	-	-	-	-	-	-
Nuclear	255	1,230	1,236	304	304	768	615	903	256	242	869	863	558	625
Win Ons	1,745	1,339	1,339	2,124	2,131	1,733	1,971	1,592	1,462	1,463	2,109	2,279	1,828	2,075
Win Offs	160	152	142	587	587	367	382	871	109	74	227	257	167	234
Biomass	492	517	517	466	461	490	465	1,013	726	726	926	1,075	863	961
PV	2,316	3,709	3,696	2,316	-	2,430	2,237	6,915	7,252	7,321	7,507	-	5,520	5,478
CSP Bas	155	1	1	5	1,920	482	503	1	48	48	80	3,157	833	803
CSP Sto	98	2	2	14	84	26	25	23	47	47	86	591	193	184
Hydro	956	594	593	748	785	680	556	290	699	716	466	457	584	670

Legend

Node	Nodes of stochastic program (0, a, b, aa, ab, ba, bb)
M	Probability weighted mean
Subscript D	for deterministic reference case
Subscript S	for stochastic setup
Subscript C	for/ across all four scenarios (aa, ab, ba, bb)
Subscript C_a	for/ across scenarios aa, ab
Subscript C_b	for/ across scenarios ba, bb

MIX
Papier aus verantwortungsvollen Quellen
Paper from responsible sources
FSC® C105338

If you have any concerns about our products,
you can contact us on
ProductSafety@springernature.com

In case Publisher is established outside the EU,
the EU authorized representative is:
Springer Nature Customer Service Center GmbH
Europaplatz 3, 69115 Heidelberg, Germany

Printed by Libri Plureos GmbH
in Hamburg, Germany